GOVERNING CHILDREN, FAMILIES, AND EDUCATION

Additional praise for
Governing Children, Families and Education:

"This collection performs important conceptual work by crossing and combining fields that are all too often kept apart: child studies, education, and social policy. Using a variety of disciplinary approaches, the authors show convincingly how policymakers in all of these domains use children as a wedge issue in efforts to reform families and restructure welfare states. By ranging across societies and over time, the articles map the impact of cross-cultural exchanges and trace the consolidation of global patterns of governance. Taken as a whole, the volume offers a fresh perspective on governmentality and the power/knowledge nexus; unique in its ambition, it has the potential to revise thinking in all of the fields it addresses."

—Sonya Michel, Professor of American Studies and History, University of Maryland, author of *Children's Interests/Mothers' Rights: The Shaping of America's Child Care Policy*

"Editors and international colleagues, Bloch, Holmlund, Moqvist and Popkewitz present in this collection a rich smorgasbord of critical views of topics all too infrequently explored. Discourses, ideologies, research methodologies and theoretical perspectives are appropriately diverse in what amounts to a comprehensive reconceptualization of education's private-public realms. Central to all contributions are thematics and relations of governing and government, of care and welfare, of reason and knowledge, of freedom and control. This is exciting reading with something for everyone who seriously considers reform."

—Lynda Stone, Professor, Philosophy of Education, University of North Carolina at Chapel Hill

GOVERNING CHILDREN, FAMILIES, AND EDUCATION

RESTRUCTURING THE WELFARE STATE

EDITED BY

Marianne N. Bloch, Kerstin Holmlund, Ingeborg Moqvist, and Thomas S. Popkewitz

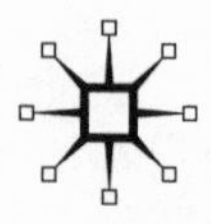

GOVERNING CHILDREN, FAMILIES, AND EDUCATION

First published 2003 by
PALGRAVE MACMILLAN™
175 FIFTH AVENUE, NEW YORK, N.Y. 10010 AND
Houndmills, Basingstoke, Hampshire, England RG21 6XS.
Companies and representatives throughout the world.

PALGRAVE MACMILLAN is the global academic imprint of the Palgrave Macmillan division of St. Martin's Press, LLC and of Palgrave Macmillan Ltd. Macmillan® is a registered trademark in the United States, United Kingdom and other countries. Palgrave is a registered trademark in the European Union and other countries.

ISBN 1-4039-6224-3 hardback
ISBN 1-4039-6225-1 paperback

Library of Congress Cataloging-in-Publication Data

Governing children, families, and education : restructuring the welfare state / editors, Marianne N. Bloch . . . [et al.].
p. cm.
Includes bibliographical references and index.
ISBN 1-4039-6224-3 — ISBN 1-4039-6225-1 (pbk.)
1. Child welfare. 2. Education. 3. Welfare state. I. Bloch, Marianne N.
HV713.R478 2003+
362.7—dc21 2003049812

A catalogue record for this book is available from the British Library.

Design by Letra Libre

First edition: December 2003
10 9 8 7 6 5 4 3 2 1

Printed in the United States of America

CONTENTS

PREFACE

We believe this book will provide a critical text for others to begin to investigate welfare state reforms and new discourses that are constituting and reconstituting the family, the child, and his/her care and education.

We also believe this volume will be a valuable addition to the field of historical and critical cultural studies in education, general educational reform literature, comparative and international education, gender and education, and to the field of education from early childhood through the secondary level.

This volume is unique in that it draws on an interdisciplinary group of social scientists whose focus is on the *theoretical* assumptions and *empirical* implications of welfare policies on children, their care, education, and schooling in (and across) numerous national contexts and state traditions. The countries represent different welfare state traditions and different conceptions of the relation of collective social goals to individual rights and responsibilities, and the relative construction of state governing systems with the associational patterns of civil society. The international group of scholars who have contributed chapters for this book have each made important investigations of changes in the welfare state in their own countries.

Our argument is that to best understand the impact of changes in a particular country's (e.g., the United States') welfare policies, it is important to examine shifting policies and their effects within and across other nations. Similarly, in Sweden, where some changes in the social democratic welfare state are being made, many argue that it is necessary to understand change within a global context that includes knowledge of what is happening in the United States, other European Union countries, Latin America, Southeast Asia, and China. In Africa, Latin America, and even in Eastern Europe and Russia, we begin to understand the changing nature of government support and policy change for the welfare of families and young citizens only in relation to an examination of the interdependence of economies, and the political, cultural discourses that "travel," in complex ways across regions and the world.

We have intentionally used multiple ways to frame the questions of "welfare" and new governing patterns, their ways of reasoning, the ways they embed power relations within their "reason," and the effects of the governing patterns on individuals as well as nations.

Finally, it was important to us as editors and contributors to engage in these analyses now—with a focus on how new governing patterns are restructuring the ways we understand childhood, families, schooling, and care in a time of increasing complexity, uncertainty, and rising cultural and economic globalization of ideas, patterns of regulation, and what we might call a social administration of "freedom."

We want to thank the contributors to this volume, our families, and our respective institutions, the Umea University in Sweden and the University of Wisconsin–Madison in the United States. In addition, we wish to thank both the Swedish authorities who have the money to support intellectual activities, and Umea University and the University of Wisconsin–School of Education that have had an ongoing exchange for the past decade. The initial symposia (one in Sweden and one in the United States) supported the discussions from which this book emerged. We are grateful for these opportunities over the past few years, as well as for continuing opportunities for collaboration. Finally, and last but definitely not least, we want to thank Sabiha Bilgi, Diane Falkner, Christine Kruger, and Amy Sousnouski at the University of Wisconsin–Madison, who provided editorial and secretarial assistance throughout the process of editing this volume. We also are most grateful for the assistance and encouragement of Amanda Johnson, Matthew Ashford, and Rick Delaney at Palgrave Press.

SECTION I

FRAMINGS

CHAPTER ONE

GLOBAL AND LOCAL PATTERNS OF GOVERNING THE CHILD, FAMILY, THEIR CARE, AND EDUCATION

AN INTRODUCTION

Marianne Bloch, Kerstin Holmlund, Ingeborg Moqvist, and Thomas Popkewitz

Introduction to the Volume

This book focuses on new patterns of governing associated with the notions of welfare, care, and education that have emerged during the late twentieth and early twenty-first century in light of historically and culturally specific, and global, relations. Recent changes in welfare state provisions have produced multiple changes in different institutions related to the structure of "care" and education for the family and the child. But when looking across nations, there is a certain similarity in the changes occurring that some suggest is related to the globalization of cultural, or political, economic patterns. We critically examine the emergence of similar ideas across different spaces through cultural and historical analyses, and by theoretically examining the ways in which local knowledge and practice travel and get translated in other places and spaces.

The volume's contributors examine power and knowledge relationships involved in the translation of ideas and practices as these "travel" from place to place, and from time to time, and take on specific meanings. In this chapter, and in many of the chapters of the volume, we examine different state traditions of welfare. We also interrogate the history of the emergence of specific nation-states as part of governing patterns that defined cultural imaginaries related to the geography of "nations," continents, and entire populations (e.g., see Anderson, 1991; Bhabha, 1994). Through the contributions by the authors in the volume, we can see the different ways in which relationships shift, and the variations in understandings of "patterns of welfare" and "governing," across nations, in the understanding of changes in family, child/childhood, care, education, and schooling as part of "welfare" and "care" for others. We examine the systems of reason through which the child, family, and nation are made into objects of scrutiny and intervention.

This chapter begins by exploring different conceptions of governing related to the welfare state, families, and children. The second section examines current conceptions of the welfare state as a problematic of providing for equity for a nation's populations, and as a critical structuring/reproduction of inequities built around gender, race, nation, and class. The third section turns to the conception of governing as the rise of new discourses associated with "the social state" (Donzelot, 1977/1997; Foucault, 1977) and new technologies that were used to "police" the well-being and security of populations within and across nation-states. We then examine what Rose (1999) and others call "the demise of the social state," and the rise of postmodern, postindustrial shifts in cultural, political, and economic patterns often linked with globalization. We discuss the new patterns of regulation related to the welfare state, conceptions of "welfare," child, family, the local in relation to "the global," and introduce the variety of issues discussed by individual contributors in relation to different theoretical and empirical understandings of these patterns of change.

Restructuring the Welfare State: Conceptions of Governing for the Welfare of Families, Children, and State

We can think about the welfare state and governing in two ways. One is to conceive of welfare policy as a way to organize legislation and institutions to "care" for populations and citizens within a sovereign nation-state. Second, governing can also be seen as a way of securing or "policing" well-being, where policing is "not understood in the limiting, repressive sense we give the term today, but according to a much broader meaning that encompassed all methods for developing the quality of the population and the strength of the nation" (Donzelot, 1979/1997, p. 7; Foucault, 1979/1991). The two different notions of "welfare" and "governing" the nation-state and its population to guarantee well-being entail different ways of interpreting and thinking about the changes that are occurring. In the following sections, two approaches to governing are discussed.

Governing the State as an Entity with Power

In one conception, welfare and governing relate to the legal/administrative, institutional, and organizational means to deliver a desired range of outcomes for citizens (and, when desired, noncitizens) within a nation. Chapters in this book, for example, explore Nordic and U.S. welfare reforms in relation to the services delivered or denied to families and child assistance through state programs and policies (see, for some examples, Holmlund's and Cannella's chapters).

This notion of governing focuses on the welfare state as an entity that has the power to provide services, or to organize institutions and policies of welfare that achieve desired goals for populations residing within (and/or outside) the territory of the nation. The family, childcare and educational institutions are central actors in this production of welfare through the provision of health services, social security, and employment/unemployment schemes. Power is viewed as something that the state has but that it can give away through decentralization or a localization of policies and practices.

The notions of *power* and the *state* as an entity that holds or distributes power are embedded within many current welfare policy formulations (e.g., Esping-Anderson, 1990, 1996; Giddens, 1998), and are elaborated within different critical analyses and theories of education. The "state" that holds power can also empower others through decentralization policies or "localization" of decision making in communities. This is typical in recent discussions of Third Way welfare policies (see, for example, Giddens, 1998, p. 78). It is also in the discussions embedded in World Bank or International Monetary Fund (IMF) documents that require many "second world" (post-Soviet) and "third world" (constructed as including selected countries in Africa, Latin America, and Asia) countries to choose neoliberal approaches of decentralization, privatization, and community participation in restructuring state social and family policies.

Using the first framing of the state as a geographically defined space and entity, and power as held and distributed, many critical theoretical analyses of the welfare state examine the history of gender, race, cultural, and class relationships in national and international laws and institutional provisions.

Critical structural analyses look toward the unequal distribution of goods and services, or the inequalities of access to different kinds of employment, schooling, income, or health services. They have also examined political economic, cultural, and ideological inequalities in economic and social institutions, including the provision of schooling, as well as the official knowledge taught in schools (e.g., Apple, 1993; Dimitriadis & McCarthy, 2001). Different authors also examine international agency policies that are ostensibly intended to equalize "opportunity," but, in contrast, appear to reproduce or produce inequities in schooling and other social provisions (for example, see Torres, 2000; Whitty, Gewirtz & Edwards, 2000). These approaches to examining the welfare state in relation to a history of colonial and postcolonial dependencies in relationships are also illustrated to some extent by Swadener's and Wachira's discussion in this volume of the effects of international agency neoliberal welfare policies on the growth of poverty and inequities in access to schooling in Kenya over the past decade.

Welfare as the Social Administration of the Welfare of Families, Children, and Nation: Knowledge/Power Relations.

A different conception of the welfare state can be considered through thinking of the notions of welfare as ways of securing or "policing" the well-being of citizens and populations through the "cultural reasoning systems" that order the individuality of the welfare person, as well as the well-being of the family, childhood, and education/care. By policing, we are drawing on Donzelot (1979/1991), who focuses on the ordering, standardizing, and making of the individual who is legible and administratable. Donzelot suggests:

> The purpose of policing is to ensure the good fortune of the state through the wisdom of its regulations, and to augment its forces and its power to the limits of its capability. The science of policing consists, therefore, in regulating everything that relates to the present condition of society, in strengthening and improving it, in seeing that all things contribute to the welfare of the members that compose it. The aim of policing is to make everything that composes the state serve to strengthen and increase its power, and likewise serve the public welfare. (Donzelot, 1979/1991, p. 7, quoting Johann von Justi, Elements generaux de police, 1768)

The "policing of the state" gives focus to, for example, the "naturalness" of child care by family/mothers, or the genealogy of "need" and "dependency," as words that emerge to describe families, women, and children within different welfare state philosophies (e.g., see Fraser & Gordon, 1994, 1997). Within education, poststructural and postmodern critical theories of *knowledge* examine the *state* as an amalgamation or circulation of diverse ideas or discursive linguistic "texts" that govern and construct identity, conduct, and the ways in which reality is experienced.

Using this approach to theory and analysis, the *state* is not an entity that holds power. Instead, power is embedded in the different texts that become what we understand as knowledge, truth, or authoritative discourse (see Foucault, 1980; also see Baker, 2001; Popkewitz, 2000). There is a focus on how discourses normalize and divide, and regulate the "mind" and action. In this second conception of governing, distinctions so important earlier such as the "state" vs. the "family," or public vs. private, and the "nature" of "freedom" are examined as effects of power/knowledge relationships that construct how we think and act, forming cultural reasoning systems that govern and regulate (Channing, 1994; Scott, 1992). This way of thinking about welfare is also to consider ways of reasoning as producing inclusions, while maintaining exclusions by defining difference, deviancy, and, particular to the subjects of this book, the *needy* mother, child, family, or even "nation" to be saved through social interventions through strategies and technologies constructed as the reason of state practice (*raison d'etat*).

The two different conceptions of governing illustrated briefly above result in different ways of thinking about reform, transformation, change, and action that are present in this book. In the first, the human actor and his/her/their agency are important in complying with, or resisting, hegemonic discourses that construct inequities and exclusions, based on poverty, and racial and gendered discrimination. By focusing on the agency of the actor in relation to the state, one can look toward transformational activities that bring or empower marginalized voices to speak or be heard. This focus also leads toward an analysis of the effects of centralized and decentralized state power, and the effects of, for example, bringing in more community participation and partners from civil society, or by including new and previously excluded knowledge into school curricula. These empowering and transformative agendas focus then on a redistribution of access to power and different types of knowledge, and a reduction in inequities based upon class, race, language, gender, or national/ethnic identity.

In contrast, the second conception of governing focuses on cultural reasoning. Change and political action are to open up a multiplicity of new possibilities for thinking, conduct, and identities that are elided through the naturalizing practices of current "thought" of the welfare family and child.

By examining governing discourses within and across different national contexts, contributors in this volume interrogate the rules and standards through which "we" know childhood, families, and the rights and characteristics of citizenship or care/education, as these are reasoned as important and necessary for different populations.

Welfare Systems: Classification and the Problematic of Equity/Inequity

In this section, we discuss different classifications of the welfare state, using the first framing of state and governing described above. This discussion is followed by a consideration of the problem of access and inequality in relation to policies concerning the family, childcare, education and schooling.

Classifications of the "State"

If we pursue the first notion of the state as a sovereign and legal-administrative entity, we can make some distinctions about the relation of the state to the family, childhood/childcare, education/schooling. Traditional distinctions between state welfare systems have a taxonomy that range from more to less welfare. The categories range from a "weak state" liberal welfare regime or model, often represented by England and the United States, to a "strong state" regime or model, often represented by the Scandinavian social democratic welfare states (Esping-Anderson, 1990, 1996; Sainsbury, 1999). These taxonomies use European and Western traditions to also focus on socialist-communist models represented by the Soviet Union, or by the Peoples' Republic of China under Mao Tse Tung, as "extremely strong (or totalitarian) state" models based upon their centrally managed economies and centralized social welfare systems.

The socialist/communist welfare policies of the Soviet bloc nations are treated as different in forging the welfare systems during the twentieth century. Until the breakup of the Soviet Union and the Soviet bloc in 1989–1990, the Communist party organized a centrally managed economy and centrally controlled political, cultural, social, and religious life. The focus was on building an imagined collective nation that would serve as a model of "family" and good citizenship. This encouraged schools, childcare centers, and families to use a unitary model of education in the context of the Soviet state. The "cradle to grave" system of social services was similar to the entitlements offered in social democracies, but differed in their explicit use of the Soviet state as the model of the "caring and moral" family/nation that was to educate children and control conduct, and in the rigorous and rigid adherence to central management of economic, social, and political life.[1]

Social democracies and the conservative-corporative "strong state" regimes inscribe welfare policies that assume that individuals and families needed the protection of a centralized "social state" to secure the collective "good." Through "social engineering," the social democratic state is described as providing universal and equal entitlements to parental leave, health care, childcare, public schooling, employment, pensions, etc. in exchange for higher, often progressive, social taxes. The conservative-corporative regime, in contrast, is described as less centralized, with both state and communities engaging in partnerships with corporations, labor, and religious and other private organizations to enact social provisions (Morgan, 2002).

If we contrast different sets of categories of the "state" as legal and administrative governing systems, the liberal "weak state" welfare regimes decentralized or "distributed" power to strengthen civil society and the individual. The liberal and neoliberal "free market" regimes relied on liberal economic theory (free market, human capital investment), democratic and social principles of freedom of speech and participation, as well as private choice, to provide equality of opportunity and social/political liberty. This portrait is used to characterize neoliberal welfare regimes in the United States, England, Canada, and Australia and is the philosophical basis for many international agencies that work with "post-socialist" ("second world") former Soviet bloc nations, as well as "third world" countries in Asia, Africa, and Latin America that are often grouped together in economic development literature (for example, see Escobar, 1995).

We can use two examples from British Third Way policies and the United States' New Federalism reforms by examining the discourses of responsibilization and autonomy that encourage countries and individual citizens to take care of themselves, their families, and communities by individually paying more for schooling, or working harder. These discourses are also used to define normal "family values," as well as normal families, defined by the existence of legal marriage and the presence of father, mother, and children. These discourses of normality are also embedded within the policies of international donor agencies (the World Bank, the U.S. Agency for International Development, OECD, and UNICEF) targeted toward Eastern/Central Europe and other countries that are part of the second or third world. In international reports and recommendations, donors use a human capital theory that encourages the expenditure of scarce resources for early childhood development programs, basic education for all, community involvement and privatization of social or state resources and subsidies (see Bloch's chapter in this volume, for specific example).

There is also a reinscription of the importance of modern knowledge systems, the rise of knowledge and information-based societies, and the need to know English as a universal marker of competitive cultural and economic globalization. The discourses of efficiency and free market are laced with notions of pluralistic and participatory "free" democratic governing processes as the natural goods and necessities within the new global and economic order of the world (Escobar, 1995; Hardt & Negri, 2000).

These heterogeneous discourses, contested by many, also parallel discourses that emphasize "structural adjustment policies" in state welfare systems that are designed to reduce social state subsidies through privatization of state services, to make economies "efficient," and to repay international loans. Pressures from international agencies, the European Union, and different governments as well as from multinational agencies encourage privatization and local community and parent involvement that involves local problem-solving, as well as the "desire" to build community schools, pay increased school fees, and compete for scarce places in the place of formerly publicly available educational services. This discursive policing of reason and practice competes with evidence of worldwide growth in poverty, decreases in parents' ability to afford schooling for all their children, and growing inequalities, particularly among children and families with young children (Bradbury, Jenkins, & Micklewright, 2001).

There are continued debates about governing and the social, political, and anomalous rights and needs of different citizens and citizenship within as well as across nations and regions around the world (Fraser, 1997; Schram, 2000). The universal rights of citizenship, or of women, or of the child are contrasted with cultural variations in needs and perspectives on rights and entitlements in various parts of the world (see Moqvist's discussion in this volume, as one example).

These discussions are viewed as critical cultural, political, and economic debates that are related to increased competitiveness on a global stage in combination with national and international perceptions of needs for new economic efficiencies. They are also, however, laced with cultural anxieties about uncertainties of work and employment, international migration, and changing cultural patterns that enhance anxieties about who the "we" of citizenship, nation, or world are, and how universal rights can be defined in an age of heterogeneity of identity, historical experience, and material and cultural differences.

The generous social benefits provided by some of the social democracies and the Western and Eastern European countries in the past appear insufficient (and, according to some, "inefficient"), when tied with national/

international budgets, economies, and uncertainties related to the universality of rights, obligations, and notions within and across nations that have often been tied to populations and citizenship. One example, in relation to education, is that of Sweden, where gaps between different social groups that the welfare system had narrowed now are widened due to neoliberal cutbacks in daycare and schooling, and also changed policy concerning salaries and taxes. The strong effects of the centralized welfare system are different compared to the early 1990s. The new discourses embedded in policy, laws, and current practices "steer from a distance" by paying more attention to individual rights, private choices, and a decentralization of decision making to communities (also see Polakow, Halskov, & Jorgenson, 2001; Rothstein & Steinmo, 2002)

The shifts in governing discourses have revived debates about the tradition of caring for the collective, when the collective now is seen as heterogeneous and racially different. Hallgren and Weiner's discussion on educational strategies in Sweden that are focused on racism, as these have become important within Sweden and the European Union over the past decade, provides one example of these debates.

Various elements of globalization are examined in this book. This includes the role of multinational corporations that spread capitalism and neoliberal market "theology" as though a natural good around the world, as well as the influences and the new governmentalities of multinational efforts that construct "human rights," "women's rights," and the "universal rights of the child" as new imaginaries of welfare for "all" (Stiglitz, 2002). As these ideas relate to the child, the family, education, and care, the analyses by Swadener & Wachira, Dussel, and Bloch, provide case studies of the omissions and imaginaries that circulate to form the reason of welfare, childhood and family that have formed different hybrid policies and practices in different locations.

Welfare Regimes: Social Equality/Inequality

In the social democratic welfare state tradition and in the Soviet communist welfare traditions, the strong state supported a notion of a living wage for all families, and equitable protection for families through universal social support in such areas as child care, education, health, employment, and pensions. The resulting social entitlements or universalist policies were to reduce hierarchical class and gender divisions, and to provide for the welfare of all citizens at reasonable standards of living. However, recent literature illustrates that class, gender, and ethnic/race-related stratification exists in the labor force, despite relatively high standards for universally provided childcare, family benefits, schooling, and health care, at least in the Scandinavian countries (Gordon, 1990; Hobson, 1993; Lewis, 1993, 1997; Orloff, 1993, 1996). These benefits have persisted, despite some reported decreases, and increased pressures from the European Union Budget/Deficit groups (Polakow, et al. 2001).

In the former Communist bloc countries, the universalist policies that provided social supports, free daycare, family subsidies, and parental leave for families have fared less well than in the Scandinavian social democracies. With increased poverty, declining fertility, rising unemployment, and small rises in fees, many kindergarten programs (for 3–6 year olds) and childcare centers (for 0–3 year olds) have closed, or have experienced a sharp reduction in enrollment. Many schools have moved toward public-private partnerships, where parental fees and other community and parental "involvement" is increasingly encouraged to support or maintain schools (buildings, textbooks, teacher salaries). Many who lost employment have been "targeted" as the new needy based on international guidelines for "targeted social assistance" modeled after the "efficient" neoliberal policies of Britain and the United States. The new discourses associated with neoliberalism relate the "free market" and "democracy" with discourses of autonomy, independence, flexibility, and choice. Those demographically targeted as "poor" or "under the poverty line" are also "dependents" who need, through employment, to become independent and responsible. At the same time, in the newly independent states of the former Soviet Union, or in East-Central Europe, there are still few opportunities to find employment, particularly employment with the social benefits that previously had been part of the "social" rights of citizenship (Bloch & Blessing, 2000; Fraser, 1997; Gal & Kligman, 2000; Michel & Mahon, 2002).

In the liberal/neoliberal welfare states, the emphasis on the need for the community to care for itself, and the autonomous, responsible (elite, male Western European) individual who could care for *his* family was embedded in new forms of "charity" that were created to work with populations that did not embody the norms of autonomy. Kindergartens, public schooling for the "masses," settlement houses, social welfare groups and, university disciplines focusing on family and poverty were developed to intervene in the lives of children and families, to make them "cosmopolitan." In England, Canada, and the United States, in the late nineteenth and early twentieth century, day nurseries, charity kindergartens, and different efforts at parent education were started as targeted interventions to provide temporary aid to families and mothers who needed to be (theoretically, temporarily) employed (Beatty, 1995; Bloch, 1987; Michel & Mahon, 2002; Polakow, 1991).

If we compare the early part and the end of the twentieth century, important changes occurred in the welfare system. In the 1930s, Aid for Families with Dependent Children (or AFDC) was developed as the United States' liberal welfare state policy to protect its needy citizens. Again, as in other liberal welfare states, targeted and categorized "needy" were defined as those from lower and working classes. They were compared implicitly to a model of the normal, self-sufficient family that embodied what in welfare literature are known as maternalist ideologies and policies supporting father employment and the mother at home with children. Pensions and benefits were tied to employment, while AFDC was tied to child-rearing by mothers. The absence of a responsible male wage-earner, or the presence of widowed, divorced, or single mothers were the most frequent demographic characteristics of the abnormal family who might need that assistance (Fraser, 1997; Fraser & Gordon, 1994; Gordon, 1994; Lazerson, 2001; Skocpol, 1992).

The United States' AFDC program by the 1980s, in contrast, focused on low-income single women ("lone parents") and their dependent children. These categories were quickly associated with race. The concept of "being on welfare" came to signify being poor, nonwhite, dependent on the state, lazy, irresponsible, and unemployable (with a particular emphasis on the number of African American mothers on AFDC) (see, Michel, 1998; O'Connor, Orloff, & Shaver, 1999; Polakow, et al., 2001; Schram, 2000; also see chapters by Cannella, Bloch, and Swadener & Wachira, this volume). By the 1990s, the discursive characterization of AFDC mothers as "dependents," unemployed, and irresponsible framed the approval of the new Welfare to Work policies, the Personal Responsibility and Work Act, and a new form of "poor relief" for families known as the Temporary Aid for Needy Families (TANF).

In contrast to the liberal welfare state models, a different democratic cluster of countries (Netherlands, Germany, Austria, Italy, France, Spain, and Portugal) are described as: "minimally concerned with either market efficiency or leveling social divisions" (Morgan, 2002; p. 145). Rather, they were "associated with conservative clericalism, a history of corporatist guild traditions, and/or reflected authoritarian bureaucrats, dictators, or conservative religious political party ideologies" (p. 145). The social policies often protected laborers and their families by combining "paternalist" (patriarchal family-father primary wage earner), "maternalist," and pronatalist ideas with male and female employment outside the home. These policies, however, produced a gendered-, class-, ethnic-, and religion-based stratification, in which parental leave policies, childcare and preschool programs, as well as public schooling, were developed as support for working-class and more elite families and for women and men.

The policies that developed over the twentieth century, nonetheless, had multiple benefits. They provided for a family wage (with strong worker support for benefits for men and women). There were social supports for parents and their children that included guarantees for employment and strong labor bargaining on wages and other social benefits, including childcare, maternity leaves, family subsidies, and free or low-cost schooling through the university level. While workers' pensions and employment were fully supported, strong wage differentials maintained a stratified gendered and classed society, reproducing family values related to maternity, social/religious policies against abortion, and inequities in access to different schools and universities. "Paternity," "Maternity," family values, *and* employment were supported, although with different wages and benefits for different groups.

In the 1990s "post-socialist" and global market, the European Union's pressure toward unification of social and economic policies put enormous pressure on this cluster of countries (e.g., France, Germany, Italy, Spain, the Netherlands, and Portugal) to reduce "social subsidies," in order to become more economically "efficient," and "in line" with 3 percent budget deficit targets for all EU countries. These pressures are now extended to new European Union members with varying responses.

The Social State: Reconsidering Governing

It would be inaccurate, however, to assume that all of the regimes discussed above, including "liberal" welfare state regimes, are not part of the conception of "welfare state." Indeed, nation-states were historical structures organized, some might say fabricated, in order to produce and organize populations "in care of the state" (de Swaan, 1988). The notion of the "social state" captures this framing of the state and of the second approach, which frames the "governing" of the individual and populations within and across "national" borders.

The rise of the "social state" across Western nations in the nineteenth and early twentieth century, was developed as a set of strategies to care for citizens. The social state, through different institutions, actors, and language embedded in policies and practices, dispersed reasoning that was to help citizens learn to take care of themselves as well as others, and to "socially administer" the freedom of self and other (see, for example, Hindess, 1996 here). Discourses of the normal child, family, and development (of child, parent, and nation) emerged to guide the reasoning of state welfare policies and practices of salvation and intervention.

The social state embedded cultural *reasoning* systems into national government, laws, and institutions, including kindergartens, schools, teacher and parent education, and social welfare organizations, which helped to construct notions of the child, the parent, families, education, and care as well as individual and national well-being. Here we examine social welfare policies as particular ways of reasoning about who and what is normal, natural, and, by comparison, who and what is abnormal, and in need of social administration, intervention, and salvation to become "civilized" and "modern." In the next section, we move historically backward and then forward to begin to understand present "state" reasoning, by examining new governmentalities that led toward the rise of the twentieth-century *social state.*

Modernity as an Era of Technological Invention, Industrialization, Scientific and Populational Reasoning

Throughout the nineteenth century, a combination of science laced with romanticism and enlightenment beliefs resulted in new discourses in which religious ideas of salvation after death shifted toward progress, modernization, and ameliorating life "on earth" through science and reason. Following revolutions in the United States and France, unification movements in Germany and elsewhere, there were shifts in work from family farms and economies toward new forms of technology and industry often requiring employment outside the home, as well as new forms of gendered divisions in the labor of reproduction at home (childcare and domestic labor). Political shifts throughout Europe resulted in a decline in the ability to rule by a sovereign or a monarch's "power," and the rise of new forms of disciplinary power, where new policies and ideas rested on new technologies of governing knowledge and on the "reason" of individuals, groups, and nations. The expansion of economies and political power also depended upon the new "possibilities" of governing elsewhere; this of course included economic and political colonization, by different European countries, the United States, the Soviet Union/Russia, and Japan (as well as other nations), of different territories in what is now Africa, Latin America, and Asia (see Chakrabarty, 2000; Young, 1990, 2001). It also involved the assumption of new colonial care and governing strategies for populations indigenous or "native" to newly colonized territories, including, for example, those within the United States, Australia and New Zealand, and Latin America. New strategies and technologies were developed to categorize and label populations, as well as to differentiate and divide normal from abnormal, the civilized from the noncivilized.

Marked by the rise of evolutionary biology and the growth of new natural and social sciences—psychology, sociology, anthropology, economics, and the mathematics of statistics—new ways arose to reason about populations, as well as to categorize and differentiate populations (Hacking, 1991). The Western European and U.S. belief in progress and the ability to find universal objective "truths" through scientific research (observation, testing, experiments) were paralleled by the emergence of new experts, who could use those methodologies to determine normality, as well as abnormality.

"Global" or abstracted discursive patterns based on "science" circulated the notions of a universal childhood, and the importance to "development" of modern schooling and "good" parenting. The natural, social, and educational "sciences," and the experts that transferred that knowledge (researchers/professors/teachers), believed that progress could be made by finding universal "truths" about childhood, families, and civilization.

The new knowledge from nineteenth-century missionary reports, travelers' reports, and anthropologists' and economists' reports were used to draw conclusions about "civilization" and "culture," and the "nature" of primitive families and childhood in exotic places (see Said, 1978). These reports, particularly when based on scientific observation of others, were used to draw conclusions about universal laws for developing children to become civilized, mature human beings; which nations were superior and which were inferior, and ways to intervene to fabricate "civilized" behavior and the cultural morals of others who were different or considered deficient. The growth in "objective" scientific research in the late nineteenth century was, of course, accompanied by a growth in social and evolutionary beliefs about the cultural survival of the fittest, and the superiority of European/Western nations and their citizens (Bhabha, 1994; Spivak, 1999; Young, 1990).

Governing the Civilized and Modern Subject through Schooling

The political landscape of countries changed, with new needs to unify skills, participation, and cultural norms that defined nationality and citizenship, difference and inclusion that also defined exclusion. The modern school developed in response to these new ideas. While kindergartens, day nurseries or childcare centers, and modern schools and their curricula were theoretically developed to modernize individuals and groups, they were also used to normalize, as well as to differentiate through different strategies of categorization and testing. In addition, a unified method of written and verbal communication was used to homogenize and unify children, and, when possible, their families, tying them to nations through a national identity and formal, similar language system. While this strategy of governing the welfare of individuals and nations was intended to homogenize and make citizens cosmopolitan, "educated" members of society, the strategies also excluded and differentiated those who could, would, or even should be allowed to adapt to or learn new customs or languages, from those who could or would not (Anderson, 1991; also see Baker's chapter, and Bloch's chapter, this volume).

The Rise of Scientific Reasoning, Expertise, and Universal Truths

Research, and potential applications from the research, were seen as critical in the rise of industry, the growth of liberal and social democratic forms of capitalism, and the development of safe and good lives for citizens. The social and educational sciences (anthropology, psychology, sociology and new fields of child development and educational psychology) that began to emerge in the late nineteenth century, appropriated ideas from natural science and assumed that reason, rationality, and the growth of information and truth would guide rational "development" (economic and human) within individuals and individual countries, as well as in nations or regions considered colonies.

The rise of scientific reasoning as an authoritative discourse was accompanied by new technologies of governing the subjectivities of citizens. Hacking (1991), for example, helps us to understand the use of statistics and populational reasoning in the development of new ways to characterize population, and to intervene into the life of "abnormal" populations. As suggested earlier, testing and measurement grew out of the reasoning associated with the idea that we could label, characterize, or group individuals as "abnormal" in intelligence, that we could locate "retardation" or intellectual abnormality, and segregate populations that were considered "at risk" biologically or because of family child-rearing. The new experts of childhood and parenthood consecrated who was considered normal, and who was not, which parents, families, or children required (temporary or long-term) intervention, and which did not (Popkewitz & Bloch, 2001; also see Popkewitz's and Baker's chapters, this volume).

With the rise of scientific discourses, new institutions emerged—as part of the state in some countries and "outside" the state in other instances—to intervene into the lives and identities of citizens. New social policy institutes emerged to develop ways to construct and govern the welfare of citizens. Professionals who were trained in science, medicine, development, and the objective observation of the child emerged as new experts (Cannella, 1997; Dahlberg, Moss, & Pence, 1997). By the early twentieth century, these new discourses of social governance of populations resulted in new technologies of governing that included preschools and kindergartens, as well as universal primary education, secondary education for children of elite families, new laws and regulations prohibiting child labor in factories, and advocating child labor in schools.

In Western Europe and the United States, the social state saw the emergence of a new expertise, technologies of science and of social administration, and new conceptions of what a modern social citizen might be like. These institutions, practices, and ideas of citizenship varied across locales, while embodying similar notions of progress through reason and ever-better science, national strategies of administration, and similar, though not identical, notions of how to use law and policy to govern individuals for individual betterment, as well as for the "collective" welfare of a country. Key to the exercise of modern social administration were new ways to govern to help normal children and families govern themselves, to learn to be autonomous citizens who could problem-solve, and to be responsible for themselves. The new citizens were to govern themselves as well as their countries by participating in the construction of a viable progressive, modern state; children, families, and citizens who were irresponsible, uneducated, or dependent were to receive different types of interventions to salvage them for society, and for themselves.

Social policies varied in their emphases on the individual and civil society as autonomous from the state (e.g., characteristic of liberal democracies such as the United States, England, Australia, Canada) or as part of a collectivity that needed careful monitoring from or by the state (e.g., the Scandinavian social democracies that began to emerge during the 1930s and particularly after World War II) (see Holmlund, 1999, 2000, and in this volume). Economic and social philosophies embodied a clear notion, however, that the economic well-being of the citizen would be fostered through the careful guidance of the social state, not only "at home," but in colonies in need of "care" through the gradual use of education, new laws, and political strategies that defined modernity and civilization.

Merging of Religious and Secular Salvation Narratives

By the late nineteenth and the twentieth century—and what Deleuze (1979/1997) termed the "rise of the social state"—the care by community for itself, a philosophy of eighteenth- and nineteenth-century Europe and the "West," was transformed into a portrait of a new individualism in social policy, based upon a normative and normalizing science and notions of modernity, that would facilitate the image and development of "free, autonomous, democratic, and problem-solving citizens." These ideas were also important as ways of governing or policing what was considered modern and progressive, or developed, and of marking the importance of Western civilization for "all" children, families, and countries. The new patterns of governing merged the eighteenth- and nineteenth-century rationalities of religious salvation of the "other" (e.g., through charitable and philanthropic activities or missionary work) with new secular technologies of governing reasoning and conduct. However, both the religious as well as the secular strategies of governing were to civilize or govern those considered different, and to "save" them from being different, less developed, or abnormal. The new secular strategies of governing were also to carry on the work of salvation, by civilizing subjectivities, desires, and the "secular" soul (see Foucault, 1977; 1991; Bloch & Popkewitz, 2000; Popkewitz & Bloch, 2001; Rose, 1990).

Global and Local: The Traveling Narratives of the Social State in Relation to Family, Childhood and National Identity

The social and secular "welfare" state had similarities, as well as distinctive differences, within and across different nations during the late nineteenth and early twentieth centuries. However, particular governing strategies were associated with the rise of social and liberal democracies that developed in the early twentieth century in Western Europe and the United States. These strategies, as they were circulated across nations and traveled outside of Western Europe and the United States to other nations, moved in different ways within different historical/cultural contexts (see, for example, Dean & Hindess, 1998, or Dussel's chapter, in this volume).

The different patterns of governing were critical to the ways in which schooling was used as a technology of governing in different places around the world, the development of preschooling or childcare, and to the variations in the ways universalized ideas of family, mother's or father's or grandparents' identities and roles were defined. Different conceptions of childhood, development, and schooling were translated into discursive practices, into laws, policies, and institutions. Perhaps what was most important were how new *governmentalities* (Foucault, 1991) came to be embedded in individual conduct and desires, as well as how these came to define what was considered reasonable, natural, and appeared as truth. The associated technologies were meant to produce an "educated," modern, developed child, parent, family and nation (see, for example, Meyer, Boli, Thomas & Ramirez, 1997). How these discursive ideas moved into different places/spaces and at different times, however, were different.

Modernity and the Universal Educated and Free Democratic Citizen

The new patterns of governing, using this framework, constructed what was reasoned as a universal set of rights and a notion of universal laws that established a truth about childhood, adulthood, and progress, within and across nations. The rationalities assumed by the notion of modernity and progress were used to inculcate "modernity" in the reasoning of individuals so that they could become "self-governing" and desire to become "modern," independent, self-problem-solving, and "civilized."

But the new governmentalities, whether rising in the United States or in Western Europe or the Soviet Union, or as they were transferred to or circulated across different countries, were also about a discursively formed and constructed "freedom" (Rose, 1999). The modern child and citizen was expected to participate actively in the formation of democratic and/or modern industrialized and scientific, progressive nations. Schooling and education in various forms were to be instruments of colonization and modernization that would help to homogenize and fabricate the cosmopolitan and modern citizen. As schooling and European or English national/international languages spread across the twentieth century, "modern" schooling came to be known as critical for the development of the "modern" child, nation, or "world development."

The Social or the Public and the Private: Social Administration of "Freedom" and "Privacy"

In Western Europe and in the United States, the liberal and social governing discourses included what is reasoned as a dichotomy between centralized governing and decentralized governing, public and the private spheres, the state, civil society, and the family. These dichotomies suggest that the social state's administration stopped at the doors of the private sector, whether civil society or family. While this seems apparent in liberal welfare state regimes, in the former Soviet Union, the Nordic social democratic states, the social state was reasoned as an important corrective to private or individual family life. The importance of the collective, and a universal entitlement to state "care" and intrusion into private family life was reasonable. But the reasoning of a separate private sphere, or of the importance or non-importance of civil society was still present.

The reason of the social state: globalization and localization of discourses of governing

During the twentieth century, the social state emerged as an amalgamation of new technologies of policing the child, family, and our reason about identity, conduct, normality/abnormality. Through a variety of interventions, including the modern public school, early childhood and kindergarten programs, health care, "scientific" truths about development and good parenting, and care for others, the social state was able to care for populations, attempting to normalize and intervene for the security of populations and a perceived or constructed well-being. But the reason of state practice included "steering from a distance." Freedom and privacy, the subjective identities and conduct of family members, as well as the reasoning of normality/abnormality were all socially administered effects of power in relation to historically defined cultural knowledge systems. The key was in how power/knowledge related to form "reason." As we move into what some call the "decline of the social state" in the late twentieth and early twenty-first century, key to our examination is still how power/knowledge relates to what we take as normal and reasonable.

Welfare and Care in a Late Modern or Postmodern Period: The Decline of the Social State and the Rise of "Community"

Welfare and Care in a Late Modern or Postmodern Period

Nikolas Rose (1996, 1999) has suggested that social policy, or the social state, has changed during the last several decades as we move increasingly into global, postmodern, and postindustrial periods. In place of the social state that tried to legislate the care of citizens, the nonsocial state is a new way to govern through the flow of ideas and knowledge that construct new cultural ways of thinking about ourselves as well as others and their care. While the social state was developed to take care of citizens, abstractly, in postindustrial "knowledge" and "networked" societies, now local communities and individuals take care of each other (see for example, Hallgren and Weiner's arguments here).

There is no longer a belief that state policy is adequate in the complex array of economic, political, and technological changes that are part of the process of global and international change. These ideas are expressed in the discursive practices related to decentralization, the private, individual autonomy and responsibility, as well as new discourses that focus on personal reflection, local action, flexibility, and choice. They are expressed in the notion of uncertainty and a flexibility and a readiness for change and instability, rather than a closure of certain care evidenced in earlier notions of social welfare. If we are to follow these ideas, welfare and care have shifted as we move into the twenty-first century. Technologies of the self, established as new ways of governing our self, our own families and communities, as well as new discourses of biological and technological control form new relations between power and knowledge. These discourses frame our own desires, actions, conduct, and the way we reason about caring for ourselves and for others. This reasoning also frames policy discourses of the World Bank, and of local/national governments. The discourses of modernity and a globalizing society continue to colonize our bodies and minds, as well as constraining the imagined possibilities.

The late-twentieth and early-twenty-first-century governmentalities continue to focus on individual choice, the natural good of the private sphere over the public, localized knowledge, and decision making over the "public" or "centralized" governing. The desire to highlight civil society, or community, to empower others, or to bring marginalized "voices" or nations into discussions also privileges this reasoning of public and private spheres and naturalizes private experience and community or individual "voice," while administering its freedom through discursively organized governing strategies.

Within this framing of welfare and care in the early twenty-first century, there are different ways of examining governing—as ways to equalize or to reproduce inequities, or as ways of reasoning that include while at the same time exclude. While the imaginations of citizenship, welfare, and care in a global democratic society (see Giddens, 1998) are available to all, or held up as ideals for many, they are governing discourses that appear inclusive, while excluding many. During the current period of uncertainty, when "empires" seem no longer (Hardt & Negri, 2000), it appears that local "community," involvement in networks of alliances, and freedom to choose are what is "needed."

But the imaginaries of equality, a local individuality, and "democracy for all" are not equally available to everyone, nor "reasonably" desired by all. Thus, while norms may be present for a Third Way (ala Giddens, 1998), or as a welfare to work policy that can fabricate autonomous, responsible, independent, and self-sufficient children, families, and nations, the cultural systems of reasoning that circulate are not possible for all, nor is an opportunity or choice available for the majority of the world. More important perhaps is the fact that current reasoning forecloses possibilities for new alternatives, new possibilities, and new ways to interrogate current governing discourses.

Using the multiple traditions of critical theory within education, one task of the contributors in this volume is to interrogate knowledge, reasoning, and power, as these are related to national and international governing practices. This volume attempts to follow such a tradition. By interrogating current practices and reason, the contributors question given assumptions, and critique current practices and the effects, to open up new possibilities for reasoning and action. We want to examine care, education, and schooling as well as new strategies of policing identity and conduct; we want to open up to new possibilities for well-being, as well as acceptance of the multiplicities of difference.

The Organization of the Volume

The sections of the book have been organized around three themes, although there are many overlapping issues that go across sections and the whole book.

The first section, "The Family and Child as an Object of Governing," includes chapters by Thomas Popkewitz, Kerstin Holmlund, and Inés Dussel that focus on cultural discourses as ways of governing how the family, woman, and child are to be constructed as normal, responsible citizens. Tom Popkewitz's chapter, "Governing the Child and Pedagogicalization of the Parent: A Historical Excursus into the Present," uses a Foucaultian cultural historical framework to question the reasoning of the present in the United States about parent and family education, involvement, and collaboration with schooling. He examines the cultural administration of the parent and child as a governing discourse he terms *pedagogicalization*. He examines the early-twentieth-century construction of the normal cosmopolitan parent, family, and child as part of governing discourses, and contrasts that with present notions of a responsible, autonomous, yet collaborating parent of today. He focuses on the inclusions/exclusions that were included in reasoning about the normal parent as having continuities from the past, but with critical ruptures in the discourses that construct the parent of today. Kerstin Holmlund's chapter, "Governing New Realities and Old Ideologies: A Gendered, Power-based, and Class-related Process," draws on a social historical framework to reconstruct the field of practice (Bourdieu, 1977/1996; 1991) in 1930s and 1940s debates by Swedish policymakers, members of unions, women's groups, and educators around childcare and early education. She uses notions of capital, habitus, language, and power to argue that the Swedish welfare state policy of childcare resulted from a struggle between different players who mobilized different discourses in effected policy. The final chapter in this section, by

Inés Dussel, focuses on "Educational Policy after Welfare: Reshaping Patterns of Governing Children and Families in Argentinean Education." It seeks to interrogate the effects of the 1990s reforms on how childhood and families are constructed. She interrelates international strategies of reform (decentralization, compensatory programs, school-based management, curriculum reform) and sets them against a homogenizing, state-centered system. Dussel illustrates how current discourses of individual poverty and "need" construct Argentina along two poles: as a diverse and pluralistic society, and one in which poverty and need are treated as individual problems rather than as effects of injustice or inequality. Discourses of the "needy," so often used in liberal and neoliberal secular and modernization discourses in other countries, are related to the ethics of care and salvation of the Catholic Church within the Argentinean context. The new discourses of governing the child and family, therefore, require careful attention as they travel, relocate, and take on meaning. She suggests that discourses about responsibilization and self-sufficiency, need, or diversity in Argentina must be carefully analyzed in terms of not only globalization of ideas, but in terms of their specific translation into the locally contingent historical and cultural space of reform within Argentina.

In the second section of the book, *The Embodied Social and Welfare State,* we examine how the current discussions of the universal rights of childhood, or the entitlements of the state, embed, often hidden, constructions of difference and abnormality. While the previous section gives reference to gender and sexuality, race, nation, and class, this section provides a more focused discussion of these categories. The chapters often merge different paradigms for their examination: the state as entity, the state as a grid of discursive reasoning, power as held and distributable or power in relation to knowledge, and systems of reasoning that appear inclusive, while also excluding.

Human Rights and Universal Entitlements

Ingeborg Moqvist in "Constructing a Parent" investigates how the current construction of the child—the child with human rights—demands a matching construction of the parent, and how this construction is made. Drawing on the notion of policing of families, Moqvist examines European policies and practices on education for parents. Particular narratives and images emerge about the good (post-) modern parent, and also a simultaneous determination to set a standard about a universal childhood and parenthood that at the same time sets standards based on a construction of historical and cultural difference. Loïc Chalmel's chapter "Early

Childhood Education: The Duty of Family or Institutions?" provides a way of historically considering the changing discursive reasoning about the normal family and child, and the rights and responsibilities of the state toward its citizens. He traces the emergence of different ideas about childhood and care and education from the Pietist roots of Comenius in the seventeenth century to the emergence of the kindergarten, crèche, and école maternelle models in the mid-nineteenth and early twentieth century in different European countries. Chalmel seeks to illustrate the common "roots" of a modern childhood that resulted, in different places, in the emergence of significantly different systems of "care" for young children. Chalmel illustrates the gender- and class-related assumptions that are embedded in the history of modern kindergartens, preschool, and childcare programs, and asks us to interrogate these systems and their embedded assumptions, as part of a search for a more unified "system" for the imagined community of "Europe."

Miriam David's "Teenage Parenthood is Bad for Parents and Children: A Feminist Critique of Family, Education and Social Welfare Policies and Practices" analyzes changing welfare discourses about teenage sexuality, pregnancy, and parenthood. She explores the challenges, complexities and contradictions in the shifts from regimes of traditional social liberalism to neoliberalism and the Third Way within Britain as well as the United States. She illustrates how new late-twentieth-century policy discourses relate to issues of social exclusion in response to changes in family life as well as global transformations. Along with others in the volume, she emphasizes the importance of the constructions of good or normal "family values" and good and bad conduct (e.g., individual personal responsibility) to frame interventions and policies targeting teenagers who are "at risk" of social exclusion.

Gaile Cannella's chapter "Child Welfare in the United States: The Construction of Gendered, Oppositional Discourse(s)" examines "oppositional (contemptuous and even punitive)" welfare discourse(s) in the United States beginning in the 1960s. This was a time period in the United States, Cannella argues, when civil rights and other social movements resulted in progress toward equity for racially and sexually marginalized groups. She uses a methodological bricolage to problematize child and family welfare policy across the twentieth century, including a variety of feminist theories to examine the history of welfare policy, and conservative political activities over the past 30 years. These policies continue to emphasize family personal responsibility and blame that is attached to welfare dependency. Cannella's chapter illuminates the activities of U.S. conservatives and "the right" in reconstructing a targeted entitlement that was to support families with dependent children, into a temporary assistance program based on class, race, and gendered portraits of the family, family values, and activities.

The themes of the final two chapters in this section, "Global/Local Analyses of the Construction of 'Family-Child Welfare'" by Marianne Bloch, and "Governing Children and Families in Kenya: Losing Ground in Neoliberal Times" by Beth Blue Swadener and Patrick Wachira, focus on the globalization and hybridization of welfare discourses within local geographical spaces. Bloch's chapter focuses on Western conceptions of the modern and universal child and family that embodies their opposite, conceptions of abnormality and difference. Using U.S., Hungarian, and Senegalese policies and practices, Bloch examines the rise of a certain imaginary of universal "rights" and "entitlements" of democracy and community welfare for all children, and for all humans in relation to the discourses of "individual, personal responsibility" and self-sufficient communities. While the global discourses of democracy and caring for others that are part of the new discursive environment appear inclusive, Bloch illustrates how the old, as well as the new, governing discourses also exclude ways of thinking and acting from the realm of possibility.

Swadener and Wachira's chapter analyzes the impacts of neoliberal or "post-socialist" global policies on children and families in Kenya. Framed in postcolonial critical theories, they link postcolonial and international structural adjustment in Kenya to policies promulgated by the International Monetary Fund, the World Bank, and others in many nations. They argue that the policies have increased inequities related to early education as well as other public services (health, public schooling) in Kenya, and poor countries more broadly. They problematize the largely Western constructions of self-sufficiency embedded within neoliberal and structural adjustment policies of welfare "reform" but argue that interdependence and cooperative economies have rich and remaining cultural legacies in Kenya and other African contexts.

In the final section of the book, *Limiting the Boundaries of Reason: New Possibilities/Impossibilities,* chapters by Gunilla Dahlberg, Bernadette Baker, and Camilla Hallgren and Gaby Weiner use different notions of *governing,* the *state,* and *power* to think of the "limitations" or "boundaries" around what is possible to think, as well as "possibilities" embedded in new governing discourses related to welfare, the child, family, and education. Dahlberg's "Pedagogy as a Loci of an Ethics of an Encounter" interrogates the notion of childhood, and pedagogical possibility as an ethics of encounter between parent, teacher, and child, as a way to deconstruct what we know of childhood. If "modern childhood" is based upon universalist, Western, and biological and scientific truths of normal child development, and pedagogies surround that knowledge, then Dahlberg asks us to abandon the known and knowable child, and open up new spaces for listening within teaching and pedagogical curricula. She draws on Deleuze and Quattari's (1999) notion of becoming, as well as Levinas's (1969) discussion of dialogue to interrogate truth and knowledge, to open up discussions of the multiplicities of possibilities that are unboundable, unknowable, and yet impossible without an interrogation of present discursive reasoning and practice. She draws on her interpretation of the Italian Reggio Emilia's early childhood program's pedagogical ideas of "listening" and "documentation" as illustrations of ways to reconceptualize thinking about pedagogical action and the ethical relationships they involve.

Bernadette Baker's chapter "*Hear Ye! Hear Ye! Language, Deaf Education, and the Governance of the Child in Historical Perspective*" shows the limits of reason about what is considered "human," educated, and literate by historically focusing on the discourses that order and produce the "hearing child" as human, and the deaf child as "different" (nonhuman). She reconstructs the historical imaginaries of normal children/humans and the education of the hearing/nonhearing child as part of a colonizing discourse related to western conceptions of humanity. By describing the limitations on reason, Baker's chapter uses a Foucaultian analysis of hearing, deafness, and disability that pulls apart present logic, and opens up new possibilities for action.

The final chapter in the volume, "The Web, Antiracism, Education, and the State in Sweden: Why Here? Why Now?" by Camilla Hällgren and Gaby Weiner returns us to a notion of the state whose practices can promote equality, or maintain inequities through its policies, and through different institutions and actors. The newly imagined "European Union" becomes the state actor that promotes new visions of social inclusion and exclusion. But, they argue, this vision involves cultural anxieties related to the changing nature of nation-state and the welfare and governing of the population of Sweden now made up of new immigrants from Eastern Europe, Africa, and the Middle East. The chapter focuses on the rise of cultural anxieties about "race" and a rising racism within schools and society through examining policies and programs that use the internet (using an illustration of "SWEDKID") to enhance a democratic pluralism, a notion of multiculturalism, and pedagogical dialogues about race and heterogeneity. The heterogeneity of cultural customs, languages, and ways of reasoning, they argue, constructs who is different and abnormal. The chapter allows us to think of the newly imagined fears associated with globalization, competitive economic pressures, population shifts, and technologies embedded in school pedagogies that normalize through assimilation, while also constructing identities associated with difference and otherness.

Note

1. Rose (1999) and Popkewitz and Brennan (1998), when speaking about governing and governmentalities, clearly would suggest that whether explicit or not, governing from a distance occurs even in the liberal welfare regimes, where "freedom" and privacy of the family is part of the philosophy and cultural reasoning systems embedded in the welfare model. The descriptions, therefore, are from the framework of the first conception of independent states, and a separate sphere for family, or the divide between public and private. The second tradition, related to cultural reasoning systems and power/knowledge relations, would break this binary apart.

References

Anderson, B. (1991). *Imagined communities: Reflections on the spread of nationalism.* London: Verso Press.

Apple, M. (1993). *Official knowledge.* New York: Routledge Press.

Baker, B. (2000). *In perpetual motion.* New York: Peter Lang Publishing Co.

Beatty, B. (1995). *Preschool education in America: The culture of young children from the colonial era to the present.* New Haven, CT: Yale University Press.

Bhabha, H. (1994). *The location of culture.* New York: Routledge.

Bloch, M. N. (1987). Becoming scientific and professional: Historical perspectives on the aims and effects of early education and child care. In Popkewitz, T. S. (Ed.), *The formation of school subjects: The struggle for an American institution,* Falmer Press.

Bloch, M. N., & Blessing, B. (2000). Restructuring the state in Eastern Europe: Women, child care, and early education. In T. S. Popkewitz (Ed.), *Educational knowledge: Changing relationships between the state, civil society, and the educational community.* (pp. 59–82).

Bloch, M.N. and Popkewitz, T. S. (2000). Constructing the Child, Parent, and Teacher: Discourses on Development. In Soto, L. D. *The Politics of Early Childhood Education.* New York: Peter Lang Publishers.

Bourdieu, P. (1977/1996). *Distinction: A Social Critique of the judgement of taste.* London: J. T. Press.

Bourdieu, P. (1991). *Language and Symbolic Power.* Cambridge: Polity Press.

Bradbury, B., Jenkins, S. P., & Micklewright, J. (Eds.). (2001). *The dynamics of child poverty in industrialized countries. UNICEF* and Cambridge, UK: Cambridge University Press.

Cannella, G. (1997). *Deconstructing early childhood education.* New York: Peter Lang.

Chakrabarty, D. (2000). *Provincializing Europe: Postcolonial thought and historical difference.* Princeton, NJ: Princeton University Press.

Channing, K. (1994). Feminist history after the linguistic turn: Historicizing discourse and experience. *Signs 19*(2): 368–404.

Dahlberg, G., Moss, P., and Pence, A. (1999). *Beyond quality in early childhood education*. London: Routledge Press.

Dean, M. & Hindess, B. (1998). *Governing Australia: Studies in contemporary rationalities of government*. London: Cambridge University Press.

Deleuze, J. (1979/1997). The rise of the social: Introduction to J. Donzelot. *The policing of families*. Baltimore, MD.: Johns Hopkins University Press.

Deleuze, G. and Guattari, F. (1999). *A thousand plateaus: Capitalism and schizophrenia*. London: The Athlone Press.

DeSwaan, A. (1988). *In care of the state*. Cambridge, UK: Polity Press.

Dimitriadis, G. and McCarthy, C. (2001). *Reading and teaching the postcolonial: From Baldwin to Basquiat and beyond*. New York: Teachers College Press.

Donzelot, J. (1979/1997). *The policing of families*. Baltimore: Johns Hopkins University Press.

Escobar, A. (1995). *Encountering development*. Princeton, NJ: Princeton University Press.

Esping-Anderson, G. (1990). *The three worlds of welfare capitalism*. NJ: Princeton University Press.

Esping-Anderson, G. (Ed.). (1996). *Welfare states in transition: National adaptations in global economies*. Thousand Oaks, CA: Sage Publication.

Foucault, M. (1977). *Discipline and punish: The birth of the prison*. (A. Sheridan, Trans.). New York: Vintage Press.

———(1980). *Power/Knowledge: Selected interviews and other writings by Michel Foucault*. (C. Gordon, Ed. and Trans.). New York: Pantheon.

———(1979/1991). Governmentality. In G. Burchell, C. Gordon, & P. Miller (Eds), *The Foucault effect: Studies in governmentality*. University of Chicago Press, 87–104.

Fraser, N., & Gordon, L. (1994). A genealogy of dependency: Tracing a keyword of the U.S. welfare state. *Signs*, 19, 309–336. (also in Fraser, N. (1997), *Justice interruptus*).

Fraser, N. (1997). *Justice interruptus: Critical reflections on the "postsocialist" condition*. London: Routledge University Press.

Gal, S. and Kligman, G. (2000). *Reproducing gender: Politics, publics, and everyday life after socialism*. Princeton, NJ: Princeton University Press.

Giddens, A. (1998). *The third way: The renewal of social democracy*. Malden, MA: Polity Press.

Gordon, L. (Ed.). (1990). (Ed.) *Women, the state, and welfare*. Madison: University of Wisconsin Press.

Gordon, L. (1994). *Pitied but not entitled: Single mothers and the history of welfare*. New York: Free Press.

Hacking, I. (1991). How should we do the history of statistics? In G. Burchell, C. Gordon, & P. Miller (Eds.), *The Foucault effect: Studies in governmentality* (pp. 181–196). Chicago: University of Chicago Press.

Hardt, M. and Negri, A. (2000). *Empire.* Cambridge, MA: Harvard University Press.

Hindess, B. (1996). Liberalism, socialism, and democracy: Variations on a governmental theme. In Barry, A., Osbourne, T., and Rose, N. *Foucault and political reason: Liberalism, Neo-liberalism and rationalities of government*. Chicago: University of Chicago Press.

Hobson, B. (1993). Feminist strategies and gendered discourses in welfare states: Married women's right to work in the United States and Sweden. In S. Koven & S. Michel (Eds.). *Mothers of a new world: Maternalist politics and the origins of welfare states* (pp. 396–430). New York: Routledge.

Holmlund, K. (1999). Cribs for the poor and kindergartens for the rich: two directions for early childhood institutions in Sweden (1854–1930). *History of Education, 28*(2): 143–155.

Holmlund, K. (2000). Don't ask for too much!: Swedish pre-school teachers, the state and the union, 1906–1965. *History of Education Review, 29*(1): 48–64.

Lazerson, M. (2001). *The price of citizenship.* New York: Metropolitan Books.

Levinas, E. (1969). *Totality and infinity.* Pittsburgh, PA: Duquesne University Press.

Lewis, J. (1993). Introduction: Women, work, family, and social policies in Europe. In J. Lewis (Ed.). *Women and social policies in Europe: Work, family, and the state* (pp. 1–24). Aldershot, UK: Edward Elgar.

Lewis, J. (1997). Gender and welfare regimes: Further thoughts. *Social Politics: International studies in gender, state, and society, 4*(2), 160–177.

Meyer, J., Boli, J., Thomas, & Ramirez, F. (1997). World society and the nation-state. *American Journal of Sociology* 103 (1), 144–81.

Michel, S. (1998). *Mother's Interests/Children's Rights.* New Haven, CT: Yale University Press.

Michel, S. & Mahon, R. (2002) (Eds.). *Child care policy at the crossroads: Gender and welfare state restructuring.* New York: Routledge Press.

Morgan, K. (2002). Does anyone have a "libre choix"? Subversive liberalism and the politics of French child care policy. In S. Michel, & R. Mahon, (Eds.). *Child care policy at the crossroads: Gender and welfare state restructuring.* New York: Routledge Press.

O'Connor, J., Orloff, A., & Shaver, S. (1999*). States, markets and families: Gender, Liberalism and social policy in Australia, Canada, GB and USA*. Cambridge, UK: Cambridge University Press.

Orloff, A.S. (1993). Gender and the social rights of citizenship: The comparative analysis of gender relations and welfare states. *American Sociological Review, 58,* 303–328.

———. (1996*). Gender and the welfare state.* Institute for Research on Poverty Discussion paper no. 1082–96. Madison: University of Wisconsin–Madison.

Polakow, V. (1991). *Lives on the edge: Single mothers in the other America.* Chicago: University of Chicago Press.

Polakow, V., Halskov, T. & Jorgensen, P.S. (2001). *Diminished rights: Danish lone mother families in international context.* Bristol, UK: The Policy Press.

Popkewitz, T. S. (2000) (Ed.), *Educational knowledge: Changing relationships between the state, civil society, and the educational community* (pp. 59–82). Albany: State University of New York Press.

Popkewitz, T. S., & Brennan, M. (1998). Restructuring of social and political theory in education: Foucault and a social epistemology of school practices. In Popkewitz, T. S. & M. Brennan (Eds.). *Foucault's challenge: Discourse, knowledge, and power in education* (pp. 3–38). New York: Teachers College Press.

Popkewitz, T. S., & Bloch, M. N. (2001). Administering freedom: A history of the present: Rescuing the parent to rescue the child for society. In K. Hultqvist & G. Dahlberg (Eds.), *Governing the child in the new millenium* (pp. 85–118). New York: RoutledgeFalmer.

Poulantzas, N. (1977). *Fascism and dictatorship. The Third International and the problem of fascism*. London: NLB .

Rothstein, B. and Steinmo, S. (Ed.) (2002). *Restructuring the welfare state: political institutions and policy change.* New York: Palgrave Press.

Rose, N. (1990). *Governing the soul: The shaping of the private self.* London: Routledge.

———(1996). The death of the social: Refiguring the territory of government. *Economy and Society, 25:* 327–366.

———(1999). *The power of freedom*. London: Cambridge University Press.

Said, E. (1978). *Orientalism*. New York: Random House.

Sainsbury, D. (1999). Gender, policy regimes, and politics. In D. Sainsbury (Ed.). *Gender and welfare state regimes* (pp. 246–275). New York: Oxford University Press.

Schram, S. (2000). *After welfare: The culture of postindustrial social policy.* New York: New York University Press.

Scott, J.W. (1992). Experience. In Butler, J. and Scott, J. W. (Eds.). *Feminists theorize the political.* New York: Routledge.

Skocpol, T. (1992). *Protecting soldiers and mothers.* Cambridge, MA: Harvard University Press.

Spivak, G. C. (1999). *A critique of postcolonial reason: Toward a history of the vanishing present.* Cambridge, MA: Harvard University Press.

Stiglitz, J. (2002). *Globalization and its discontents.* New York: W.W. Norton Press.

Torres, C. (2000). Public education, teacher's organizations, and the state in Latin America. . In Popkewitz, T. S.(Ed.) *Educational knowledge: Changing relationships between the state, civil society, and the educational community.* Albany: State University of New York Press.

Whitty, G., Gewirtz, S. and Edwards, T. (2000). New schools for new times? Notes toward a sociology of recent education reform. In Popkewitz, T. S. (Ed.) *Educational knowledge: Changing relationships between the state, civil society, and the educational community.* Albany: State University of New York Press.

Young, R. J. C. (1990). *White mythologies: History writing and the west.* London and New York: Routledge Press.

Young, R. J. C. (2001). *Postcolonialism: An historical introduction.* Oxford, UK: Blackwell.

SECTION II

THE FAMILY AND CHILD AS AN OBJECT OF GOVERNING

CHAPTER TWO

GOVERNING THE CHILD AND PEDAGOGICALIZATION OF THE PARENT

A HISTORICAL EXCURSUS INTO THE PRESENT[1]

Thomas S. Popkewitz

Introduction

The child, family, and community are sacred sites of modern politics and social welfare systems. Family values are what modern social reformers retort will bring national consensus and fight cultural disintegration. The health and the sanctity of the child are said to be pivotal for national preservation, social regeneration, and the progress of humankind. But the family and child are not just there to beguile and to recoup a paradise lost. The family, the school, and the community are historical sites of governing. This governing is not only the institutional procedures or organizational practices that provide the welfare "nets" for the family or to enable the education of the child. From the late eighteenth century, the child, family, and community have been subjects of regulating the intimate relations interests and aspirations as an instrument of regulating populations. "[T]he family becomes an instrument rather than a model: the privileged instrument for the government of the population and not the chimerical model of good government" (Foucault, 1978/1991, p. 100). The governing is embodied in linking of the development of the rationally ordered life of the child and family with the "political will" and progress of the nation.[2]

Pedagogy is the inscription device of governing. Its intellectual techniques map the interior of the child and parent to render them visible and amenable to government. The site of inscription is the mapping of the "reason" that orders the child's action and participation—the principles that order a child's problem solving, thought, and decision making. But the "system of reason" classified by pedagogy is not something of logic, or a psychology that describes what is innate and natural to "thought." The maps of thought are fabricated to regulate how judgments are made, conclusions drawn, rectification proposed, and the fields of existence made manageable and predictable.

This chapter examines the changing historical configuration of thought related to a cosmopolitanism in fabricating the "reason" of the child and family. Cosmopolitanism is related to eighteenth-century European Enlightenment and Protestant elite notions that a universal reason will bring progress and harmony to humanity.[3] Cosmopolitanism brought together the social projects of the scientific ordering of reason with the production of an individual who reasoned through science. Cosmopolitan universalized reason as a subject of government made possible the conditions of the modern state, its citizens, and the pedagogy.[4] The projects to calculate and govern the action of the child through cosmopolitan principles traveled in uneven ways from John Locke, John Dewey, Lev Vygotsky, among others.[5] But cosmopolitanism was never universal. The universal inscriptions of reason to bring progress embody distinctions that normalized and divided the qualities of the child who is "civilized" from those who are not.[6]

The first section focuses on the broad contours of the rules and standards of cosmopolitan reason inscribed in the fabrication of the child and parent in eighteenth- and nineteenth-century American schooling. The principles of pedagogy, I argue, are related to the governing conditions of the "welfare" state. By welfare state, I focus on the conditions of government that link the child and family actions to a salvation narrative about the cosmopolitanism of the nation, an American Exceptionalism. I consider the new sociologies and psychologies of pedagogy as inscription devices to govern the qualities of reason. The sciences of pedagogy order the principles of action and thus tame and regularize the uncertainties of the world of individual liberty and freedom. But the inscriptions that order thought also produce divisions between the *civilized* child and the anthropological "Other," the child without the qualities of the cosmopolitan "reason." The second section moves to contemporary narratives and images of the child, family, and community. Today's patterns of governing inscribe a system of reason of a cosmopolitanism as a lifelong learner who is problem-solving, active, flexible and self-managed. The parent is pedagogicalized as a surrogate teacher whose child-rearing practices follow the didactic principles of teaching children. Contemporary distinctions and differentiations about the qualities of the child who is cosmopolitan do not talk about civilized and uncivilized, but phrased differences in a language about the qualities of the child in need of remediation and rescue.

My focus on cosmopolitan reason is then a strategy to trace changing contours of ideas, institutions, and technologies that produce the distinctions that make the child and family in schooling over time. My concern is how thought plays a part in holding the contingency of social arrangements together and contesting the givenness of social life.

The Calculated Family and Child: Governing Reason in the Eighteenth & Nineteenth Centuries

This section focuses on the cosmopolitan reason that orders the child and family in pedagogy in the late eighteenth to the nineteenth century. One element of the calculation of reason is to order action in a world that no longer has certainty. Liberal modes involve a new calculus of governing through the liberty and freedom of the individual. Harmony and stability are in the rules of reflection and action as the external world is bound by an uncertainty, as the future is not foretold. Second, the social and educational sciences provide inscription devices (theories and methods of child and family studies) to specify, measure, and administer the characteristics of the "reason" of action. Third, notions of "reason" and the "reasonable" person embody redemptive discourses about freedom that separates those who are constituted as "civilized" from those who are "uncivilized."

The Cosmopolitan Child, The Responsible Parent, and the Disciplining of Reason

The narratives that connected the child, the family, and the home have an interesting genealogy. In the Middle Ages and ancient Greece and Rome, for example, parents could conduct themselves spontaneously. Parents were influenced more by their own interests than by what their thoughts and actions meant to children. Children lived with parents, with no architecture in the household to separate them from adults, and no idea of constraints and responsibilities for parents to "develop" children through calculated methods of child-rearing.

The home and the family provided analogies to connect the local and immediate with the broader society in which the individual "belonged." Authority and liberty flowed from the structure of personal relations and not from the political organization of the society. The family was the model for describing most political and social relationships, not only between the king and subjects but also between superiors and subordinates. The American early colonial situation, for example, defined the family or household as the basic institution in society in which all rights and obligations were centered.[7] Puritans called the family the little commonwealth. It was the fundamental source of community and continuity, the place where most work was done, and the primary institution for teaching the young, disciplining the wayward, and caring for the poor and insane. The notion of the family reached even into the courts. Criminals were treated as wayward children (Wood, 1991, p.72).

The family ceased to be simply an institution for the transmission of a name and an estate from the eighteenth century onward. At first, elite families "saw" the child no longer simply as a little adult but as having a life distinct from parents that required regulation and development. Parents had now formal obligations in the development and civilization of the child (see Ariès, 1960/62; also see Elias, 1980/1998).[8] Children no longer mixed freely in adult society but were segregated to prevent premature contact with servants and other corrupting influences. The child was regarded as a person with distinct attributes—impressionability, vulnerability, innocence—requiring a warm, protected, and prolonged period of nurture (Steedman, 1995). A new expertise emerged to advise parents on child-rearing (Wood, 1991 pp. 148–149). Educators and moralists in the years around the American Revolution stressed the child's need for play, for love and understanding, and for the gradual, gentle unfolding of the child's nature. Parents were criticized in the literature for placing family pride and wealth ahead of the desires and integrity of their children. Children became moral beings to be cared for, and educated to have agency.

The new theories of the "home" and childhood embodied cosmopolitan values. The child was to personify the universal values of reason that enabled the freedom and liberty of the individual.[9] The values of childhood presupposed, at one level, a global humanity in which the child has inalienable rights that transcend nations but are protected through particular communities. Child-rearing became more demanding as a result, and emotional ties between parents and children grew more intense (Lasch, 1977, pp. 5–6).[10]

The child-rearing practices were to provide a child who developed and ordered life through a cosmopolitanism that was tied to the nation. The child's freedom was to enable the promise of America to be realized through the agency of the individual. Drawing on a reading of Locke into American cultural projects from the mid–eighteenth century, coercion was viewed as sometimes necessary, but not preferred, for effective long-lasting parental authority. Liberty was to come with the increase of years, and the child had to be gradually trusted with his own conduct. The parental relation was to win respect and esteem of their children through reason, benevolence, and understanding. Affection rather than force was to bind parents and children together (Wood, 1991, p.151).

Parental child-rearing and pedagogy were to govern the *soul.* While many might object to the interjection of the idea of the soul, as the European Enlightenment, for example, sought to discredit religion, a controlling impulse in the American Revolution and the early Republic was the merging of religion and the state (Ferguson, 1997, p. 21). The state was evidence of "divine sanction" that was inscribed in Puritan images of the commonwealth (Bercovitch, 1978). The administration of the child and family in the school was to manage conduct and "mold" character for social ends that embodied this divine sanction. School, as expressed by Superintendents of Schools for the Office of Education in 1874,

> is a phase of education lying between the earliest period of family-nurture, which is still a concomitant and powerful auxiliary, on the one hand, and the necessary initiation into the specialities of a vocation in practical life, on the other. In America, the peculiarities of civil society and the political organization draw on the child out of the influence of family-nurture earlier than is common in other countries. The frequent separation of the younger branches of the family from the old stock renders family-influence less powerful in molding character. The consequence of this is the increased importance of the school in an ethical point of view. (Government Printing Office, 1874, p.13)

The studies of the child took the imaginary of the nation as a governing principle of pedagogy. Edward Ross, one of the early sociologists, argued in the 1920 book *The Principles of Sociology* that no institution is as important as the school in the taking of diverse populations and making them into one social entity that would comprise the nation: "Thoroughly to nationalize a multitudinous people calls for institutions to disseminate certain ideas and ideals. The Tsars relied on the blue-domed Orthodox church in every peasant village to Russiafy their heterogeneous subjects, while we Americans rely for unity on the 'little red school house'" (cited in Franklin, 1986, p. 83).

The pedagogical practices were to domesticate the child in an image of virtue related to seemingly unified American culture. Romantic and religious moral visions combined with science to calculate the processes of the mind in which virtue could be produced through education. G. Stanley Hall (1905/1969) viewed the new discipline of psychology and child study as replacing moral philosophy in the challenge put forth by the materialism of Darwin. He spoke of *the soul* in describing the focus of psychology and the relation of psychology to pedagogy. For Hall, psychology was a way to reconciling faith and reason, Christian belief and "Enlightenment empiricism." Embodied in the notion of human nature was the soul of the child to be nurtured and directed in the world of liberty and social contingencies. The cosmopolitan child into the twentieth century was a *socialized* individual who expressed the cosmopolitanism of the nation in the universalized laws of child development and the child-rearing practices of the scientifically organized family.

Educational Sciences, Taming Chance, and the Calculated Family

The care of the soul of the child for social ends was embodied in the projects of social and educational sciences that emerged in the nineteenth century. British and American social sciences, for example, combined Christian ethics and notions of salvation with strategies to understand, govern, and sometimes to missionize populations in the name of freedom and cosmopolitan values. Research and social planning would rescue immigrants and marginalized people for "civilization" through remaking their sense of belonging and collective "home."[11]

The family embodied the metaphor of *the home* as a redemption and salvation narrative of the nation in the new social sciences.[12] The family became the earliest and most immediate place for socialization and the paradigm of the self-abridgment of culture, and thus a way of expressing fears and anxiety of the threat to a national identity. The connections between child, family, and community in social policy, health, social science, and schooling joined the metaphorical "American family" of the nation with the family and the child in the construction of identity (Wald, 1995).

While family was placed as a private rather than public sphere, the family was an administrative instrument for regulating populations. Lasch (1977) argues that as rule of force gave way to rule of law, social relations took on increasing importance, and, in the United States, were tied with the notion of civil society as something distinct from the state. But the rise of a concept of civil society mediated a pattern of governing as agencies outside of the family expropriated and reformulated parental functions through an abstract, impersonal, evolutionary process known as the "transfer of functions" (Lasch, 1977 p. 25).

Parents, under the guidance of the new social theories of health, would develop altruistic instincts that expressed self-obligation and self-responsibility in their children (see Lasch, 1977). The school was narrated in the image of the family, yet had to supercede its norms and cultural values in order to produce the citizen that would guarantee the future of American progress. The teacher was analogous to the mother, but where the tasks of the teacher were to fabricate the child who embodied the rules and standards of a cosmopolitan "reason." Teacher education, for example, was to select self-motivated and morally devoted candidates who would shape the character of the child. That child was to embody the norms and dispositions that embodied the American exceptionalism that was cosmopolitan in orientation and detached from local, provincial, and ethnic interests.

The government of freedom, Rose (1999) argues, involved the invention of a range of technologies that enabled the family to inscribe the norms of public duty while not destroying its private authority.

> The government of freedom, here, may be analyzed in terms of the deployment of technologies of *responsibilization*. The home was to be transformed into a purified, cleansed, moralized, domestic space. It was to undertake the moral training of its children. It was to domesticate and familiarize the dangerous passions of adults, tearing them away from public vice, the gin palace and the gambling hall, imposing a duty of responsibility to each other, to home, and to children, and a wish to better their own condition. The family, from then on, has a key role in strategies for the government of freedom. It links public objectives for good health and good order of the social body with the desire of individuals for personal health and well-being. A "private" ethic of good health and morality can thus be articulated on to a "public" ethic of social order and public hygiene, yet without destroying the autonomy of the family—indeed by promising to enhance it. (p.74, italics in original)

The notion of community was an important site in connecting the intimate relations and inner capabilities of the child and family with cosmopolitan images and narratives of the nation. The "Community Sociology," developed at the University of Chicago during the early decades of the twentieth century, for example, connected the family, marriage, and urban conditions. Working with civil service bureaucracies, the labor movement, state agencies, the new philanthropic agencies, feminist movements of the times, and the faculty at the University of Chicago (most notable was George Herbert Mead and John Dewey), community studies gave direction and a decidable structure to the cultural practices that related the child and family to a collective belonging and home.

The civic ideal of community mobilizes the sublime and secular as progress. The sublime is the home of God (Cronon, 1996). In the early nineteenth century, the landscapes of nature were the places where one had a better chance than elsewhere to glimpse the face of God's creations. The installations of ordered gardens and parks of nineteenth-century urban planning were to make communities function in relation to God. But the notion of community also expressed the return to the conditions in which neighbors form the democratic institutions viewed as predating modernity. The reinstalling of face-to-face interactions through theories of the family and community were to produce democracy and its images of the sublime as an American identity among diverse ethnic communities of immigrants. Dewey's notion of community provided "a civilizing process" to make the dispositions and sensitivities of the individual "intrinsic and constructive" so as "not to submerge individuality in mass ideas and creeds to produce a reverence for mediocrity" (Dewey, 1922/1929, p.479; 1916/1929). Science was central to the making of democracy through the progressive school, in which the classes were "grouped for social purposes and with diversity of ability and experience is prized rather than uniformity" (Dewey, 1928, p.199).

The "primary group" was an intellectual technique to render the characteristics and relationships of the child, community, and family as visible and amenable to government. The primary group secularizes religious notions of community and salvation through calculating the close relations that fabricated the civilized person (Greek, 1992). The work of Charles Horton Cooley, one of the leaders of the Chicago School of Sociology, for example, saw the family and the neighborhood as providing the proper socialization through which the child could lose the greed, lust, and pride of power that was innate to the infant, and thus become fit for civilization. The communication systems of the family would, according to Cooley, establish the family on Christian principles that stressed a moral imperative to life, and self-sacrifice for the good of the group. Parents, under the guidance of the new social theories of health, would develop altruistic instincts that expressed self-obligation and self-responsibility in their children (Lasch, 1977).

The narratives of the family, the child, and the school were embodied in modern pedagogy. The legal doctrine of *parens patriae,* for example, is introduced in the early twentieth century to establish a regulatory relationship in the law between schooling, the child, and the parent (see Grubb & Lazerson, 1981; Richardson, 1989, pp.77–78). Pedagogy was a "civilizing" process to prevent the child from straying into "barbarism," to quote Horace Mann, the first secretary of a state's (Massachusetts) Board of Education and a leader of the public school movement in the nineteenth century. Mann saw the duty of education to guarantee the harmony and stability of the republic by making the child who would ensure the future of the civilization embodied in the American Republic and democracy. The administration the child in the family was to inscribe rules and standards of reason that ensured a liberal democracy bound to participation in which the future was always conditional. But the ordering of the family as a primary group and "community" also took "the ideals of mutual aid and democratic freedom as related to . . . the source of American institutions and racial inheritance" that were to shape a particular national character through focusing on the psychological qualities of a human nature (Ross, 1991, pp. 264–265).

The shaping and inculcation of "private" responsibility assign a key role to experts. Science was to provide methods to fabricate the freedom of modern individuality through the child and family (see, Franklin, 1986). One of the founders of American sociology involved in the problem of education, Frank Lester Ward, argued that government and the social administration of individuality was necessary to harness liberty through directing the individual "to act as desired" (Ward, 1883, p. 233). Ward was concerned with bringing order and regularity to the conditions of a democracy that produced uncertainty. To be an active member of society is to participate and thus make decisions about the direction and progress of the civilization itself. This action is not random but based on identifying the fundamental laws of human action that direct freedom. Science, Ward continues, provides the relation between discipline and freedom—laws and indeterminacy. Science provides the ordering of the contemplative "man." Education is to provide for the ordering of society and modifies it by allowing for the artificial construction of evolution.[13] He rejected what he called "the direct method" of controlling natural forces, as it "has accomplished comparatively very little" (Ward, 1883, pp.159–160). In its place is "the indirect method," which is to produce a civilized cosmopolitan individual. Progress in government was not simply an education to accommodate society but "must be in the direction of acquainting every member of society more thoroughly with the special nature of the institution, and awakening him to a more vivid conception of his personal interest in its management"(Ward, 1883, p. 243).

Ward's indirect method was to produce a desiring agent who joined a cosmopolitanism with a particular American exceptionalism. The American exceptionalism embodied an ongoing process of interaction that included a self-realization of personal fulfillment through pragmatic, scientific solutions to social problems (for a general discussion of freedom in the American context, see, e.g., Foner, 1998). The Exceptionalism told a saga of a generalized American race whose nation was the natural and progressive evolution of humanity and that separated the people of the nation from the savage (Wald, 1995).

Redemption Through Governing the Same/"Other"

Embodied in American Exceptionalism is the Other. This is oddly evident if the changing focus of the idea of liberty as the "pursuit of happiness" is considered as an object of the organization of government and society. The Enlightenment ideal of the cosmopolitan embodied a universal individual whose freedom is to "pursue happiness," which government was to both enable and not restrict. This notion gave emphasis to the strength of civil society and the limitations on government restricting freedom. By the end of the nineteenth century, however, the ideal of the pursuit of happiness was reversed to focus on those who were deemed unhappy, almost as though government had taken up the mantle of means-testing of populations before it became fashionable in welfare policy. The "unhappiness" of populations stood as an oppositional category of groups residing outside the norms of civility and civilization.

The turn to populations not able to participate in the pursuit of happiness, "the unhappy" populations, is found in the shift in the notions of poverty. The initial focus on environmental conditions that produced poverty and thus prevented participation was transmogrified into the personal and psychological characteristics of the child and family that made for unhappiness. That soul required pastoral care. Measurement theories, psychological testing, and notions of child development took the notion of the Great Chain of Being, a term that had circulated in the nineteenth century, and made into norms of development that established a hierarchy of an evolutionary racial history (Lesko, 2001). For some of the early social scientists and social reformers, the Old World cultures had to be destroyed through resocializing the immigrant family and child. But for others, salvation narratives were constructed with the city and the immigrant groups producing the ideal of the future. A text for Protestant missionaries in New York (Grose, 1906), for example, wrote about the problem:

> *Unguarded Gates*
> Wide open and unguarded stand our gates,
> And through them presses a wild, motley throng—

> Men from the Volga and the Tartar steppes,
> Featureless figures of the Hoang-Ho,
> Mayalan, Scythian, Teuton, Celt, and Slav,
> Flying the old world's poverty and scorn;
> These bringing with them unknown gods and rites,
> In street and alley what strange tongues are these,
> Accents of menace alien to our air,
> Voices that once the Tower of Babel knew! (p. 3).

The fear of unguarded gates evokes a military language of protecting against the invasion of the uncivilized hordes that crossed over from Europe. There was the evangelistic hope of bringing the Christian gospel to the heathen in America. That gospel mixed religious notions of salvation (a generalized Protestant Christianity) with the notions of an American exceptionalism. "How long will American Christianity allow this process of degeneracy to go on, before realizing the peril of it, and providing the counteracting agencies of good?" (Grose, 1906, p. 224). For some there was an acceptance of the immigrants in the New World, a fear that the laws of equitable conditions for the new immigrants were not enforced (Grose, 1906, p. 66) and of intolerance toward racism as some groups "brutally abuse the immigrants" (Grose, 1906, p. 10).

The uncivilized were not fixed and stable notions of populations or individuals but were distinctions of the "Other" that continually changed. At different points and with some overlapping, the "uncivilized were the immigrants from Southern and Eastern Europe; African American migrants from the South after the American Civil War; the poor populations of the cities; and the Rockefeller's and Mellon's, as "the Robber Barons" of the turn of the twentieth century whose excessive individualism of capitalism seemed to undermine Christian values among the new professionals of the social sciences.

The family was the most immediate place for expressing the fears and anxieties of the making of the immigrant into a civilized individual, as illustrated in the community studies of sociology. History, fiction, or pedagogy dwelt on issues of familial responsibility and warned against the evils of parental tyranny through harsh and arbitrary modes of child-rearing of an older, more savage age. Some of the early political scientists, for example, spoke about "race suicide" as a national problem produced through immigration but understood and represented the threat as a challenge, literally and metaphorically, to the American family (Wald, 1995). MIT President Francis Amasa Walker, a statistician involved in the censuses in 1870 and 1880, explained how an outcome of immigration is the destruction of the American family. John R. Commons, a prominent progressive economist, raised the question of race suicide through the analogy of the family and gender, heeding Teddie Roosevelt's warning that "if the men of the nation are not anxious . . . to be fathers of families, and if the women do not recognize that the greatest thing for any woman is to be a good wife and mother . . . that nation has cause to be alarmed about its future" (in Wald, 1995, p. 245).

Social and educational theories about the family and the child classified and divided the poor and immigrant families who were outside the psychological and social conditions of normalcy.[14] Studies abounded to define the disintegration of the immigrant family. The domestic science movements, for example, were directed to the rationalization of the home of the urban family that lived in unhealthy social and psychological conditions. The "muckraking" reporting of the urban conditions and the community sociology of immigrants focused on the environmental impediments for the pursuit of happiness.

The formation of childhood and the family are central to Dewey's concern with a cosmopolitanism threatened by immigration at the turn of the twentieth century. Dewey (1902) decried the too rapid "de-nationalization" of immigrant children: "They lose the positive and conservative value of their own native traditions, their one's own native music, art, and literature. They do not get complete initiation into the customs of their new country, and so are frequently left floating and unstable between the two. They even learn to despise the dress, bearing, habits, language, and beliefs of their parents—many of which have more substance and worth than the superficial putting on of the newly adopted habits" (in Wald, 1995, p. 247). Respect of the immigrant cultures would produce a healthy American that pragmatically worked toward the future.

To this point, I have explored a particular image of the cosmopolitan as a governing principle that overlaps with that of an American exceptionalism. The future of democracy and the republic were guaranteed through regulating the inner qualities of the child.[15] Reason was the inscription device. It tamed change and the chance produced through regimes of a participatory politics that inscribed particular universal norms of humanity—a cosmopolitan individual. The cosmopolitan child into the twentieth century was a *socialized* individual who performed according to universalized laws of child development, the child-rearing practices of the scientifically organized family, and the legislative expertise of the teacher who used "indirect methods" to fabricate national imaginaries and a collective sense of belonging. But in the notions of exceptionalism and its civilizing practices were also the principles of regulating the populations who embodied the conditions that produced "unhappiness." The personal and family characteristics of "unhappiness" become sites for the cultural practices of the governing of conduct. Today that unhappiness is universalized, as every family and child is "at-risk" and potentially a site for counseling their "unhappiness," although some deserve more attention than others.

Reconstituting the Child, Parent, Community, and the Teacher: Reconstituting Freedom at the Turn of the Twenty-first Century

Today's cosmopolitanism in the reason of the parent and child embody different internments and enclosures than those discussed in the previous section. I discuss these first, through the new salvation narratives of democracy and freedom as a child is prepared for the future in a *knowledge society* and *globalization*. Reason, as earlier, is to govern and tame that future. The second section places the narratives of democracy with those of the problem-solving, lifelong learner. Whereas the cosmopolitanism of the child of the turn of the twentieth century was to live as a *socialized* individual who embodied the national exceptionalism, today there is little talk of socialization. The cosmopolitan child lives in networks of communicative norms that order the classroom and family through a problem-solving, active, flexible, and self-managed lifelong learner. The third and fourth sections trace the narratives of the new cosmopolitanism of the lifelong learner with other cultural practices. The regulatory ideals of *community* and the *pedagogicalization of the parent,* for example, reclassify the principles of the home to connect the scope and aspirations of public powers with those of the personal and subjective capabilities of the ethical conduct of life of the child. But the new global cosmopolitanism brings into focus different contours of "the Other," with the new barbarian as one who does not embody the norms of the lifelong learner. The new boundaries of action make resistance and revolt more distant and less plausible as the individual is in continual self-development and perpetual intervention.

The Regulation of the Present for the Future Redemption of the Nation's Democracy

The pedagogical expertise is the agent of democracy that remakes the child to ensure the nation's survival. The teacher is the transforming agent. The American Council on Education's report (1999), an association of the presidents of the leading U.S. research universities, evokes the future in its title, *To touch the future: Transforming the way teachers are taught*. The risk is prophetically pronounced as a "wide-spread consensus [that] has held that the nation's schools can and must serve better the citizens of our democracy and that the quality of teaching is not what it could or should be" (p. 1). The future salvation of the nation depends on reforming the teacher by the expertise of the university and its sciences of the child and teacher.

Progress is embodied in a universal political philosophy about the natural, inalienable *birthright of the child* who now has the independent rights of the citizen. No longer are the school and family only to socialize the child to become an adult. The child stands as a fabrication that has its own rights, obligations, and responsibilities that the teacher nurtures. As one author outlined, the "reasonable" parents "express love and affection and are responsive to their child's needs and request" (Chase-Lansdale & Pittman, 2002). The unhappiness principle is now spoken of as not serving the natural *birthright* of the child, whose needs and requests stand as independent of others. The globalized individuality embodies distinctions that divide, as some children and families can never "be of the average."

Fabricating the Lifelong Learner: the Freedom of the Disciplined Autonomy

There is a new sense of choice embodied in the family, child, and teacher. Each acts as a lifelong learner. The lifelong learner is one who is "problem-solving" while actively, cooperatively, and continuously engaging the world to understand differences and to master the content of school subjects. No one escapes being a lifelong learner.

The lifelong learner embodies a new aesthetics of the ethical self that does not have a permanent "home." There is a continual remaking of oneself through the active intervention in one's life as a lifelong learner. Life becomes a continuous course of personal responsibility and self-management of one's risks and destiny—the "autonomous learners" who are continuously involved in self-improvement and are ready for the uncertainties through working actively in "communities of learning" (see, e.g., the National Council for Teachers of Mathematics, 2000).

Freedom is choice. The lifelong learner is talked about as the empowered individual who continually constructs and reconstructs one's own practice and ways of life through a perpetual intervention in one's life. The flexible, collaborative, problem-solving individual is someone who can chase desire and work as a cosmopolitan in a global world in which there is no finish line. One can refuse allegiance to any one of the infinite choices on display, except the choice of choosing.

Problem solving is the salvation practice. The notion of problem solving has a resonance with Dewey, that of focusing on the psychological dispositions and capacities for making reasonable choices and for "constructing" knowledge. But there are substantive differences that relate to the narratives and images of a lifelong learner, as the latter is not an imaginary of John Dewey. Classroom research in teaching, for example, studies how communication patterns "capture" children's thinking about problems. In one research project on mathematical thinking, the research on children's problem solving is to provide the classifications that order teaching. The teacher is to make a child who problem-solves by knowing where "connections are formed between new information and existing knowledge structures or when new information leads to cognitive conflict and therefore, to the reorganization of existing structures in order to resolve that conflict" (Warfield, 2001, p.137).

But problem solving is not some natural process found in the child through psychology and pedagogy. Rather, particular sets of distinctions and differentiations order the mind and social interactions for the social administration of the child. Research on classroom interaction and children's thinking, for example, classify and order the properties of the mind that the teacher is to work on. The effective teacher, as one research report argues, incorporates the classification systems of the research about how children think into the principles for organizing and evaluating classroom lessons.

The reforms of the lifelong learner are told as sagas of democracy, modernization, and the globalization of the nation. It is also told of individual empowerment and emancipation. The problem solving of the lifelong learners reaches into the characteristics of the dispositions and sensitivities of the teacher. The teacher works collaboratively in partnerships to rescue the child. The teacher is a decision maker who is "empowered" and given "voice" through partnerships with communities and parents.

The problem solving is a calculus of intervention and a displacement of the ethical obligation of the child. The teacher assesses the processes of learning and problem solving to calculate and supervise the making and remaking of "self" and the child's biography. The teacher observes the child's problem-solving processes from a constructivist standpoint in which there are multiple paths to attain answers. The process and choices are what is important to teaching. The teacher is also an action researcher who reworks herself and the child through a continual construction of life histories or portfolios. Revelation is in the problem solving that is therapeutic. It administers personal development, self-reflection, and the inner, self-guided moral growth of a better managed, healthier, and happier individual.

School reforms are to implement remedies that guarantee a future of the nation free of barbarians, but today's barbarians are related to a fear of nonreason and the nonreasonableness of particular children, families, and communities. The barbarians' threat to society is in not preparing the child adequately for being the citizen of a globalized world; that is, a lifelong learner who works on him or herself through self-improvement, and autonomous and "responsible" life conduct. The nation is at-risk, to quote a title of an important U.S. Department of Education report. Democracy is held in the balance. Pedagogical practices are to administer the soul as an individual, and a "lifelong" learner.

Lifelong Learning in Community

The revelation of problem solving is, as in the nineteenth century, in a community. The metaphor of community is evoked in the U.S. reforms of site-based management, home-school collaboration, and curriculum discussions of the classroom as a "community of learning," again not part of the mix of the previous reforms. The current notion of community brings together multiple salvation themes of progress. Community seems to have a universal value of goodness and progress through a promise of multiple values of individual and collective progress. Community emphasizes the idea of the autonomous citizen, whose collective actions constitute the moral good. Diversity, difference, and multiculturalism are embraced within this notion of community. Teaching standards in the U.S. reforms, for example, relate educating "all children" and "diverse populations" with involving parents through collaborating with communities. Teachers are asked to go into the "community," to become part of communities to "better know" their pupils and their families, to become trusted, or to "know" what they should include from "community knowledge" (Borman, 1998; Delpit, 1995).

If I can return to a distinction about the notion of community from the earlier discussion, the turn of the century notion of community was to socialize the individual into a universal social whole. All social classes were to embody the norms of a universal *cosmopolitan self.* The hyphenated Italian, Polish, Irish-American emphasized the unified "home" and commitment of "self" to the end of the phrase, the American. Today's communities embody different distinctions, divisions, and principles of a home. The community spoken about today is one where the problem-solving, lifelong learner forges emotional bonds and self-responsibility circumscribed through networks that produce a *belonging.*

This version of democracy and freedom in community seems to have no collective social identity except in the collection of different communities themselves. Where the self-management of the early part of the twentieth century linked the child and family rules to the universal collective national sagas, there seem no externally validated social morals and obligations. The empowerment of the lifelong learner is to construct and reconstruct one's own "practice," participate in collaborative problem solving, and to be self-managing in the autonomous ethical conduct of life (see, e.g., Rose, 1999; Rose & Miller, 1992). But it would be incorrect to talk of the demise of the social as the demise of collective identities. Current reforms of the child, the family, and the school reconfigure the ethical and cultural imaginaries in which political aspirations are fabricated. The principles in governing conduct make resistance and revolt more distant and less plausible as it is the *self's* capacities and potentialities of a lifelong learner that is the site of perpetual intervention (see, e.g., Boltanski, 1999; Rose, 1999).

The Pedagogicalization of the Parent and Communities of Partnerships: Fashioning Territories of Nonmembers

The salvation themes of the lifelong learner, partnership, and community fashion a double map. One is a grid of the cosmopolitan child and family discussed above—the lifelong learner. This grid of values and characteristics is found in the media and in the popular handbooks for child-rearing on the shelves of the bookstores. The second is a map of the child and family who are not cosmopolitan, in need of remediation and, as I argue below, mental health counseling. The second map is of rescuing the child of poverty whose failure is determined by psychological and family characteristics. The sets of distinctions and divisions are about the child (and parents) whose capabilities and capacities do not "fit" as problem solving, reflection, and life long learning.[16]

There is a universal language about justice and equity through the remaking of the parent who develops that children as a lifelong learner. A report calling for national teacher standards, *What matters most: Teaching for America's future,* for example, decries the failure of the past in its commitments to a democracy and offers rectification through the reform of the teacher who works with parents. The David and Lucille Packard Foundation journal, *The future of children* (2002), asserts that the future of an egalitarian society requires parent involvement in education to prevent children's failures from inadequate parenting (Shields & Behrman, 2002).

While the language is inclusive and about *all* children, the distinctions and differentiations are about children and families that fail (see, e.g., Hidalgo, Siu, Bright, Swap, & Epstein, 1995; U.S. Department of Education,

2001). There is talk about reforms as programs where *all children learn* and there is to be *no child left behind*. The distinction about *all children* however involves a textual division that names the qualities and characteristics of those who do not qualify as among *all* the children. It is the children who are not in *the all* that are the targeted populations for social policy to rescue from failure.

Who are these children who are outside the norms that constitute *all children?* Finer distinctions are made to build profiles of the failing child and family that lie outside the values that classify the *all children* (for discussion of making humankinds, see Hacking, 1995). The categories of the child who does not "fit" are ones "who lives in poverty, students who are not native speakers of English, students with disabilities, females, and many nonwhite students [who] have traditionally been far more likely than their counterparts in other demographic groups to be the victims of low expectations. . . . [This child needs further assistance, for example,] to meet high mathematics expectations, such as non-native speakers of English and students of disabilities who need more time to complete assignments" (The National Council for Teachers of Mathematics, 2000, p.13).

The categories of the child intersect with the characteristics of the family that is differentiated as non-functioning with labels such as the "fragile family," and the "vulnerable families" (Hildago, Siu, Bright, & Epstein, 1995, p. 500). The parents have a lower level of education and socioeconomic status; are immigrants (according to the length of time living in the country); live in poor areas of residence; and are ethnically defined (living or not living in ethnic enclaves), among others (Hildago, Siu, Bright, Swap & Epstein, 1995, p. 501). The social and economic classifications of the child and family are linked to structural relations of gender, such as whether the mother is a single or teen parent (Hildago, Siu, Bright, Swap, & Epstein, 1995, p. 501, David and Lucille Packard Foundation, 2002).

A digest of personal facts fashions territories of membership and nonmembers in the *all* children. The categories of the different populations who do not "fit" the norms of success become a particular human kind. The various categories of the child and family are a determinate classification of deviance that has succinct chronological, cultural, physiological, and psychological characteristics. The aggregate of the "fragile" and "vulnerable" family acquires the abstraction of science or the impersonal management to reason about group and personal capabilities and capacities.

The problem of failure is placed in the psychological and communicative interactions of the child and parent of children who are classified as minority or poor. These parents do not have "high expectations" for their children. The parents lack the work norms and expectations needed for school success. The patterns of *proper* conduct and interaction seem to appear as if taken straight out of a teaching manual. Parents receive training in one program for helping their children develop critical thinking skills, evaluating their children's educational progress, and helping with homework and project assignments.

The successful parent is a pedagogical one. The parent is one who "learned, for example, how to use a list of common words to help children make sentences, learn grammar, and sharpen their reading skills; they also learned how to use a 'number line' manipulative to help children practice adding and subtracting" (U.S. Department of Education, 2001, p. 11).

Whereas the turn of the past century was to provide parents with technologies to calculate proper child development, contemporary research and policy *pedagogicalizes the parent.* Parenting is a surrogate to schooling. The cultural patterns of the classroom are the model of action for the parent. Parents are to produce the cognitive and affective learning skills of the school to ensure children's learning and development (Bloch & Popkewitz, 2000). Parent participation is to produce social progress through making better readers, more positive attitudes about school, improved attendance, and better homework habits (see, e.g., Eldridge, 2001; U.S. Department of Education, 2001).

These networks involve partnerships. The partnership is to bring parents and school together as a seeming expression of empowerment and social solidarity. Partnerships stand as a strategy to form the intimate ties of community to patterns of belonging and a home where individuals can be self-actualized as lifelong learners. The greater harmonization brought by educational standards and public partnerships, for example, is to create a lifelong learning that, Giddens (1998) argues, is the friend rather than the enemy of diversity.

The U.S. Department of Education (2001), for example, published exemplars of programs that are successful parent/school partnerships. The twenty programs illustrate training parents for participation, teaching children at home, and volunteering. The child will be reformed through parents who work in partnerships.

At first, the collaboration and participation seem merely as procedures to bridge differences between groups and to improve "the sharing of information." But the school/parent partnerships are more than producing a sense of belonging. Instead of talking about primary groups, the governing of the child is within "the theory of overlapping spheres of influence" and through "(t)he restructuring of both environments of home and schools to make each more aware of, responsive to, appreciative of, and in partnership with the other to support children as students" (Hidalgo, Siu, Bright, Swap, & Epstein, 1995, p. 17). The parent, the family, and the teacher are in overlapping "learning networks" in which the flow of knowledge occurs. Programs of parent education and teaching jointly treat the psychological qualities and patterns of communications that prevent children's achievement.

One conclusion of the research on parents and children's achievement is to direct social policy away from the welfare and employment system and toward those with mental illness or learning disabilities treated and monitored by the mental health system. Family-professional partnerships are to provide "the link between parental mental health and the ability to parent effectively (Chase-Lansdale & Pittman, 2002, p. 171).

The transmogrification of *all children* to that of the child who lies outside of normalcy is analogous to a museum exhibit on *art and primitivism*. The *art and primitivism* exhibit might show the influence of African images on modern art by juxtaposing the painting of Picasso, among other artists, with African masks and other cultural objects. The African objects would stand on museum pedestals for the spectators to look at and to marvel about the lines, textures, and design perspectives of the masks that were thought to inspire the European modernists. But to move the African objects into the museum display cases reclassifies and recontextualizes the objects into something that defines its difference in relation to the rules of the European cultural artifacts. While appearing as African masks, the masks no longer functioned in relation to the myths and everyday life in which they were fabricated. The masks perform as images within a particular museum space of an "art" world that revisions them through modernist discourses of "culture," aesthetics, and commodification.

So it is with notions of family and children that are positioned in the spaces of school reform. The museum exhibition moves the cultural and social spaces of the African into a set of categories that defines differences in relation to a sameness. So it is with the inclusionary move to direct reforms to include *all children*. The child and family are brought closer to the pedagogical theories of the lifelong learner discussed earlier. But that closeness is not one of equivalencies but difference from a normalized image of the child. The pedagogicalized parent is placed outside of the qualities of normalcy—the active, collaborative, and problem-solving individual. Families of the deviant child are studied, communication patterns identified to order expectations and disposition. The process of rescue and salvation is one that transports and translates distinctions and divisions that place the child and family as outside of normalcy and not capable of ever being "of the average."

Reason and Governing the Child, Family, and Community

Our notions of the child, family, and schooling cannot be left as unmediated practice and experience of realities. I used the notion of cosmopolitanism to focus on fields of cultural practices about ordered and divided objects of reflection and action. Theories and school programs, I argued, fashion and shape a cosmopolitan whose freedom is bound by the practices of reason. But that reason of freedom is not a thing of logic but historically fabricated in fields of cultural practices to govern who the child and family are, should be, and who embodies the characteristics that lie outside of that normalcy. I moved between the eighteenth and nineteenth centuries and the turn of the twenty-first century to consider the internments and enclosures that substantively alter social relations, individual and collective responsibilities and, simultaneously, anxieties, displacements, and exclusion.

The new keywords of governing the child and family are "lifelong learners," "community," "partnership," "problem-solving," and what I call the pedagogization of the family. The keywords, however, are not merely words but an amalgamation of practices that order, classify, and normalize that qualify and disqualify individuals to act and participate. The inclusions and exclusions are embodied in the inner characteristics and qualities of the child. The excluded are those who do not have the capabilities of problem solving, lifelong learning, and collaboration. The capabilities of problem solving, as I argued, are not "merely" normative principles about a higher good, but are historically inscribed through fields of cultural practices and effects of power.

In proceeding in this historical way to think about the present, I placed pedagogy as a particular normalizing project related to modernity. As Wagner (1994) has argued, "Modern institutions do not merely enhance liberty but offer a specific relation of enablement and constraint, the substance and the distribution of enablements and constraints become important. The history of modernity cannot simply be written in terms of increasing autonomy and democracy, but rather in terms of changing notions of the substantive foundations of a self-realization and of shifting emphases between individualized enablements and public/collective capabilities" (p. xiv).

Further, the "enabling" practices for the autonomous conduct of life have multiple effects. One of these I spoke about as the new territories of the individual make resistance and revolt, as I argued, more distant and less plausible. Life becomes a continuous course of personal responsibility and self-management of one's risks and destiny. Freedom is talked about as the empowered individual who continually constructs and reconstructs one's own practice, and the ways of life. But the greater participation and collaboration in the self-management of choice and the autonomous conduct of life has no mooring and thus makes it less likely to assess the rules by which truth is told about the self acting on the world.

The problematic of this essay is diagnostic, to show the contingency of the arrangement that we live and thus open up the possibilities of other ways of living. Interrogating historically the systems of reason that have played a part in holding those arrangements together is to contest the strategies that govern human possibilities. Rose (1999) expresses this thought about inquiry: it is "to disturb that which forms the very groundwork of our present, to make the given once more strange and to cause us to wonder how it came to appear so natural" (p. 58).

The strategies are to consider the historical and anthropological dimensions in the construction of the child and family and to open the critical possibility of breaking from the boundaries of thinking in terms of the society that we already know. To disturb the groundwork that makes the present possible is a form of resistance that makes possible other alternatives.[17] My argument, however, should not be read as a critique of the attitude embodied in a cosmopolitan reason. The discussion is, I believe, framed in the Enlightenment's faith in reason and, in that sense, I am a "child" of the Enlightenment. But the discussion recognizes, to borrow from Foucault, that reason is not a universal and that different modes of reason need to be called forth to engage in the present and its systems of internments and enclosures, although none are ever outside of the historical practices that make them possible and thus always dangerous.

Notes

1. Excursus: "a digression" from the chronology of curriculum history. I would like to thank Mimi Bloch, Kenneth Hultqvist, Julie McCloud, Dar Weyenberg, and Amy Sosnouski for their comments as I worked on this paper. The discussion is part of a broader project of understanding the present that I currently call, "The Reason of Reason: Cosmopolitanism and Governing Social Inclusion and Exclusion."
2. My argument is to focus on knowledge as a material practice that concerns how differentiations and distinctions are concretely produced that are typically either taken for granted in structural analyses of the welfare state or

reduced to an epiphenomenon of "forces" of history. I also recognize the importance of structural and institutional studies and this discussion should be viewed as complementary to those inquiries.

3. This is not to say that it miraculously appears in the Enlightenment or that there were not multiple cosmopolitanisms (see, e.g., Breckenridge, Pollock, Bhabha, & Chakrabarty, 2002).
4. The notion of modern is used hesitantly. My concern is not to engage in the periodization but to explore the slow and uneven changes in the categories, epistemologies, and distinctions from the late seventeenth to the nineteenth centuries that make possible the institutional developments of the state, industrialization, and urbanization, among others, that are often associated with modernity.
5. While Vygotsky sought a psychology related to the Marxist notions of society, the Communist ideology maintained the French liberal notions of the citizens as a practical matter in the transition to communism.
6. Civilized is used historically to focus on the inscriptions of divisions about those who "possess" reason as differentiated from those who do not. It is deployed under different words that differentiate and divide (Popkewitz & Bloch, 2001, Bloch & Popkewitz, 2000).
7. I use the notion of American as it is one used most frequently in the literature, while recognizing that it refers to North America and particularly to historical formations of the United States.
8. Wood (1991) suggests that the changes in the family were related and part of the republicanization that occurred in the eighteenth century that included, among other changes, the giving of legal rights of women for divorce (p. 147).
9. There are distinctions between the republicanism that embodied these values in many European contexts and the liberalism of the United States, but with different emphases, the values related to the cosmopolitan individual.
10. The translation and transportation of Locke and Rousseau as related to child-rearing can be read in relation to this field of cultural practices of the home in North America.
11. There is also the development of the social sciences that related to the missionizing and pacifying of colonial populations, particularly anthropology. This history, however, is not necessarily related to freedom and cosmopolitanism but to people being considered savages, although the cosmopolitan values were also used in resistance.
12. "Natio" meant initially the local community, domicile, family, and the conditions of belonging.
13. Evolution and freedom is a strange couple; natural selection is a blind biological process with no end and adaptive fitness is contingent, while freedom embodies a notion of socially formed relationship. The two though were joined at same time as natural *philosophie* in late eighteenth and early nineteenth centuries to give a prestructured meaning to an

inherent development process. At the same time, teleological theories of history such as those of Comte and Hegel gave qualitative development to the human mind, such as the latter's view of history as the development of self-consciousness and freedom (Hirst, 1994, p.48).

14. Happiness was spoken about in radical English political theory of the Reformation as concerned with the end of government to protect and uphold unconstrained enjoyments of a number of specific civil rights (see, e.g., Skinner, 1998). The end of government is the good, as it ensures people's securing the enjoyment of their rights, without pressure and oppression from rules or fellow citizens. Liberty is embodied in the system of laws that enables the pursuit of happiness that constitutes freedom.
15. The Republican ideas and instantiation of democracy overlaps with multiple historical practices that include capitalism. My focus is on the cultural patterns that travel along and make possible the overlapping institutional practices.
16. If I can use an example of this in mathematics education, the constructivist literature continually cites the work of Walkerdine (1988) to register some limitations. Walkerdine studied how progressive, child-centered pedagogies were presented as universal rules of thinking and reasoning but were related to particular gendered and bourgeois mentalities that produced distinctions and divisions among children. The mathematical educational literature ignores this critique of universalizing reason in its presentation of constructivism as a universal.
17. There is a need for a self-reflexivity embodied in this statement that requires an intellectual strategy that rejects its own hubris and its own notion of the intellectual as a legislator. I think that embodied in the notion of historicizing is one approach for that reflexivity as it continually inserts knowledge as a product of its time/space and thus needs to be scrutinized.

References

American Council on Education. (1999). *To touch the future: Transforming the way teachers are taught: An action agenda for college and university presidents.* Washington, DC: American Council on Education.

Ariès, P. (1960/1962). R. Baldick, trans. *Centuries of childhood: A social history of family life.* New York: Vintage Books.

Bloch, M. and T. S. Popkewitz (2000). Constructing the parent, teacher, and child: Discourses of development. In *The politics of early childhood education.* L. D. Soto, (Ed.) New York: Peter Lang Publishers: pp. 7–32.

Boltanski, L. (1993/1999). *Distant suffering, morality, media, and politics.* G. Burchell, trans. New York: Cambridge Press.

Borman, K. (Ed.) (1998). *Ethnic diversity in community and schools.* Norwood, NJ: Ablex.

Breckenridge, C., Pollock, S., Bhabha, H., & Chakrabarty, D. (Eds.) (2002). *Cosmopolitanism.* Durham, NC: Duke University Press.

Chase-Lansdale, P. & Pittman, L. (2002). Welfare reform and parenting: Reasonable expectations. Special Issue "Welfare Reforms and Children *The Future of Children* 12: 1. Palo Alto, CA: The David and Lucille Packard Foundation: 167–187.

Cronon, W. (1996). The trouble with wildernesses or getting back to the wrong nature. W. Cronon (Ed.). *Uncommon ground: Rethinking the human place in nature.* New York: W. W. Norton, pp. 69–90.

Curti, M. E. (1959). *The social ideas of American educators, with new chapter on the last 25 years.* Paterson, NJ: Pageant Books.

David and Lucille Packard Foundation Journal. (2002). *The future of children.*

Delpit, L. (1995). *Other people's children.* New York: New Press.

Dewey, John (1916/1929a). American education and culture. Joseph Ratner, ed. *Character and Events; Popular essays in social and political philosophy; Volume II.* New York: Henry Holt and Company, pp. 498–503

———(1916/1929b) Our educational ideal. Joseph Ratner (Ed.). *Character and Events; Popular essays in social and poltiical philosophy; Volume II.* New York: Henry Holt and Company, pp. 498–503.

———(1922/1929). Mediocrity and individuality. In J. Ratner (Ed.). *Character and events: Popular essays in social and political philosophy* (II). New York: Henry Holt and Co., pp. 479–85.

———(1928). Progressive education and the science of education. *Progressive Education, 5:* pp. 197–204.

Eldridge, D. (2001). Parent involvement: It's worth the effort. *Young children, 56*/3: 65–69.

Elias, N. (1939/1978). *The History of manners: The civilizing process. Vol. 1.* E. Jephcott (Trans.). New York: Pantheon Books.

Ferguson, R. A. (1997). *The American enlightenment, 1750–1820.* Cambridge, MA: Harvard University Press.

Foner, E. (1998). *The story of American freedom.* New York: W. W. Norton & Company.

Foucault, M. (1979). Governmentality. *Ideology and Consciousness, 6,* 5–22.

Franklin, B. (1986). The first crusade for learning disabilities: The movement for the education of backward children. In T. Popkewitz (Ed.), *The formation of school subjects: The struggle for creating an American institution.* New York: Falmer, pp. 190–209.

Giddens, A. (1998). *The third way: The renewal of social democracy.* Malden, MA: Polity Press.

Glaude, Jr., E. (2000). *Exodus! Religion, Race, and Nation in Early Nineteenth-Century Black America.* Chicago: University of Chicago Press.

Government Printing Office. (1874). *A statement of the theory of education in the United States of America as approved by many leading educators.* Washington, DC: Author.

Greek, C. (1992). *The religious roots of American sociology.* New York: Garland.

Grose, H. (1906). *Alien or Americans? Forward mission study courses, edited under the auspices of the Young People's Missionary Movement.* New York: Easton & Mains.

Hacking, I. (1995). The looping effects of human kinds. In D. Sperber, D. Premack and A. J. Premack (Eds.), *Causal cognition: A multidisciplinary debate.* Oxford, UK: Clarendon Press, pp. 351–94.

Hall, G. (1905/1969). *Adolescence: Its psychology and its relation to physiology, anthropology, sociology, sex, crime, religion, and education.* New York: Arno Press and *The New York Times.*

Hidalgo, N., Siu, S., Bright, J., Swap, S., & Epstein, J. (1995). Research on families, schools, and communities: a multicultural perspective. In J. Banks (Ed.), *Handbook of research on multicultural education.* New York: Macmillan, pp. 498–524.

Hirst, P. (1994). The evolution of consciousness: Identity and personality in historical perspective. *Economy and Society 23*/1, 47–65.

Lasch, C. (1977). *Haven in a heartless world: The family besieged.* New York: Basic Books.

Lesko, N. (2001). *Act your age: A cultural construction of adolescence.* New York: Routledge.

McEneaney, E. (2003). Elements of a contemporary primary school science. In G. S. Drori, J. W. Meyer, F. O. Ramirez and E. Schofer (Eds.), *Science in the modern world polity: Institutionalization and globalization.* Stanford, CA: Stanford University Press, pp. 136–154.

National Commission on Teaching & America's Future. (1996). *What matters most: Teaching for America's future.* Washington, DC: National Commission on Teaching and America's Future.

National Council of Teachers of Mathematics. (2000). *Principles and standards for school mathematics.* Reston, VA: Author.

Popkewitz, T. and M. Bloch (2001). Administering freedom: A history of the present—rescuing the parent to rescue the child for society. In K. Hultqvist and G. Dahlberg (Eds.), *Governing the child in the new millennium.* New York: RoutledgeFalmer, pp. 85–118.

Popkewitz, T. (2002). How the alchemy makes inquiry, evidence, and exclusion. *Journal of Teacher Education, 53*/3, 262–7.

Rose, N. (1999). *Powers of freedom: Reframing political thought.* London: Cambridge University Press.

Rose, N. and P. Miller (1992). Political power beyond the state: Problematics of government. *British Journal of Sociology, 43*/2, 173–205.

Skinner, Q. (1998). *Liberty before liberalism.* Cambridge, UK: Cambridge University Press.

Steedman, C. (1995). *Strange dislocations: Childhood and the idea of human interiority, 1780–1930.* Cambridge, MA: Harvard University Press.

U.S. Department of Education (2001). *Family involvement in children's education. Successful local approaches. An idea book. Abridged version.* Washington, D.C.: Office of Educational Research and Improvement, U.S. Department of Educational Research and Improvement.

Wald, P. (1995). *Constituting Americans: Cultural anxiety and narrative form*. Durham, NC: Duke University.

Walkerdine, V. (1988). *The mastery of reason: Cognitive development and the production of rationality.* London: Routledge.

Ward, L. F. (1883). *Dynamic sociology, or applied social science, as based upon statistical sociology and the less complex sciences.* New York: D. Appleton and Co.

Warfield, J. (2001). Where mathematics content knowledge matters: Learning about and building on children's mathematical thinking. In T. Wood, B. S. Nelson and J. Warfield (Eds.), *Beyond classical pedagogy: Teaching elementary school mathematics.* Mahwah, NJ: Lawrence Erlbaum, pp. 135–55.

Wood, G. S. (1991). *The radicalism of the American Revolution.* New York: Vintage Books.

CHAPTER THREE

GOVERNING NEW REALITIES AND OLD IDEOLOGIES

A GENDERED, POWER-BASED, AND CLASS-RELATED PROCESS

THE PUBLIC DEBATE CONCERNING MOTHERHOOD AND EARLY CHILDCARE INSTITUTIONS IN SWEDEN, 1938–1950

Kerstin Holmlund

Introduction

During the first half of the twentieth century, the generally accepted and disseminated views in Sweden concerning women's rights and their position in the social structure were based on relatively stereotyped gender concepts. A woman's biological ability to give birth was presumed to endow her with an obvious suitability and an unequivocal responsibility for childcare.[1] The home became the private place of work for the married woman's unremunerated labor, and her husband was her protector and sponsor. The mothers themselves were hemmed in by rules and regulations, which they had little opportunity to influence. But in the public debate, they were considered responsible for what they did or did not do and for any consequences this may have for the family and society.

Bearing in mind the fact that the labor market in Sweden was divided into different sectors for women and men, gender differences have been accentuated and have contributed to the creation and survival of collective myths. The labor market's obvious need for female workers during the 1940s can be seen as a shift regarding the position of the family and society and a sign of a move of the mother from home to work. Researches have found that issues concerned with who should take care of the children when the mother was at work became a true social issue when the decision-making process began to include the home.[2] This social alteration had a strong impact on the political discourse. The question of what type of institution would be best to care for the nation's youth was emerging as a problem for the nation, the schools, and the family.[3]

A gender-based situation is also considered in international literature and we can identify some crucial areas concerning mothers and motherhood. In the first place it has been of significance if a woman is construed either as a mother or as an individual.[4] The definition of a woman as a mother has had the result that men have been presumed to be the breadwinners and that dependence has been considered the normal destiny for wives.[5] Government decisions have also sometimes meant freedom and independence while, simultaneously, making women dependent on their husbands.[6] In the second place, motherhood has been fundamental for the improvement of power relations. Mothers have been subordinated and patriarchs have exercised power in the family.[7] Since the family, to a great extent, has been decisive for women's livelihood, this view has been extremely determining for mothers. The opportunity to make one's own living by going out to work or by state allowances is interpreted as being of great significance to the situation of mothers.[8] In the third place, it is important to consider the target group for social policies. In countries where parents, rather than individuals—not male breadwinners nor dependent mothers—have been in focus, justice is seen as a question of equalization between families regardless of parental occupation, marital status, or class.[9]

In this chapter there will be an examination of how decision makers have interpreted and transformed changing social patterns into the state system of daycare institutions, of the values that shaped the arguments for and against full-day care and half-day care, and of how gender-based identities were constructed during that process. The source material consists of government inquiries, parliamentary documents, archive materials, and journals. The theoretical tools of analysis are drawn from theories pertaining to capital resources, power, and gender. A study of how different interest groups have looked upon mothers provides us with understanding of how position, disposition, and habitus have affected different value judgements.[10] How can the language used to describe wage-earning mothers be perceived and what images are used to describe their relationship to their children? Bourdieu uses the concept of symbolic power to describe a covert exercising of power where those involved do not wish to admit to their participation in it.[11] It is by means of symbolic violence that dominant groups try and make their lifestyle seem to be superior and make the others inferior.[12] Language is an important aspect of exercising power. Language is used to define reality. The description of a situation is presented as if it were derived from generally applicable principles, i.e., it is the only possible way of seeing it. But the definition of a reality can not be derived from generally applicable principles, neither for those participating nor for those who are outside the process.[13] The basis for this definition of reality varies, in that it also embraces difference of power. It holds true because of the power relationships between groups in society.

Changing social patterns affecting government decisions

From the beginning of the 1930s policies were directed toward the establishment of a welfare state with the home in focus.[14] Swedish welfare policies have been directed toward a high level of general welfare whereby state subsidies have conformed to rather general regulations. In principle these have purported to reach the entire population at different stages of their life.[15] These policies have been characterized by attention toward public services, a standardized system of allowances, and means-tested contributions. Institutions for the care and supervision of children and infants have been part of the general thrust of these welfare policies. However, at the beginning of the 1940s these policies had not yet crystallized. Rather members of Parliament, supervising authorities, and professional groups with vested interests in the area of welfare, were all engaged in finding out how these directives should be formed.

In several government inquiries and proposals, as well as in parliamentary debates at the end of the 1930s and during the entire decade of the 1940s, the advantages of half-day care—kindergartens—were given prominence compared with care provided in day nurseries.

In the Official Report of the Population Commission in 1938 it was implied that during good times it was both desirable and possible to support the kindergartens and that it was, in fact, solely this pedagogically based half-day care form that deserved the support of society.[16] The report however, did not result in a parliamentary decision because of the outbreak of the war.

Still, in 1943 the Population Commission emphasized the advantages of the kindergarten without taking into consideration whether or not the mother was a wage earner.[17] A differentiation was made as to which institution should be given priority in the future and which one should receive government subsidies in the short term. In the Peace Program and other postwar initiatives the kindergarten was given a prominent position. However, the researchers and their report to the Government implied that state subsidies should be given to the day nurseries in view of the prevailing times of crisis. This proposal may be viewed as being tactical. Rhetorically, the commission supported the prevailing view of women. However, through its decision a social and cultural change was initiated toward bringing welfare to all. This meant greater equality and a public commitment to childcare.

In 1945 the issue of regulations for government subsidies was again raised in Parliament. There were new arguments that pointed toward the need for change. These arguments became a vehicle for criticism of the social and cultural changes brought about by government decisions. In the first place, criticism was voiced against the fact that full-day care, which was seen as a temporary solution for working mothers, had received a better economic position compared with desirable, pedagogically based half-day care. This was especially unfortunate since half-day care, in the general consciousness, had achieved increased recognition and "had occupied a position similar to that of school education."[18] In the second place, the kindergarten constituted a necessary and important assistance for mothers who were working part-time and for families who were living in small, cramped apartments. In the third place, the regulations for government subsidies had been an obstacle in places where, for economic reasons, the people wished to invest in childcare institutions by building kindergartens. It may be that, at this particular point in time, it was not viewed as being expedient to continue with centrally controlled change. The government decided, in fact, that day nurseries and kindergartens should be given equal status in matters pertaining to government subsidies.

The representative of the government in the committee of 1946 for half-day care was of the opinion that reality limited decision making.[19] The export industry was in dire need of female workers and many of these were mothers with small children. Provisional suggestions were, therefore, presented regarding government commitment to childcare. These were not seen as diverging from the fundamental ideas proposed in the general plan, which eventually would be presented in the final version of the Official Report. The provisional suggestions meant that local councils should be recommended to set up daycare activities (family daycares), to use school canteens as afternoon centers and create temporary day nurseries. In 1949 Parliament was, thus, given the opportunity to weigh the pros and cons of "what was inherently desirable and what was possible in practice in the given situation."[20] In its zeal to resolve the acute problems, the government decided to sell off excess surplus property (military huts) in order to facilitate the building of day nurseries by the local authorities.[21] The government ignored centrally based solutions and preferred local alternatives. This was a strategy that reflected a real need, which, in the long run, could create the foundation for more radical government decisions regarding the development of a welfare state.

The kindergartens were, of course, badly suited to the prevalent economic and social reality. State investigators, ministers, and other MPs were forced, time and again, to reiterate that other solutions than those that were desirable for childcare and supervision should be given priority for the time being. The government subsidies that were initially granted to the day nurseries were provisional, but, in practice, the decisions contributed to establishing and regulating the activities of the day nurseries and, by extension, making them permanent (barely a century before primary schools had been introduced with the same arguments).[22] They were also considered a temporary measure, created to deal with an imperfect reality.

From the scenario I have presented it can be seen that the decisions made concerning the day nurseries and kindergartens were considered neutral, well balanced, and logical. Decisions appear that way when one obscures the decision-making process, value judgements, and power relationships that are a part of the context in which the decisions are taken. In the following part of the chapter I shall examine the value judgements and power constellations and give power, gender, and class a decisive role in the analysis.

The decision-making process

Government innovations have been of great importance for the development of Swedish society, not least for women. But women's participation in development was limited for a long time by their lack of opportunity for involvement in the decision-making process. Men's perspectives have, thus, been crucial for the shaping of reforms that have touched women's lives. These decisions have affected the practical situation of women and have varied in significance depending on the nature of the decision and the women's groups that were in the fore.

Motherhood and childhood interrelated with an idea of a future society

Thanks to the Population Commission, family and women's issues became political questions.[23] The Commission comprised both men and women. One of these was Disa Västberg, who, two years earlier, had become leader of the Social Democrats' women's association.

The Commission had half-day care as the starting point for their assumptions. It was considered pedagogically desirable and suited to all classes within society. The Commission related the proposal to children's welfare. It was proposed with "the child's best interest" in focus. The mothers' situation did not come first. If the proposals brought about an improvement for the mothers it should be regarded as an effect of the children's increased welfare. Problems of cramped housing, children without siblings, and children with behavioral problems could be solved by the kindergartens, while, at the same time, providing the mother with more time for herself. The care time in the kindergarten facilitated, in addition, the mother's housework and gave her the opportunity to calm her nerves and develop her personality, which, according to the Commission, was the requirement made of her by modern marriage and society.[24]

The Population Commission had its focus on the children of the future; the future society, the man and woman of the future. They imagined something new, something different and more equal. A future member of society was imagined as mobile, malleable, and able to adapt to different environments, and this vision put new demands on bringing up children. In this respect the home was not sufficient and needed to be complemented. The security of the home should be combined with a more adaptable environment. Public responsibility for childcare was even deemed to increase mothers' opportunities to influence society.[25] The picture of the new mother was of one who should be at home but functioning in a more modern way than previously. The man should be a wage-earner as before, but should make other demands of his wife.

The state representatives regarded their view of future society, the future citizen, family, and mother to be universally applicable. The child was described as genderless, without class, with characteristics that were generally applicable and visible. The imagined child should be creative, emancipated, independent, and socially adaptable.

Suggestions for state intervention were based on mothers who were housewives. This meant there were no reasons for the Commission to provide day nurseries with economic support over and above that given to half-day care since it was solely this institution that was warranted from the mother's point of view. The economic subsidies should be the same, in spite of the fact that full-day care was much more expensive. An extended time for care should not be paid for by state funds but rather by parents, local authorities, foundations, and others. Support for mothers who had their children in full-day care could, therefore, only be done if the mother made economic demands on society in excess of the costs of half-day care. Those mothers who could afford child supervision for the whole day should not be criticized. But economical assets are very much a question of social class.

Working mothers constituted a small part of the entire group of mothers. Most mothers were housewives and could recognize themselves in the description of the good mother. One's position within a group is of significance for the value that is attributed to the primary characteristics of a group of individuals.[26] The position of working mothers among their peers was affected by the judgements placed on their professions, but also on secondary characteristics such as educational capital, economic resources, and age. To speak about half-day care as the sole suitable institution for political, population-oriented measures was encased in a context that could also decide which resources were valued and which type of mothers were considered desirable, exemplary, and superior. For working mothers, their economic assets were crucial to the kind of mothering and childcare they would choose.

The Commission was keen to point out that they had not forgotten impoverished working mothers. But many of these mothers were found in incomplete families—unmarried mothers, divorced mothers, and widows. For the single mothers, housing allowances were recommended in the first place so that the mother and children could establish a family.[27] As the ideology was placed before the reality, these measures can be seen as a political attempt to normalize and discipline these mothers.

At the same time, and independent of the attitudes of the Commission, the number of unmarried mothers decreased and the percentage of married women in the age group 20–64 increased dramatically. During the first three decades of the twentieth century, the female population growth was divided more or less equally across the different age groups.[28] After that, the growth of the young women's group ceased because of falling birth rates, while the older women's group was greater than previously. The male workforce had decreased because of military service, at the same time as industries with links to the military had increased production. In this situation women constituted a reserve workforce that had to be employed. The percentage of married women employed within the female workforce rose from 16 percent to 29 percent during 1940—mainly due to the war industry.[29] In keeping with these developments, the number of domestic maids decreased.[30] Half-day care could not alone solve the supervision needs of workers who were working full-time. Could the priorities afforded to the kindergartens be comprehended as being "the best for all children?" The pressure on the government to seek other solutions for childcare became ever greater.

The relationship between the state and expertise in creating the mother and the child through childcare

The Population Commission, comprising male experts and a male secretary, sets up a special delegation for home and family issues consisting solely of female members and a female secretary. Both groups participated in the inquiry and proposals concerning government subsidies for kindergartens and day nurseries, and, in certain areas, there was even consultation with representatives for the kindergarten teachers.[31] When the Population Commission was set up in 1940 it described the government's financial state as being in crisis. The labor market was in need of women workers and the birth-rate was, at the beginning of the 1930s, the lowest in the world.[32] From the perspective of the birth rate it was considered important that a wage-earning mother was assured that she could continue to work after giving birth. The investigators' opinion was that the mother should remain at home and take care of her child, but the risk that women might choose work instead of giving birth was alarming. At this point, it became legitimate to make it attractive for women to combine motherhood with earning a living. It was felt that subsidies to day nurseries should be higher than those to kindergartens, bearing in mind the fact that the children were fed at the full-day care institution. This was an attitude that aroused strong feelings in many people.

The general view held by these female investigators was, nevertheless, that children who were four or older were in the greatest need of some hours of educative activity. However, for working mothers, social motives were the strongest ones—that is, care and supervision for the whole day so that the mother could work. However, when the mother was a wage earner, then the day nursery was seen as a better alternative than temporary solutions. What constitutes "temporary solutions" is not stated, but the document talks of preventing "children from drifting around and being subjected to negligence."[33]

Experts never appeared to waver in their judgement as to which kind of institution was generally best for the children. Such an institution, however, made special demands on the mother. Not all mothers were able to live up to these demands and some mothers viewed them as being wrong. For those women who did not match up to the normative view of motherhood, the women-only staff of the nurseries could go on acting as mothers.

On account of the crisis at that time the investigators were thus forced to support the day nurseries in spite of the fact that they did not consider this to be a fully desirable development.

The way in which Parliament tied motherhood to childhood

The day nursery challenged established symbols, normative concepts, and the subjective gender-identity of many members of Parliament. During the 1940's the Swedish Parliament was almost exclusively a male institution.[34] This gender distinction meant that men also were allotted power over issues that, first and foremost, concerned women. Government decisions concerning institutions for young children can be considered over-explicit examples of how gender-related exclusive interests (male), to a greater extent, governed decisions about women's everyday life and professional work. The ideas surrounding the married women's role as wife and mother were built on a collective myth that had a bastion in the male-dominated Parliament. In parliamentary debates support was expressed for the traditional nuclear family, with a mother who was a housewife. Attitudes like these were easy to combine with the half-day activities of the kindergarten.

The capacity of these debates to create new identities was weak as the view of gender relationships was, as previously, built up on the subordination of the woman and male dominance, with a division between the work of reproduction and wage-earning.

Institutions for young children were, by no means, a high priority area. In spite of the fact that those in power considered half-day care to be a good thing, they never reached a decision as to how the kindergartens could be integrated into the welfare state and the remaining school organization. The state subsidies were kept at such a low level, thus implanting into the state machinery those economic problems which, in the initial phase, characterized both kinds of institutions. When the subsidies were raised, the previous subsidies were taken as the starting point. This created a vicious circle that kept the status quo. The economic subsidies for these children's institutions were never compared to those of other activities, as, for example, in the compulsory school.[35] Revision of priorities between the different sectors was never carried out. The proposal from the kindergarten teachers to reduce military expenditure in favor of young children was never given a hearing.

Activities for young children were established in their own sphere where they could compete against each other over the tiny state allocations. The needs of different groups of women, but also social groups, were put at loggerheads. Statistics show that from 1944 to 1945, parents from Social Group 1 had extremely low numbers of children in day nurseries. They made use of the kindergartens. For parents from Social Groups 2 and 3 the pattern was the opposite—they made use of the day nurseries for supervision of their children.[36] This complicated the situation for women to perceive and, thus, to react against this gender- and class-based exercise of power.[37] Thus, housewives and working mothers have symbolically competed against each other over government resources and status.[38]

Whoever has power over the economic capital also has power over disciplining the groups that are subordinated to them. The assumptions about, and experiences of, the female sex's importance for social tasks related to childbirth and reproduction meant that many MPs were concerned about the changes taking place. We shall now attempt to get better acquainted with two different opinions about motherhood prevalent in parliamentary debate during the 1940s.

The language of duties as the Parliament's ideal

When the subject of institutions for young children was broached in Parliament, the main discussions dealt with women's duties as mothers and difficulties as wage-earners. In the parliamentary debate of 1943 concerned with state subsidies to kindergartens and day nurseries, it was considered that a woman's duty was to take care of her home, husband, and children.[39] The argument was that those mothers who could not stay at home should be able to combine part-time work and motherhood. In that case the government should not pledge itself to long-term economic support for day nurseries but rather wait to see the results of the first year's subsidies before making these measures permanent.[40] Perhaps the government could be involved instead with providing part-time work for women.[41] "That is how the issue was seen among the rank and file. Their view was not, as some ladies would have it, a question of providing assistance to be able to send the children away, but rather the problem for these [working] mothers was how they keep their children with them."[42] This quotation shows how gender power structures were used to maintain the prevailing gender order. In the contributions to the debates, three concepts were used to differentiate between different women groups; wives, ladies, and mothers. How these concepts related to women in reality was never explained, but these women groups were attributed with varying attitudes to motherhood. The power to assign different attributes is what Bourdieu refers to as the power to execute symbolic violence.[43]

The consequences for the health situation of working mothers who ignored their duties were also raised in parliamentary debates. A disastrous result of married women with children who went out to work was that they could become " . . . nervous and worn-out and estranged, so that they would have to be furnished with precious health care and, perhaps, they may not even be able to take care of their children when they were incapacitated from contributing to production."[44]

The day nursery was deemed, quite simply, to entice women into acting against their own good and that of their children, family, society, and, perhaps most important, of their husbands. Day nurseries should exist for unmarried mothers; "who have no husband to take care of them and have no one to contribute to their maintenance in the same way as a proper husband can."[45]

The day nursery was a true threat to the ideal of how a woman should be. However, this ideal began to be called into question in spite of the fact that the number of women and mothers in Parliament was so low.

The language of rights as a struggle to redefine the Gender Contract

Discussions about the suitability of mothers as wage-earners were constantly raised in Parliament. The Secretary of State Gustaf Möller's view in 1943 was that this debate was mainly of an academic character.[46] His view was that one had to have as a starting point the fact that trade and industry required the services of the mothers and that it was society's duty to build day nurseries so that these women could also bring children into the world and raise them.

There were also those in Parliament during the 1940s who were prepared to make a more extensive and radical redefinition of motherhood. A woman's rights discourse began to emerge alongside the duty discourse. The women Social Democrats, for example, thought that it was not only the husband's duty to take responsibility for the economy of the home. Neither should women have exclusive duties regarding their home and children.[47] If both husband and wife were wage earners then they should help out with the housework. The woman should have the same rights as the man.

Women are also independent individuals in a modern society and not merely appendages to their home and children. They should also possess the right to choose the task to which they are best suited. In my view society's role is to facilitate and not hinder their choices. We demand that we should avoid the risk of being thrown in and out of the production line as if we were inanimate objects.[48]

Women were considered independent individuals who ought to be respected for their choices. Moreover, there were people who reasoned that women were needed for the labor market both at that time and in the future.[49] The demographic development with all the more elderly people spoke clearly in favor of this.

But there was a tension between advocates who defended the idea of the mother's duty and those who fought for their rights, and Parliament was not the only place for that struggle. How other public figures viewed the day nurseries and wage-earning mothers will now be examined.

Civil-based associations and economic entities interrelate with the state to form policy and governing programs

In 1943 Parliament took a positive stance for industrial companies to start day nurseries and preschools.[50] It was assumed that they were able to guarantee the consistency and continuity necessary to ensure that state subsidies could be granted.[51]

Swedish Chocolate Factory Cloetta Ltd.'s (located in Ljungsbro south of Stockholm) commitment to the building of day nurseries is an example of how companies took part in the development of society. The company had built modern housing, fixed water and sanitation, street-lighting, premises for a doctor's surgery, as well as dental-care, and telephone communications and even cheese production had come to the area.[52] The daycare center that at the end of the 1940s stood ready to be opened was designed by Cloetta's architect Henry Fraenkel and was extremely modern in regard to form and equipment.[53] A public debate was held at the opening of the daycare center concerning mothers, wage earning, children's welfare, and day nurseries' benefit to society.[54]

Eight women and two men contributed to the discussion. The men were representative of authority, while the women, from several aspects, constituted a heterogeneous group. All were wage-earners, but with different social and educational backgrounds. Their experience of motherhood was also varying.

The way in which motherhood and childhood were created by people in authority

The General Director of the National Board of Health and Welfare, Ernst Bexelius, claimed, on his part, that individual care of children was superior to collective care. According to Bexelius, the mother should endeavour to remain at home until the child is three years old as she has her motherly duties to attend to. To grow up in a "natural home" environment with a natural relationship between the mother and child is a good thing for the child.[55] There were, Bexelius admitted, women who strayed from what was natural, and in this case it was good that others could take care of the child's upbringing.

Bexelius argued that women were working and that subsequently would continue to do so, since social changes had worsened the financial opportunities for women and children to be supported at home.

The day nursery was not seen, however, as a satisfactory solution to the issue of child supervision, according to the director-general. He felt sorry for those children who had attended day nursery from infancy until school age. But since the mothers were out working, the day nursery could prevent the children from ending up in worse "situations." "In other words, it is not our expressed wish that we should have day-care centres."[56] When he spoke about the mothers he used a technical language that heightened the unnaturalness of the situation. The children were dragged out, according to him, in the mornings and delivered or thrown into the day nursery. His own personal opinions about the mother's relationship to her child were presented as if they were objective knowledge about real-life situations. "Then she fetches the child after her day's work is done and fixed their dinner and consequently there cannot be very much time spent together between the mother and her child."[57] In his statement he combines a description of unavoidable events—the act of fetching the child when the working day is over—with value judgements concerning the quality of the time spent together. This statement about the quality of their time together became somewhat of an objective truth, a natural consequence of the previous description of the train of events that were based on reality.

The negative consequences of the separation between mother and child were transferred to the activities of the day nursery, according to Bexelius. The best alternative to care outside the home was, in his view, the family daycare combined with a kindergarten. The family daycare was a home where a woman, a mother, could leave her child with another mother who would happily look after several children during the day. "We are wont to believe that the individual care of children is superior to full-time care in a collective [day nursery]."[58] But due to the great demand for day nurseries, the representatives of the Swedish national board of health and welfare were ready to support an increase in building these institutions.

This was, otherwise, the prevailing standpoint of the Swedish National Board of Health and Welfare, which was expressed into the early 1960s.[59]

The Headmistress and Doctor of Philosophy Carin Ulin expressed similar views. She took the good home as a role model. In the natural home the mother went to the market and the shops and did her shopping with her child, so that the child could see the fish from the market to the cooking pot. In the home, the child could comfort mommy when she had a headache. These common experiences created "root-ties" between mother and child. Against this role model was presented the picture of a mother who was working in an office, a shop, or a factory and hardly knew her child—a child who was away from the mother the whole day and who became like a stranger to her. The day nursery would miss out on providing many "root-ties" that would be of significance for the future mother-child relationship. These children missed all the natural contexts of the natural home. Another negative issue concerning the day nurseries was that they were too tied up with children, furniture, toys, and the "ladies," whose job was to look after the children. Even the kindergarten was child-centered, but since the time spent by the children there was only relatively short, there was no serious damage.[60] The day nursery gave children an unreal picture of an unnatural life, according to Ulin.

These descriptions of mother and child are grounded in the psychology that influenced the realm of the preschool during the 1930s and 40s. Ph.D. Ulin was an authority in this area; she made her statements from an academic viewpoint.[61]

Career women think that working mothers should be respected

It is during the 1930s and 40s that working women joined forces to raise and pursue common issues. Working-Mothers' Co-operative Council (YSF) was founded in 1944 and was an amalgamation of the committee for Swedish Association of Women's Councils, formed in 1934 and followed in 1937 by the National Association of Working-Mothers.[62] YSF was a national organization, a national executive for working women, employees, those with their own businesses, and those who exercised a profession. The aim of the organization was to safeguard women's professional, social, and economic interests.[63] The organization was, therefore, not a union and, as such, lacked, for example, negotiation rights.[64]

In 1947 the Swedish Trade Union Confederation had founded a women's council to get women more involved in the work of the union. In the same year the union nominated its first and, for a long time, only woman ombudsman.[65] Sigrid Ekendahl herself was of the opinion that the Swedish Trade Union Confederation was working actively to raise opinions about the extension of the day nursery program, and that during the 1940s theirs was the only voice to be heard pursuing the issue of the day nurseries.[66] Prior to this, the issues of child supervision had not been considered important and the Women's Council had even met opposition within the Swedish Trade Union Confederation.[67]

In this situation Sigrid Ekendahl represented the Swedish Confederation of Trade Unions and the issue was now a problem at least for the Women's Council. The Working-Mothers' Co-operative Council was represented by Getrud Julin (who was herself a mother with a child who had been to day nursery). What areas of interest did these two women and the union represent in the issue of day nursery?

The obvious relationship between mothers, children, and issues of supervision that certain speakers referred to in their contributions to the debate were scrutinized by Julin and Ekendal. They had another perspective on these issues and questioned those value judgements that others had presented as truths. Gertrud Julin considered that there were many different reasons why women worked, including, for example, those related to finances, education, and marriage. Sigrid Ekendahl's view was that the situation of those members she represented was very different to the one described in the general debate. Working mothers were forced to have their children in a day nursery for financial reasons.[68]

Sigrid Ekendahl did not wish to assert that all the mothers affiliated with the Swedish Trade Union Confederation wanted places in day nurseries for their children. However, she did consider that the lack of day nursery places was a real problem for the women's group that she represented.

Gertrud Julin questioned the statement that the mother needed to be at home for the child's welfare. "It is not so, that just because one has given birth to a child that one is immediately suitable to take care of it."[69] Her experience of having children in day nursery was that the children liked being there and sometimes did not want to go home when their parents came to collect them. This was not because the children had a miserable time at home, but rather that the day nursery was their playground. Both of these representatives of working mothers also argued that it was psychologically taxing for mothers with children in day nurseries to constantly hear that they were such lost souls. If one wanted to support the working mothers it was possible to regulate the long period the children had to spend in the day nursery, argued Ekendahl. If the working day was to be shortened for both parents—not only for the mother—then the time spent together at home with the parents could be lengthened.[70]

Gertrud Julin had also heard different versions on the theme of how children were dragged to day nursery. She had never done so and in her opinion placing the day nursery within the housing area would reduce the problem of getting there. Even Carin Ulin's description of the educating mother in the home was called into question.

"I must state, I do not believe that there are so many of those kinds of mothers. I have seen quite a lot of mothers, and when they are doing their shopping there aren't so many of them who discuss with the children what they are buying. But they do say: 'Stand still,' 'don't climb on that,' 'wait a moment,' 'go out and wait for me outside,' etc."[71]

Julin seems to construct a more supervising mother. She also reacted against the statements claiming that if a mother went out to work this would automatically mean bad relationships with her children. Why did no one have thoughts about the relationship between the working father and his children, she wondered?

"Then there is all this talk of the mothers' alienation from their children. I just think that it is so strange that among all these men who are fathers, that you applaud this so much, but don't you also feel alienation towards your children since you are also wage-earners?"[72]

These public women challenged the status quo regarding men and women's standards and pointed out the arbitrariness of these value judgements. They called into question the standards of that time for women and mothers and demanded that women be respected as wage-earners.

The opinions of some working mothers about daycare and motherhood

Elvira Wedberg-Larsson was a member of the City Council in Stockholm, a Social Democrat and active in the women's confederation and the Swedish Trade Union Confederation. She had worked in the industrial sector for 30 years and her children had attended day nursery. Karin Kroon was a working mother with children in a day nursery. Britta Åkerström also claims she represents the mothers. For Åkerström's family's first child, she had employed a nanny and the necessary servants. At that time her attitude was as follows: "So then, just this sentimental spirit: The mother is obviously the best person to take care of her own child."[73] Her second son, who had been extremely troublesome, became a "new child" after a year attending the day nursery. "Of course, you might respond by saying that his parents were incapable of taking care of the kid."[74] But she decided that such a conclusion was untenable. Britta Åkerström established the fact that attending the day nursery had been a positive experience for her son.

Wedberg-Larsson stated that her objections to the day nursery lay in the long day that children had to endure. A woman industrial worker had an

8-hour day. Adding her lunch break of an hour meant a total time of 9 hours. If traveling time were further included, the working day would become 10 hours. With her wages, the working woman could not afford cooked food, assistance with the laundry, or other home help, so when she came home she had to do all these chores. This was the daily reality also described by Karin Kroon.

As I start my working day at 7 o'clock I have to wake my little daughter before 6. It's in fact too early to expect a child to get up. It would be ideal if I were able to start at 8 o'clock and have Saturdays free, but that's unthinkable, so it's not much point to expect that. I will finish by stating that, in my present situation, I am a warm supporter of the day nursery system.[75]

But what was a normal working day for a wage-earning mother? The Population Commission carried out in 1950 a series of 390 research surveys with mothers working in the textile, clothing, and chocolate industry, with the identical kind of jobs to those performed in the Cloetta chocolate factory.[76] The average woman in the survey was described as follows:

She wakes up at 5.30 A.M. and does her chores for almost an hour: When the working day is over she takes care of the housework for approximately 4 hours and 20 minutes. Leisure time every day is reduced to 48 minutes and her night's rest starts at about 10 P.M. This means that her night's sleep is 7 hours in duration.[77]

Among the married women roughly one-third said that they had "help every day" from their husbands. However, just as many stated that they never received help with the housework. The results of this survey clearly demonstrate that these working mothers had an extremely heavy workload, and it meant that they took responsibility for taking care of their children, homes, and making a living.

Wedberg-Larson also pointed out that the women working in industry had problems combining their roles as mother with their vocation. Many women were overworked, worried about the future of their children, and had a bad conscience because they did not have the strength nor the time to perform the duties placed upon them as women. There was no certainty that they would have been competent mothers in the home situation, but, as they lacked the opportunity to be with the children except for in the evenings, they assumed that everything should be better devoting themselves wholly to their children. These demands upon themselves meant that working mothers were constantly unhappy, dissatisfied, and, thus, they deteriorated on all fronts—as mothers, wives, and workers. It was the reality—the finances, the traveling, and the working day—that formed their experiences.[78]

Wage-earning mothers wanted to give the best to their children, but they could always judge themselves as failing in relation to the symbols and discourses of motherhood produced and presented in the public debate. The society of which they were a part highlighted values of motherhood that were almost impossible for working mothers to reach. For example, the rules of the labor marked forced them to go to work early in the morning while the so-called good mother should stay at home with her child. Using Scott's concept we can state that the working mothers' subjective identification with that of the "good mother" gave them a bad conscience.[79] The mothers took responsibility as citizens but at the same time they judged themselves for not acting in the way that the idea of the good mother and power structures insisted.

However, their dissatisfaction had nothing to do with the day nurseries. Wedberg-Larsson had nothing but positive judgements to make about them. "I really like the day nursery and especially having seen such an example as this one is, and several of those built recently in Stockholm are extraordinarily good."[80]

The kindergarten teachers' perspective on day nurseries and kindergartens

The kindergarten teachers had vested interests in the day nurseries and kindergartens. In order to comprehend their views it is necessary to take a short historical look at the situation.

At the turn of the twentieth century bourgeois women opened kindergartens, and started training courses and associations.[81] They underlined the fact that the kindergarten was a complement to the home. There should not exist any competition between the kindergarten teacher and the mother. The mother was considered to be the most suitable person to take care of her children. This attitude facilitated entry into the labor market and had consequences as to which institution the leadership of the union would advocate. The kindergarten was considered a pedagogically motivated institution and, thus, desirable for society. However, the day nursery, which received children for the whole day, was a problematic institution. Nevertheless, the issue was difficult to resolve and turned out to be a stumbling block for the teachers' status, legitimacy, and professional success. During the 1940s when these institutions for small children became incorporated into the welfare state, the problematic relationship between wage-earning and motherhood was further accentuated.

Housewives and half-day care—the best for children

The kindergarten idea was based on a premise that everyone, regardless of gender, class, marital status—in short the entire society—had a common interest in child issues. If this were the case, then the labor market would only employ women who had no small children or organize part-time work for mothers. Another unargued assumption underlying the logic of choice was that the women would choose what was best for their children—namely to be housewives—and the state would give its support to half-day care. On the other side collective whole-day care for small children involved risks in the psychological perspective, argued the kindergarten teachers. They managed to gain the support of doctors, who claimed that the day nursery meant considerable risks for child development.[82] The day nursery was a care institution, which should only be applied in emergency situations.[83] Their view was that, primarily, women without children or mothers with older children should be wage-earners. Should mothers with young children go out to work, part-time work was a possible solution. This point of view was tacitly accepted with reference to what was "best for the child."[84]

The kindergarten teachers based their arguments on the historical mother and child. This involved a retrospective view of the middle-class home and its children. The nuclear family, the established mother, and the kindergarten were all important for the child's development and seen as something natural not related to questions of class. In a similar fashion to the Population Commission they wanted to maintain the gender-order and pointed out the splendid qualities of the home and kindergarten contra the imperfections of the day nursery. This retrospective view on history was also used by many MPs, the director of the Swedish National Board of Health and Welfare, and even the researcher Carin Ulin. The headmistress and Social Democrat Alva Myrdal, who had been a member of the governing body of the kindergarten teachers, lost her right to be a member in 1944 when the charter of the association was changed. She had not been a kindergarten teacher and her commitment toward the future was far too loosely rooted and critical of history.[85]

A union in trouble

However, more and more kindergarten teachers had salaries by being on the staff of day nurseries, and during the 1940s they started to make their voices heard in the public debate. Their view was that even if there had been a general consensus that the best thing for a child should be a harmonious home, a mother who was a housewife, and access to a kindergarten, the body of teachers could not, in good faith, ignore the development of society and the needs of the labor market. One should therefore attempt to reduce the deleterious effects of the day nurseries.[86]

The short length of stay at the kindergartens had previously been used as proof of their superiority compared with the day nurseries. But when the daycare activities started to be reevaluated, longer daycare time was presented as being an advantage.[87] If the staff could train the parents to limit the care time to a reasonable level and if the group of children was suitably small, "then the day nursery would be at the same level as the kindergarten."[88]

Stina Sandels, who as chairwoman of the congress had continuously spoken of the advantages of the kindergarten compared to the day nursery, created quite a stir at the Union of Scandinavian Kindergartens' AGM in 1947, when, in an introductory speech, she was self-critical about what had been going on. Her opinion was that the kindergarten teachers had all too easily perceived the exclusive advantages of half-day care and the disadvantages of full-day care. But, meanwhile, changes had taken place in society. Mothers went out to work, parents were tired and stressed out, they were devoid of knowledge about child psychology and nutritional physiology. Homes such as these were not helped by half-day care in a kindergarten. "This type of institution is far too clearly psychological and educational in the old sense of the word—it takes too little consideration of the whole child."[89]

New research findings had made Stina Sandels doubt her conviction as to the superiority of the kindergarten.[90] She left her post in 1949 as chairwoman of the union.

At the end of 1948 the union leadership had discussed a direct question from one of their members, enquiring whether or not the union had changed its fundamental attitude to the day nurseries. "During the course of the discussion it was strongly emphasised that this had not occurred."[91] The union leadership continued to give priority to the kindergarten.

Working mothers on to the agenda

The active, progressive, and noble-minded struggle that we have been able to follow was the beginning of a tenacious process that eventually led to putting issues concerning working mothers and day nurseries onto the public agenda. But in the period that I have studied, resistance was very strong. It is a well-known fact that real change requires a long time. To trace concrete changes in issues such as these that are connected to power, gender, and class can, therefore, be difficult.

During the decade in question, there was strong support for the good mother. The myth of motherhood gives the impression that a gender division between the roles of women and men in the home is biologically rooted, generally applicable, and permanent. This should be eternal—a type of social magic. Using Bourdieu's concept, this notion can be described as an official truth grounded on specific interests and, if taken too seriously, can receive the status of a norm born out of practical experience, in this case, from marriage.[92] As the article shows gender relationships do not affect only the relationships between husband and wife but are also construed as economic assumptions and political decisions.[93]

By means of abstractions, generalized descriptions of mothers decrease the number of concrete examples and the opportunity to generalize increases. Abstractions are found in ideologies and where power is found. The woman or mother is found in real life and always in relation to other characteristics, for example class or ethnicity or in a more complex relationship to vocation, housework, unemployment, living conditions, or education. It is only on the basis of abstractions that one can evaluate different kinds of women and mothers with the same standards. To tone down the separate reality of different women and make statements about general standards for motherhood strengthened the gender and class situation of that time. Generalized descriptions of women and mothers can be viewed as spreading the value judgement of a certain group. This act constitutes a symbolic show of strength and is placed upon other power relationships empowering them and increasing them.[94]

The situation for mothers, fathers, and children is different today than it was in the 1940s. The conditions of motherhood vary, and the fact that social differences exist is evident from the statistics.[95] The question focusing on the extent to which welfare-state decisions are established on ideology or reality needs to be asked just as much today as it was in the 1940s. This is especially the case, since value judgements seem to relate to different realities and be affected by position, disposition, and habitus. To scrutinize the focus for social policies concerning parents and children is important in a society that endeavors to attain democracy and equivalence among its citizens.

Notes

1. Holm, U. (1993). *Modrande & praxis. En feministfilosofisk undersökning.* Göteborg. Daidalos, p. 209.
2. Hirdman, Y. (1989). *Att lägga livet till rätta: studier i svensk folkhemspolitik.* Stockholm: Carlssons.; Rothstein, Bo. (1994). *Vad bör staten göra? Om välfärdsstatens moraliska och politiska logik,* Stockholm: SNS Förlag.
3. Holmlund, K. (1996). *Låt barnen komma till oss: förskollärarna och kampen om småbarns-institutionerna 1854–1968.* Umeå: Borea Bokförlag.

4. Berkovitch, Nitza. (1997). "Motherhood as a national mission: the construction of womanhood in the legal discourse in Israel." *Women's Studies International Forum* 20, 5–6: 605–19. In this study concerning the construction of womanhood based on the legal discourse in Israel around two laws passed in 1949 and 1951 respectively, the analysis shows that a Jewish-Israeli woman is construed as a mother and not as an individual.
5. Pedersen, S. (1993). *Family, dependence, and the origins of the welfare state: Britain and France 1914–1945.* Cambridge: Cambridge University Press, p. 12.
6. Compare: Lister, R. (1997). *Citizenship: Feminist perspectives.* London: MacMillan, p. 168.; David, M. E. (1993). *Parents Gender & Education Reform.* Cambridge: Polity Press.
7. Gordon, L. (1994). *Pitied but not entitled: Mothers and the history of welfare.* New York: The Free Press, p. 13.
8. O´Connor, J. (1993). "Gender, class and citizenship in the comparative analysis of welfare state regimes; theoretical and methodological issues." *British Journal of Sociology* 44 (3): 501–18; Orloff, A. (1993 June). "Gender and the social rights of citizenship, the comparative analysis of gender relations and the welfare states," *American Sociological Review* 58: 303–28; Wilson, E. (1985). "Feminism and Social Policy." In Loney, M. Boswell D., & Clarke, J. (Eds.). *Social policy & Social Welfare.* Milton Keynes, UK: Open University Press, p. 34.
9. Pedersen, S. (1993), pp. 17–18.
10. Bourdieu, P. (1995). *Praktiskt förnuft: Bidrag till en handlingsteori*. Göteborg. Daidalos, pp. 11–29.
11. Bourdieu, P. (1991). *Language and symbolic power.* Cambridge: Polity Press, p. 164.
12. Bourdieu, P. (1996). *Distinction: A social critique of the judgement of taste.* London: T. J. Press, p. 511.
13. Callewaert, S. (1992). *Kultur, Paedagogik og videnskap.* Viborg. Akademisk Forlag, pp. 116–17.
14. Compare Yvonne Hirdman's description of the welfare state. Hirdman, Y. (1989), pp. 128–158. "Social engineers" is a term used during recent years to describe politicians who characterized Sweden during the 1930s. Ibid., p. 97.
15. Rothstein, B. (1994), pp. 25–27.
16. Official Reports of the Swedish Government (SOU). (1938:20). *Betänkande angående barnkrubbor och sommarkolonier m.m.,* p. 41.
17. SOU. (1943:9). *Utredning och förslag angående statsbidrag till daghem och lekskolor,* p. 49.
18. Official parliamentary publications (RT). (1945). P 1:5, point 91, p. 113.
19. "Half-open child care" was the title given to the form of care where the child was only attending the institution in question part of the day and, for the remaining time, was at home. RT. (1943), pp. 332. 55.
20. RT. (1949). The Government Parliamentary bill (KMP). 46, p. 5.

21. RT. (1949). KMP. 46, pp. 7–8.
22. Johansson, U. (1988). Uppfostrans förstatligande. *Tvärsnitt* nr 1.
23. Karlsson, G. (1996). *Från broderskap till systerskap: Det socialdemokratiska kvinnoförbundets kamp för inflytande och makt i SAP.* Lund. Arkiv Förlag, p. 95.
24. SOU. (1938:20), pp. 15 and17.
25. Lindholm, M. (1994). "En modern samhällsmoderlighet? Concerning Elin Wägners and Alva Myrdals ideas during the 1930s." In Baude, A., & Runnström, C. (Eds.). *Kvinnans plats i det tidiga välfärdssamhället.* Stockholm: Carlssons. 11–9.
26. Bourdieu, P. (1987). *Kultursociologiska texter. I urval av Donald Broady och Mikael Palme.* Stockholm: Salamander. 249; Pierre Bourdieu, P. (1996). 100.
27. Here is mentioned the large group of maids who had children living with their employer. SOU. (1938:20). 3.
28. Qvist, G. (1974). *Statistik och politik. Landsorganisationen och kvinnorna på arbetsmarknade.* Stockholm: Prisma. 17.
29. Ibid. 1.
30. Gough, R. (1994). Från hembiträden till social hemtjänst. In Baude, A. & Runnström, C. (Eds.). 39–56.
31. There were five representatives of the kindergarten, among others Principal Mrs. Alva Myrdal and Ph.D. Miss Carin Ulin.
32. Löfgren, B. (1994). "Hemarbetsfrågan i svensk politik under 1930- och 1940-talen." In Baude, A., & Runnström. C. (Eds.). 21–37.
33. SOU. (1943:15). 40 and 43.
34. Carlsson, S. (1966). *Den sociala omgrupperingen i Sverige efter 1866, Samhälle och riksdag del I.* Stockholm: Almqvist & Wiksell. 479. In the first Chamber in 1945 there was roughly 1 percent of women and in the Second Chamber there were approximately 7 percent women.
35. Lirén, G. (1989). *Facklärarna i skolans och arbetsmarknadens perspektiv. En kamp för jämlika villkor 1950–75.* Stockholm: Svenska Facklärarförbundet. 178. During the 1950s, after the end of the war, government subsidies to day nurseries and preschools constituted merely 0.5 percent of government expenditure for compulsory schooling.
36. SOU. (1951:15). 50.
37. Holmlund, K. (1996). 195–204; Karlsson, G. (1996). 233.
38. Compare Ferguson, E. (1991). "The child-care crisis: Realities of women´s caring." In *Womens Caring.* Baines, C., Evans, P., & Neysmith, S. (Eds.). Toronto: McClelland & Steward Inc. 81 ff.
39. RT. (1943). Official record of proceedings of the second chamber in the Swedish Parliament (AKP). 32:51. 26. Ebon Andersson (right-winger, r).; Jonas Erikssson i Frägsta (r). Ibid. 12–13; Harald Hallén (social democrat, s). Ibid. 18.
40. RT. (1943). Official record of proceedings of the first chamber in the Swedish Parliament (FKP). 32:12. 54. Gustaf Björkman (r).; See also RT. (1943). AKP. 32:51. 26–27.

41. RT. (1943). AKP. 32:51. 14–15. Wiktor Mårtensson (s).; Harald Hallén (s). Ibid. 17–18; Johansson, J-E. (1992). *Metodikämnet i förskollärarutbildningen. Bidrag till en traditionsbestämning.* Gothenburg: University of Gothenburg. 105.
42. RT. (1943). AKP. 32:51. 18. Harald Hallén (s).
43. Bourdieu, P. (1990). *Reproduction in education, society and culture.* London. Sage.
44. RT. (1949). FKP. 14:43. 50. Axel Mannerskantz (r). Similar arguments are presented by Sture Henriksson (s). RT. (1949). AKP. 14:97. 102.
45. RT. (1949). FKP. 14:43. 50.
46. RT. (1943). Governmental Bill (KMP). 339. 55.
47. RT. (1949). FKP. 14:43. 52. Anna Sjöström-Bengtsson (s).
48. RT. (1949). AKP. 14:97. 108. Disa Västberg (s).
49. Ibid. 99. Sture Henriksson (s).
50. RT. (1943). KMP. 339. 60. Concerning allowances to day nurseries, etc. (Gustav Möller).
51. The day nurseries' responsible authorities were, in October 1950, divided as follows: local authority 68, association 99, industries 31, private persons 3, and regional authority 1. SOU.(1951:15). 108.
52. Swedish Chocolate Factory Cloetta Ltd. (1948). *Theobroma Cacao.* Stockholm: Esselte AB. 90 ff.
53. The day nursery had cost 226,000 Swedish crowns to build and 39,000 crowns had been received in state subsidies. *Arbetaren* (1949 Monday 21st March). 66.
54. National Archives (RA). Committee number 1672, "The importance of the day nursery for society." The meeting was on Saturday nineteenth March 1949.The debate was broadcast via microphone and was taped.
55. RA. Committee number 1672 (The importance of the day nursery for society". 27. Bexelius.
56. Ibid. 3.
57. Ibid.
58. Ibid. 7.
59. Holmlund, K. (1996). 229–46.
60. Bourdieu, P. (1990). *The logic of practice.* Cambridge: Polity Press. 52–65; Jenkins, R. (1992). *Key Sociologist Pierre Bourdieu.* London and New York: Routledge. 75.; Callewaert, S. (1992). 136.
61. Hultqvist, K. (1990). *Förskolebarnet: En konstruktion för gemenskapen och den individuella frigörelsen.* Stockholm. Symposion. 12–13, 141–4.
62. Rössel, J. (1950). *Kvinnorna och kvinnorörelsen i Sverige.* Stockholm: YSF:s Förlag. 111.
63. The archive of the National Association of Swedish Pre-School Teachers (SFRA). (May 1944). Box 51–318. To the National Association of Kindergarten Teachers; SFRA. (January 1962). Box 53–318. To YSF:s membership groups. Confidential, Urgent. Letter from the Swedish Co-operative Board of Women's Associations 31st.

64. Rössel, J (1950). 112.
65. Sigrid Ekendahl was chairperson of the Women's Council between 1947 and 1967. Englund, E. (1989). "Arbetsmarknadens kvinnonämnd." In *Saltsjöbadsavtalet 50 år. Forskare och partner begrundar en epok 1938–1988*. Edlund, S. et al. (Eds.). Stockholm: Arbetslivscentrum. 178–179.
66. Axelsson, K. (1980). *Den första i sitt slag, Stockholm. Tidens Förlag.* 55.
67. Waldemarson, Y. (1998). Women's Council of Swedish Trade Union Confederation: En paradoxal historia om framsteg och motgång. *Arbetarhistoria.* 22: 1/2: 54–58.
68. RA. Committee number 1672. "The importance of the day nursery for society". 19. Sigrid Ekendahl.
69. Ibid. 9. Gertrud Julin.
70. Ibid. 20. Sigrid Ekendahl.
71. Ibid. 23. Gertrud Julin.
72. Ibid. 23. Gertrud Julin.
73. Ibid. 11. Brita Åkerström.
74. Ibid.
75. Ibid. 24. Karin Kroon.
76. RA. Committe number 1672 file 10. This research was carried out between August 7 and 19, 1950.
77. Ibid. 4.
78. Bourdieu, P. (1996). 175.
79. Scott, J. W. (1966). *Feminism & History.* Oxford, New York: Oxford Readings in history.
80. RA. Committe number 1672, "The importance of the day nursery for society". 13. Wedberg-Larsson.
81. Holmlund, K. (June 1999). "Kindergartens for the poor and kindergartens for the rich: two directions for early childhood institutions in Sweden (1854–1930)." *History of Education* 28 no 2.
82. See for example RA. (December 31, 1948). Committee number 1672 volume 9, report by Dr. Per J Nordenfelt, "*Daghemmen kan inte undvaras*" He stated that children should grow up in complete families and be looked after within the home. His view was that childcare was not suitable for children under 3–4 years of age.
83. SFRA. (June 9, 1944). Box 103. SBR's statement about normal regulations, to the Royal national board of health an welfare.
84. *Barnträdgården.* (1946). nr 3. 36–41. Report from AGM in Uppsala.
85. She is also critical of the old principles of child-rearing. Myrdal, A. (1935). *Stadsbarn: en bok om deras fostran i storbarnkammare.* Stockholm. 96.
86. "Diskussion om daghem." *Barnträdgården* (1944). nr 5. 92–93.
87. Ericsson, A. M. "Se positivt på daghemmen!" (1946). *Barnträdgården* 7. 118–121; Andersson, R. (1946). "Mera deltidsarbete i allmän tjänst." *Barnträdgården* 9. 159.
88. Ericsson, A-M. "Se positivt på daghemmen!," *Barnträdgården* (1946). 7. 118.

89. *Barnträdgården* (1947). 6. 100.
90. Longitudinal comparisons had been carried out in America among several hundred children; full-day care, half-day care, and children in the home. According to Stina Sandels, all of the studies showed that the children who had attended kindergarten had developed better than the children at home. The best results were measured among children in full-day care. Among other things, their IQ was tested.
91. SFRA. (October 11, 1948). Box 105. Minutes from SBR.
92. Bourdieu, P. (1990). 172.
93. Skeggs, B. (1997). *Formations of class & gender: becoming respectable.* London: Sage. 161.
94. Callewaert, S. (1992). 116–117.
95. The Swedish National Board of Health and Welfare. (1994:4) *The condition of children in an age of change.* Stockholm: The Swedish National Board of Health and Welfare; Barnombudsmannen. (1997). *Barndom sätter spår* (Childhood makes a mark). Stockholm: Barnombudsmannen.

References

A. Unpublished Source material

The Teachers Building

The Archive of the National Association of Swedish Pre-School Teachers 1918–1968.

The National Swedish Board of Health and Welfare

The Archive of the National Swedish Board of Health and Welfare 1940–1968.

The National Archive

National Archives (RA). Committee number 1672, "The importance of the day nursery for society."

B. Published Source material

Official parliamentary publications (RT) 1940–68.

Official Reports of the Swedish Government (SOU). (1938:20). *Betänkande angående barnkrubbor och sommarkolonier m.m. Avgivet av befolkningskommissionen.* (Report on child cribs and Summer camps. Produced by the Population Commission). Stockholm: Isaac Marcus Boktryckeri-Aktiebolag.

SOU (1943:9). *Utredning angående statsbidrag till daghem och lekskolor m.m.* [Report on state subsidies to day nurseries and play schools]. Stockholm: K. L. Beckmans boktryckeri.

Newspapers and periodicals

Barnträdgården [The Kindergarten]1919–1968

Arbetaren [The Worker] (1949 Monday 21st March). 66.

C. Literature

Axelsson, K. (1980). *Den första i sitt slag.* [The first of its kind]. Stockholm: Tidens Förlag.

Berkovitch, Nitza. (1997). Motherhood as a national mission: the construction of womanhood in the legal discourse in Israel. *Women's Studies International Forum, 20,* Issues 5–6, 605–19.

Bourdieu, P. (1987). *Kultursociologiska texter.* [Cultural Sociological Texts]. I urval av Donald Broady och Mikael Palme. Stockholm: Salamander.

———. (1990). *Reproduction in education, society and culture.* London: Sage.

———. (1991). *Language and Symbolic Power.* Cambridge: Polity Press.

———. (1995). *Praktiskt förnuft: Bidrag till en handlingsteori* (Practical Reason: Contribution to a Theory of Action). Göteborg: Daidalos.

———. (1996). *Distinction: A social critique of the judgement of taste.* London: T. J. Press.

Callewaert, S. (1992). *Kultur, Paedagogik og videnskap* [Culture, Pedagogy, and Science]. Viborg: Akademisk Forlag.

Carlsson, S. (1966). *Den sociala omgrupperingen i Sverige efter 1866, Samhälle och riksdag del I* [The social regrouping in Sweden after 1866: Society and parliament Part I]. Stockholm: Almqvist & Wiksell. David, M. E. (1993). *Parents, Gender & Education Reform.* Cambridge: Polity Press.

Englund, E. (1989). Arbetsmarknadens kvinnonämnd. In S. Edlund, et al. (Eds.), *Saltsjöbadsavtalet 50 år. Forskare och partner begrundar en epok 1938–1988* [The labor market's women's committee: The agreement of Saltsjöbaden 50 years. Researchers and partners reflect on an epoch 1938–1988]. Stockholm: Arbetslivscentrum.

Ferguson, E. (1991) The child-care crisis: Realities of women's caring. In C. Baines, P. Evans, & S. Neysmith (Eds.), *Women's Caring.* Toronto: McClelland & Steward Inc.

Gordon, L. (1994). *Pitied but not entitled. Mothers and the history of welfare.* New York: The Free Press.

Gough, R. (1994). Från hembiträden till social hemtjänst. In Baude, A. & Runnström, C. (Eds.). *Kvinnans plats i det tidiga välfärdssamhället* [The woman's place in the early welfare society]. Stockholm: Carlssons.

Hirdman, Y. (1989). *Att lägga livet till rätta: studier i svensk folkhemspolitik* [To lay the life in order: Studies in the politics of the Swedish welfare state]. Stockholm: Carlssons.

Holm, U. (1993). *Modrande & praxis. En feministfilosofisk undersökning* [Mothering and practice: A feminist philosophic investigation]. Göteborg: Daidalos.

Holmlund, K. (1996). *Låt barnen komma till oss: förskollärarna och kampen om småbarnsinstitutionerna 1854–1968* [Let the children come to us: Preschool teachers and their struggle for the childcare institutions]. Umeå: Borea Bokförlag.

———(June 1999). Kindergartens for the poor and kindergartens for the rich: two directions for early childhood institutions in Sweden (1854–1930). *History of Education, 28,* no 2.

Hultqvist, K. (1990). *Förskolebarnet: En konstruktion för gemenskapen och den individuella frigörelsen* [The preschool child: A construction for the solidarity and the individual liberation]. Stockholm: Symposion.

Jenkins, R. (1992). *Key Sociologist Pierre Bourdieu*. London and New York: Routledge.

Johansson, J-E. (1992). *Metodikämnet i förskollärarutbildningen. Bidrag till en traditionsbestämning* [The methods in subject teaching in the school of pre-school education: Contribution to a determination of tradition]. Gothenburg: University of Gothenburg.

Johansson, U. (1988). *Uppfostrans förstatligande* [The nationalization of upbringing]. Tvärsnitt nr 1.

Karlsson, G. (1996). *Från broderskap till systerskap: Det socialdemokratiska kvinnoförbundets kamp för inflytande och makt i SAP* [From brotherhood to sisterhood: The social democratic women's association's struggle for influence and power in the social democratic party ([SAP)]. Lund: Arkiv Förlag.

Lindholm, M. (1994). En modern samhällsmoderlighet? Concerning Elin Wägners and Alva Myrdals ideas during the 1930's [A modern motherliness of society? Concerning Elin Wägner's and Alva Myrdal's ideas during the 1930s]. In A. Baude, & C. Runnström (Eds.), *Kvinnans plats i det tidiga välfärdssamhället* [The woman's place in the early welfare society]. Stockholm: Carlssons.

Lirén, G. (1989). *Facklärarna i skolans och arbetsmarknadens perspektiv. En kamp för jämlika villkor 1950–75* [The specialist teachers in the perspective of the school and labor market: A struggle for equal conditions 1950–75]. Stockholm: Svenska Facklärarförbundet.

Lister, R. (1997). *Citizenship: Feminist perspectives.* London: MacMillan.

Löfgren, B. (1994). *Hemarbetsfrågan i svensk politik under 1930- och 1940-talen* [The question of housework in Swedish politics 1930–1940]. In A. Baude, & C. Runnström (Eds.), *Kvinnans plats i det tidiga välfärdssamhället* [The woman's place in the early welfare society]. Stockholm: Carlssons.

O´Connor, J. (1993). Gender, class and citizenship in the comparative analysis of welfare state regimes; theoretical and methodological issues. *British Journal of Sociology, 44* (3), 501–18.

Orloff, A. (1993 June). Gender and the social rights of citizenship, the comparative analysis of gender relations and the welfare states. *American Sociological Review, 58,* 303–28.

Pedersen, S. (1993). *Family, dependence, and the origins of the welfare state: Britain and France 1914–1945.* Cambridge: Cambridge University Press.

Qvist, G. (1974). *Statistik och politik. Landsorganisationen och kvinnorna på arbetsmarknade* [Statistics and politics. The confederation of trade union and the women in the labor market]. Stockholm: Prisma.

Rothstein, Bo. (1994). *Vad bör staten göra? Om välfärdsstatens moraliska och politiska logik* [What should the state do? About the moral and political logic in the welfare state]. Stockholm: SNS Förlag.

Rössel, J. (1950). *Kvinnorna och kvinnorörelsen i Sverige* [The women and the women's lib in Sweden]. Stockholm: YSF:s Förlag

Scott, J. W. (1966). *Feminism & history.* Oxford and New York: Oxford Readings in History.

Skeggs, B. (1997). *Formations of class & gender: Becoming respectable.* London: Sage.

Swedish Chocolate Factory Cloetta Ltd. (1948). *Theobroma Cacao.* Stockholm: Esselte AB.

Waldemarson, Y. (1998). Lo:s Kvinnorad: en paradoxal historia om Framsteg och motgång [Women's Council of Swedish Trade Union Confederation: A paradoxical history about progress and adversity]. En paradoxal historia om framsteg och motgång. *Arbetarhistoria* 22: 1/2: 54–58.

Wilson, E. (1985). Feminism and Social Policy. In M. Loney, D. Boswell, & J. Clarke (Eds.), In *Social policy & Social Welfare.* Milton Keynes, UK: Open University Press.

CHAPTER FOUR

EDUCATIONAL POLICY AFTER WELFARE

RESHAPING PATTERNS OF GOVERNING CHILDREN AND FAMILIES IN ARGENTINEAN EDUCATION

Inés Dussel

Introduction

The title of this chapter is based on Sanford Schram's book *After Welfare* (2000). Analyzing social policy in the contemporary United States, Schram points out that it is important to avoid the trap of welfare advocates who only see exclusion and dismantling of structures of power in recent transformations. Instead, he claims that welfare and its aftermath need to be understood as " . . . a 'technology of citizenship' that empowers people to be citizens but in ways that also disable them. At this point, welfare advocacy may need to begin emphasizing that the problem it confronts is not so much how the liberal order has excluded welfare recipients as citizens, but rather how it has included them" (Schram, 2000, p. 25).

Schram's opponents are also present in current debates about educational reforms and social policies in Argentina. Many people, including scholars, union leaders, politicians, and journalists, share a certain *nostalgia* about the past. When Argentina had a powerful welfare state, we are told, things were better; there was no social and political crisis, no educational crisis, teachers were better paid, students learned more. The problem is that this system did not include enough people, or was not strong enough to counterbalance the works of neoliberalism and globalization. Its failure was more related to external forces than to its own dynamic and productivity.

It is difficult not to be nostalgic in present-day Argentina. Half of the country's population has dropped beyond the poverty line, at a pace of 700,000 persons per month, in the last 6 months. Schools have to attend the plights of parents and families that are unemployed, homeless, and hungry on a daily basis. To talk about reforming the curriculum or even replacing a broken window seems out of place in today's schools.

However, nostalgia is not a good counselor, not only because a return to the past is impractical but also because it is not desirable. My aim is to introduce a discussion that is at the same time political and theoretical. As I will attempt to show in this chapter, the inequalities and injustices of the present have not emerged out of the blue. Moreover, if we insist in dealing with them in the old ways, then it is very likely that we will not produce a more just or equal situation.

Inspired by Schram's arguments, I will debate against/with this nostalgic view, and propose instead that the restructuring of the state and the educational system have to be thought of within a different framework than the one assumed by words such as "neo-liberalism" and "globalization," and by rhetorics of restoration. I will propose that recent reforms have to be understood as part of government technologies that intend to shape the way people are to act, think, and feel about the world, that combine the old and the new in unique ways. These technologies of government are not the expression of a single "will to power" (Rose, 1999a), but rather have to be considered as a "combinatory repertoire" (Hunt, 1999) that adapts and relocates different discourses and strategies in the process of governing teachers. I will ground my analysis on the works of Michel Foucault, Nikolas Rose, and on other studies about governmentality, and especially on Tom Popkewitz's readings of this school of thought from the educational field (Foucault, 1980; Rose, 1989,1999a; Popkewitz, 1993,1998).

The chapter has two sections. In the first, I will discuss current efforts of reform related to the construction of new patterns of governance effected on families and parents that are based on personal responsibility and self-government, and will focus on the production of a new figure, the "needy," to contain those who are incapable of those conducts. In the second, I will analyze the changes in the regimes of truth and the new role of the pedagogical intellectuals and how they have changed the way in which children are spoken of and thought about in schools. In both cases, I will show how the new patterns of government combine heterogeneous discourses, and cannot be enclosed in "neoliberal" or "neoconservative" labels. Also, I will argue against those who see the current situation as the sole product of globalization, and emphasize instead the translation and recontextualization processes that are taking place in the educational field. I hope that the unfolding of these combinations will illuminate the complex works of power, thus providing new grounds for the development of an "inventive politics" (Rose, 1999b).

The current wave of reform, or how diversity became the opposite of responsibility

Argentina did not stand still to the wave of reforms that affected many Western educational systems in the 1990s. "Reform" spread as a "contagious discourse" (Schram, 2000) that brought into national debates categories that were not neutral nor innocuous. Decentralization, accountability, managerialism, professionalization, national standards, were all topics of the rhetoric of educational reform that became popular. They depicted an old-fashioned system, caught between struggles of interests and conservative tendencies. The rhetoric promised to put Argentina "in the twenty-first century" and in the first world, and an imagery of computers, English, and highly dynamic teachers and principals networking collaboratively was offered to picture the "school of the future."

These discourses involved ways of seeing the world, senses of selfhood and otherness, disciplinary knowledges, and power relations that were introduced in the local arenas along with the words used to describe the situations and prescribe the solutions. They were part of "regimes of truth" (Foucault, 1980) that established criteria for judgement and validation processes that had political and ethical implications. They affected how teachers and parents are considered and the kind of actions that they are supposed to perform.

These reform discourses, generally produced in the United States and Western Europe, were quickly adopted in Argentina. They were propagated by state-centered agencies, while the universities remained largely in the opposition.[1] Proposals such as professionalization of teachers, unified curricular contents, site-based management of schools, and decentralization became the cornerstone of change and innovation. New legal and administrative frames were established. A compulsory general education of 10 years replaced the traditional 7-year-long primary school, and a vocationally oriented 3-year "*Polimodal*" education (with several orientations) took the place of secondary schools, traditionally structured, despite the technical or administration schools, by a humanist curriculum (cf. Dussel, 1997).

The institutions that were in charge of teacher training were transferred to the provincial governments. The transference was accompanied by a deep restructuring of teacher education, including the reshaping of the teacher training system into a network with principal and subsidiary members, new standards for accreditation, and new content areas for the curriculum.

While the administrations that held office during the 1990s have diverged in their pedagogical orientations, some being more driven by the goals of participation and autonomy and others by managerial discourses, professionalization has remained a common thread among them. The notion that the teacher has to transform herself/himself into a different kind of practitionner, more academically oriented (i.e., with an updated knowledge purveyed by research findings) and more accountable in terms of her/his results, has received wide support.[2] The reforms have been oriented to give more weight to content areas and to "institutional knowledge" about how schools work. The notion of "institutional knowledge" has implied a new look at the life of schools in teacher education, which was significantly absent in the Normalist tradition—only requiring a small practicum at the end of the mostly theoretical training. But it has also implied the introduction of managerial language into pedagogy and curriculum. Schools are to be thought of as organizations that have to be managed, balancing inputs and outputs, controlling the flux of communications, and putting into numbers the daily interchanges and processes that take place in schools. These numbers will make teaching more accountable and thus, it is presumed, better.

The discourses of reform are related to shifts in the technologies and targets of power, which are described by Nikolas Rose (1999a) in the following way:

> Today, perhaps, the problem is not so much the government of society as the governability of the passions of self-identified individual and collectivities: individuals and pluralities shaped not by the citizen-forming devices of church, school and public broadcasting, but by commercial consumption regimes and the politics of lifestyle, the individual identified by allegiance with one of a plurality of cultural communities. Hence the problem posed by contemporary neoconservatives and communitarians alike: how can one govern virtue in a free society? It is here that we can locate our contemporary "wars of subjectivity." (Rose, 1999a, p. 46)

The education of "passions" through the intervention on people's lifestyles is a substantial change in the way power is effected and the technologies it brings into play. The very idea of "lifestyle" points in a different direction than the liberal technologies of the self, and introduces new practices of identity formation through "the active and practical shaping by individuals of the daily practices of their own lives in the name of their own pleasures, contentments, or fulfilments" (1999a, pp. 178–179). These techniques, on the other hand, are homologous to the ones involved in the reorganization of the workplace and the military, which focus on competency, flexibility, adaptability, and a reeducation of the will ("the entrepreneurial self," as Popkewitz [cf. his chapter in this book] defines it). The individual is responsible for self-actualization, in a continual work upon the self in order to fully develop their potentialities.

In this new configuration of the social, "local" or "private" endeavors are seen as the most democratic and dynamic practices. The discourse on individual and community responsibility involves local institutions and individuals assuming what used to be done by the welfare state, and it links decentralization and democratization in a way in which each educational institution is the primary constructor of the new social agenda rather than the educational system as a whole. One central province, San Luis, has started six charter schools, although they have remained as a confined experiment until now.[3] But responsibility and accountability have permeated other areas as well. In what follows, I would like to briefly explore the links between this discourse on personal responsibility in the school context, and broader social discourses about personal responsibility, and how they have shaped the notions about which families and parents can participate in the life of schools and how they should do it.

Sanford Schram, portraying how welfare reforms have focused on a notion of "responsibility" that is defined in an economic as well as in a therapeutic register,[4] stresses that this kind of responsibility has reinscribed liberal ideals about the subject in terms of new racial, gender, and class relations. For Schram, a discourse on personal responsibility defines citizenship and rights in relation to the ability to work and to provide an income for oneself and one's family. Not being able to work is a consequence of bad habits (addictions, early pregnancies, inability to constitute a two-parent family) and of an incapacity to discipline oneself (Schram, 2000, ch.1). The "welfare queen," an artifact produced by this discourse (Schram, 2000; Cruikshank, 1999), is the epitome of this deficiency: women, and particularly women of color, are seen as especially incapable of controlling their impulses, governing their selves; in short, incapable of living as independent beings. This formulation encourages the idea that, as Schram puts it, "women are more likely to be poor and the poor are more likely to act in a female-like fashion. . . . In this process, not only does poverty get feminized but personal responsibility is again reinscribed as a male phenomenon that women lack" (Schram, 2000, p. 41).

The French sociologist Alain Ehrenberg provides another take on the categories and distinctions introduced by this therapeutic-economic discourse. In a compelling book on depression and society, Ehrenberg points out that the new patterns of governance stress performance and success as the "normal" outcomes of conduct, and produce a new pathology of insufficiency and incapacity (Ehrenberg, 2000). As with Schram´s welfare queen, all those who are incapable or unwilling to discipline themselves in line with the kind of performance needed to succeed will be considered a failure, and will enter into the realm of therapeutic technologies to redeem the self. These changes have deeply transformed the experiences and practices of the self. Ehrenberg points out that the psychological illness of our time is no longer Freud's neurosis but depression, "being tired of being oneself," of lacking sufficient initiative or responsibility. Pharmacology and the sciences of conduct will come to the rescue, helping the individual to cope with this incapacity—although, Ehrenberg argues, they have been complicit in the production of the pathology. Ehrenberg's work is important because it shows to what extent the new patterns of governance are centered on the self, and imply new languages, new categories, and new actions to be taken for/against the self.

In Argentina, most of these discourses have circulated intensely, although with heterogeneous effects. Their dissemination poses interesting questions to think about the dynamics of the global and the local in the circulation and recontextualization of educational discourses. For example, the discourse on personal responsibility has been a central piece in educational reforms, providing arguments for changing the work of head teachers and supervisors, but it has crashed against a firmly entrenched teaching ethos that defines teaching as a wage-based work, subordinated to the central state. One example of this has been the unsuccessful attempts to reform the Teachers' Statute, a protective law passed in 1957 that grants stability and special privileges to teachers. Despite aggressive campaigns in the media, accusing teachers of being lazy and self-compliant, even corrupt, none of the three administrations that held office from 1990 to 2001 have been able to change it.[5] It should be noted that, after the generalized political and social crisis that is taking place since December 2001, it is fairly obvious that these discourses on personal responsibility have very few chances to be efficacious when corruption and ungovernability are widespread, and it would be highly unpopular to blame the unemployed for their situation. However, the discourses on personal responsibility and autonomy did produce some effects on how school principals and supervisors conduct their schools, having become more liable both in juridical and social terms. Several districts have implemented meetings with parents on a regular basis, more in the spirit of client/consumer than on the traditional way of relating to parents, which could be described as paternalistic and "civilizing," oriented by Enlightenment conceptions about schooling and the social (cf. Popkewitz, in this book; Sarti, 2003 about the modern family).

Another example of how these discourses have been recontextualized locally is the case of "diversity," which is read in terms of social deficiency or deprivation. Many compensatory programs have been developed that are called "Attention to diversity," understanding by this mostly the education of the poor, but also the disabled or recent immigrants. Interestingly, poverty is to be considered a sign of a diverse, pluralistic society, and not the effect of injustice or inequality.[6]

As I have attempted to show elsewhere (cf. Dussel, 2000), this absence of racial, ethnic, or gender inflections to discourses of difference has to be explained historically and politically. Simplifying a complex history for the sake of this argument, I would say that in the United States, difference has been primarily articulated in terms of race (Winant, 1994). The "arboreal imagination" of multiculturalism posits the homology between the nation and a tree with a central trunk of common mores rooted in white Christianity (Connolly, 1996). The very notion of minorities, which on the one hand has been so productive socially and politically for the Civil Rights movement, has nonetheless tended to perpetuate the affirmation of a center-majority that is White, male, and Anglo-Saxon, and also to essentialize racial affiliation as a "natural" line (Grant and Ladson-Billings, 1997; McCarthy, 1998).

On the other hand, in Latin America, one could sketch two broad types: the Andean countries, which have been sensitive to ethnic issues (due to the fact that a larger Indian culture has survived the genocide of conquest), and the Southern Cone societies (like Argentina), with weaker Native groups that were quickly exterminated. These latter countries have tended to be more European-like in their patterns of inclusion and exclusion, organized around class lines. The Argentine nation was thought of as a homogeneous and continuous entity, and the only "visible" difference was in terms of social class. In both regions, multicultural discourses have entered the educational field only in recent years, mostly related to the translation of Anglo-speaking authors. But as can be seen in the previous example, this translation relocates the new categories in terms of the old classifications, and has not produced, at least not until now, the kind of ehtnic-centered and self-contained identities that are typical of U.S. multiculturalism.

In the case of compensatory programs, "diverse" has become a euphemism for "difficult," "irrecoverable," "irredeemable." It is assumed that to work in the poorer neighborhoods is to undertake a task that is condemned to frustration. In these environments, the constructivist pedagogies are not useful, because the child's resources are considered to be scarce or inexistent, the patterns of family discipline different, and the background sinful or amoral. It can be speculated that, given this introduction, the discourses of diversity will hardly produce an openness or new awareness of cultural difference among teachers and students, and will reinforce the national imaginary of homogeneity and cultural uniformity.

Thus, diversity stands in fact in opposition to responsibility. Those who are responsible, capable of self-government, of conducting their lives in the right way, are not in need of programs that attend "diversity." On the contrary, those who are deficient in these capacities are to be included in that category. This can be analyzed through the figure of "the needy," the subject that is to receive compensatory programs and who lacks the ability to govern her/himself. While the poor were previously ascribed certain positive characteristics (i.e., they were decent, working people), they were later classified as "dangerous people" (Gonzalez, 1993) or "at-risk populations" (Castel, 1991). The present administration in Argentina has produced a new rhetorical figure, "the needy," turning their demands into a natural incapacity to provide for their own basic needs and abstracting them from the realm of daily politics. The "needy" receive "gifts" and "donations" from "those who care": a paternalistic relationship is reinstalled, which coexists with new discourses and technologies of government in other spheres (and may even be directed at the same population).[7]

The "needy" is the individual subject who is identified by the welfare offices as the object of their policies, due to her/his incapacity to provide for her/his basic needs. The "needy" is portrayed as an individual "voice" whose representation should no longer be assumed by articulated ways of representation, but through individual negotiation with the state agencies. Following this rationale, democracy's duty is, first and foremost, to take care of the basic needs of the population; rules and debates are not as important. In that respect, current reforms involve significant changes in the social organization, as these "voices" emerge as an alternative to organized, unionized, and other traditional means of representation.[8] This is particularly important in countries like Argentina where, unlike the United States, social action was always conceived of in terms of groups or classes, either supportive of or opposite to the state, as with trade unions, mass demonstrations, and multitudinous demands (Rama, 1997, p. 55).

In the case of the National Ministry of Education of Argentina, the "needy" has become the privileged subject/recipient of educational policies, at least until 2001. This entails the displacement of the traditional state agencies involved in the mediation of demands and the construction of a direct, radial connection between the central state and individuals. This displacement is visible in the ambitious compensatory program called the "Plan Social" started in 1992. This program ranked all the schools in the nation according to their infrastructure, the social backgrounds of their students, and drop-out rates, and designated the 1,000 schools that ranked the lowest on the scale to be the recipients of textbooks, new buildings, and in-service teacher training. Along the same lines, the national system of evaluation included the social background of the school population within the "independent variables" to be taken into account (here, again, it can be seen to what extent issues of class have been the central ones). In 1996, the schools with low-income populations that ranked higher were rewarded with special resources. In all of these instances, a complex set of meanings is being attached to "democratic," including redistribution, poverty, and equality of conditions. At the same time, the discussion of procedures, challenges to authority, and public accountability of expenses—meanings that could also be mobilized around the term "democratic"—have been excluded.

In different provincial settings, the discourse of "the needy" is linked to an "ethics of caring" more related to the social doctrine of the Catholic Church than to secular modernization theories and the dominant discourses on efficiency promoted by the national government. For example, in the province of Buenos Aires, the provincial administration has implemented dozens of compensatory programs that range from nutritional supplements to vacations at the beach. Its slogan has been to make available for everyone, especially the "needy," basic goods, in addition to "leisure goods" such as soccer championships and end-of-the-year school trips paid for by the government.

It is essential to point out that the construction of the "needy" is related to a particular discourse on gender. Many of the provincial compensatory programs have been organized by the Woman's Council (in singular in Spanish), formerly the Department of Social Welfare. This council was conceived of as one of the main ways of decentralizing the educational system. It gives money to the parents' associations for lunches and snacks, and distributes free school uniforms, supplies, and sport shoes to those designated as "needy" according to socioeconomic indicators. Intermediate bureaucratic organizations that formerly distributed basic goods, such as the supervisors and school boards, are displaced by this new structure of power. The Woman's Council becomes a centralized agency that relates directly to each school and parents' association, establishing radial connections to the center.

The parallels between the concept of a Woman's Council and particular conceptions of womanhood and nurturing are evident. The council is in charge of distributing food to the "needy" and taking care of the population. It is structured in district committees, run by women that promote an organization by blocks. In each of these committees, a woman is in charge of transmitting information and local demands to the district leaders. Most of these demands are made by the women in the families, who reportedly feel more comfortable talking to other women about their problems. This approach involves a sort of "invasion" of the public sphere by women that has few antecedents in Argentine history. However, these acts of "giving" and "chatting" are considered to be independent from politics: politics involves negotiation and dirty issues that are supposedly not at stake in assisting the "needy." Assisting others has become "a women's issue," while "politics" remains a masculine domain (Auyero, 1997).

As has been said, the "needy" hardly ever intersects discourses on personal responsibility and autonomy.[9] Mostly, it appears as its discursive opposite: one can be responsible, or else enter into the "diverse" populations. It is intended to include those who will not be able to be responsible for themselves, reestablishing a paternalistic relationship with heavy patriarchal overtones between the central state and the poor. While it is too early to see its effects, the current crisis that started in December 2001 has done nothing but intensify that movement, given that the urgency and drama of the social situation has strengthened the hierarchies and populist dynamics of most compensatory programs.

Before moving to the next section, I would like to summarize my argument so far. Throughout this section, I have tried to unfold the many nuances and inflections that the discourses of reform have taken in the Argentinean educational field and that have changed the ways in which parents and families are governed. The figure of the "needy" is a powerful example of the assemblage of several discourses and strategies that constitute the patterns of governance of families. Mitchell Dean has argued that regimes of government always put together features that are heterogeneous—some paternalistic, some liberal, or simply authoritarian (Dean, 1999). As I have said before, these regimes are not the expression of a single "will to power" (Rose, 1999a) but rather have to be considered as a "combinatory repertoire" (Hunt, 1999) that adapts and relocates different discourses and strategies in the process of governing schools, teachers, and students.

The new "regime of truth" in the educational field and the role of intellectuals: The child as a psychopedagogical construct

In the past section, I have argued that the new patterns of governance are related to discursive categories that establish distinctions and authorize particular ways of reasoning and of acting. In this section, I would like to move from the social policy discourse to curricular and educational discourses centered on teaching and learning, in order to analyze the scope and extent to which reforms have reshaped the categories and languages we use to think about and act toward the child.

The reforms have implied a reconfiguration of the rules and procedures by which educational knowledge is produced and validated in the educational system. Knowledge does not only refer to an organized and disciplined corpus of information but also to the ways in which people perceive their social experiences and relate to each other. These ways are inscribed in what Foucault calls regimes of truth, that is, the types of discourses, mechanisms, and means by which particular statements are sanctioned as truth, the techniques and procedures that make this possible, and the status of those who are charged with the sanctioning function (Foucault, 1980). In introducing this notion, it is important to stress that current reforms also involve issues of cultural authorization (Bhabha, 1994); in other words, how discourses and representations are articulated to produce particular regimes of truth, particular visibilities and invisibilities, and how they authorize certain speeches and dis-authorize others.

The regime of truth that is emerging from the recent changes is one that confers a new role to "expert knowledge." In doing that, it goes against a longstanding tradition of social science defined otherwise. Since the Spanish colonization, intellectuals have always assumed the function of the "literate prophets," trying to discipline their societies into the norms and hierarchies of the written word and becoming major allies of the state (Rama, 1997). The emergence of the intellectual field, thus, has been pregnant with this statist perspective, and it could be said that, although many intellectuals have engaged in alternative views of society, they still have not given up their role as "prophets."

It is interesting to trace these changes in relation to how childhood was thought about, and with which discourses. For almost the entire twentieth century, in Argentinean schools childhood was defined as immaturity, incapacity, instinctivity that needed to be governed and protected, both from others and from itself. There were many associations between childhood and delinquency, for example in the laws concerning minority and infancy dating from 1918, but also in the writings of educational psychologists of that time. Two historians of psychology claim that:

> From the academic discourse, medicalized and biologized, the notion of the child gets severed from the figure of the innocent angel, free of sins. In the context of a Haeckelian biogenetic law, according to which the development of the individual recapitulates the stages of development of the species, the fact that the child has outbursts of violence from his primitive and less developed personality in his adaptation to the environment acquires the category of evidence. It constituted an analogy between the child-like stage of humanity and the child-like stages of the future adult. (Ríos and Talak, 1999, p. 142)

The child, then, could not be abandoned to his or her instinctive tendencies but he/she had to be molded through education, an education that was conceived as the outline of natural, evolutionary laws that defined normality. The educational psychology that emerged out of this framework was not, then, one that valued "child development" but one that tried to steer it into a very defined pattern. In this respect, and paraphrasing Nikolas Rose, I would say that the technologies of the self that were effected through the prevalent pedagogy in Argentina were based on a calculable individual, an individual whose performance was to be judged and measured against a norm, but that this individual was not seen as calculating him/herself, as a self-regulatory being that had to enact these calculations on his/her own (Rose, 1999a, p. 133). One can take, for example, the history of physical education in Argentina and compare it to Britain or the United States. While in these two countries sport practices played a pivotal role in the training of the body (see for example Kirk, 1992), in Argentina the practice of sports was explicitly prohibited in 1908 because they were thought to promote too much competitiveness and no solidarity. The privileged form of physical education became the drills and marching. One can speculate that the body imagined by Argentinean educators was one that had to respond to commands but that was not required to calculate its autonomous contribution to a collective endeavor.

The discourses on personal responsibility and diversity that have been described in the previous sections illustrate the changes and continuities that have occurred in the discursive field in education. Traditionally, educational sciences have centered on broad visions of society, either philosophical or sociological, and not on psychological notions of the individual. It could be said that partiality and engagement in social projects have been highly prized as ingredients of good intellectual work

(Schwartzman, 1991). Until very recently, the notion of expertise that portrays the role of intellectuals as giving objective, impartial knowledge (Larson, 1977) has not had the same impact in shaping the intellectual work in Argentina as in other countries. The introduction of the discourses on expertism, thus, can be analyzed as another example of cultural adaptation or translation.

Children are now discussed in terms of reflective individuals whose development has to be guided and oriented by teachers. The reform erases some corpus of knowledge and recognizes others as the most valid and useful. The disciplines that have been put at the center of the field are no longer the philosophies or sociologies of knowledge, or psychologies of deviance, but didactics and pedagogies applied to fields of knowledge (Feldman, 1996). Also, economists and business managers have been hired as consultants in order to bring into the educational field the new discourses of flexible management and efficiency. Teachers are conceived of as "post-professionals" who have to work collaboratively and reflectively, acting as community leaders with increased technological expertise (Aguerrondo, 2000), in order to better understand how to steer children's development.

These new discourses on the child as a psychopedagogical entity are entangled in compensatory programs as well, producing a particular mixture of social, pedagogical, and psychological arguments, as can be seen in the implementation of remedial programs developed by the provincial administration of Mendoza since 1992. For example, a compensatory program for "at-risk" populations focuses on reducing drop-out rates. This program is based on a strategy of "reverse discrimination," which is also at the base of the discursive construction of the "needy." According to Charles Taylor, populations and institutions that share an unequal differentiation are identified, and they benefit from a temporary "reverse discrimination" that intends to gradually "level the playing field and allow the old 'blind' rules to come back into force in a way that doesn't disadvantage anyone" (Taylor, 1994, 40). Differences are temporarily highlighted so that equality will be possible eventually.

The program in Mendoza was designed to combat school failure, and retention rates became the central concern of the administration. The schools that performed the lowest on a ranking determined by the combination of drop-out and retention rates, performance on the provincial standardized tests, and a socioeconomic indicator, were given textbooks, special in-service teacher training, lunch for the children, and specially trained teachers to support children and adolescents "with special difficulties" or who were considered "at-risk." Also, this program has introduced the widespread use of statistics, which has certainly changed the social geography of schools.[10]

A successful school is one in which performance rates are high, disregarding other considerations. Children are ranked in relation to their results in tests, and failure to achieve high scores is a child's problem, not the school's. The links to discourses on personal responsibility and autonomy described above are clear.

The construction of a new regime of truth is made visible through the analysis of this program and similar ones that structure homogenizing tendencies. The programs introduce a new language and techniques to identify and address school problems. Indeed, as happened with the efficientist movements in U.S. education (Kliebard, 1986), the problems themselves are configured as problems of achievement and performance that can be expressed through "objective" scales and charts. Innovation and increased school performance become the focus of the efforts and the key to success and to acquiring more resources. Change is understood as a manageable process that depends on an adequate formulation of goals and a tight control of the execution and produces univocal results. New technologies and agents are authorized through this reconfiguration of the educational field.

This way of reasoning also shapes the uses that schools can make of the autonomy they are supposedly given, because the menu of options from which they can choose is limited. Innovation is understood as the introduction of certain formulas and techniques and equated to remedial programs. The "needs of the community" also tend to be predetermined by what the system has defined, with prewritten templates on how to conduct surveys and questionnaires to parents and families.

The effects of the program are also visible in their breaking apart the old classification systems of the republican school and introducing a new set of rankings that refines and deepens the differences in achievements. Focusing the programs on particular "target-populations" has been a major change in a system that, since the old republican ethics, has always posited the homogenous policies as the most democratic ones. Students are ranked according to standard deviations. The appearance of the "at-risk" language is pervading the way educators think about "learning problems," and the notion of "danger" is influencing how teachers relate to youth.

Statistical and technical knowledge has come to play a new role in the field, and this implies changes for the agents themselves. While educational scientists were previously expected to act as intellectuals, prescribing goals and criticizing policies, they are now supposed to perform a technical role, bringing the "know how" to deal with these new languages and techniques, and the formulas for success. The notion of "public servant," specialized technician, or even that of technocrat, denote these changes in the production of educational knowledge.

The changes have had heterogeneous effects. For example, one of the changes that has faced the biggest opposition is related to the restructuration of the public expenditure. In the province of Buenos Aires, the so-called rational allocation of resources mandates that each classroom and school fulfill a minimum quota of students (35 in the case of classrooms). That means that, if a classroom has 50 students, another section cannot be opened until they reach a total of 70 (2 groups of 35). In some cases, the rationalization caused several teachers and administrative employees to lose their jobs as their schools did not reach the fixed ratios. In others, many schools rounded up their numbers to be able to keep their personnel. Numbers became the central site for struggles. The results were both a police-like control of the statistics by the central administration, with sudden inspections of schools, and generalized chaos in the quantitative data. Although it is not clear how widespread this action is, its emergence nonetheless undermines the claims of validity of statistics as the objective representation of reality.[11]

Again, one can see here the combinatory repertoires that are being mobilized to configure the new patterns of governance. The shift from the critical intellectual to the specialized technician has supposed losses and gains for the individuals who occupy that space in the intellectual field. While they have become more modest in their pretensions, nonetheless they have remained a key player in the field, providing a language that is supposedly objective, neutral, measurable, and keeping the monopoly of legitimate knowledge to speak about classrooms and learning. The managerial discourse has to coexist with the republican tradition of "common education for all," and this has been the source of much contradiction and resistance (cf. Dussel, Tiramonti and Birgin, 2000).

Concluding remarks

> Contracting America underscores how one discourse must of necessity invoke another, given the pervasive impossibility of getting beyond intertextuality. In the quest to use discourse as an attempt to make coherent the incoherences of public life or of life generally, one discourse trades on another, borrowing metaphors for justification, creating an inevitable layering of meaning. . . . The "metaphors of contract" highlight the "contracts of metaphor." Meaning becomes contingent upon the deferred promise of representation. (Schram, 2000, p. 9)

My aim in this chapter has been to propose a theoretical and political reading on the educational reform in Argentina that goes beyond restorative or nostalgic alternatives. I have attempted to show that we miss a lot if we define this reform as simply "neoliberal," "neoconservative," or "globalization effect." I have focused on the multiple discourses that are being mobilized to structure new patterns of governing the child and the families, discourses that are not easily reduced to a single entity, be it ideology, political party, or policy. I have argued that they combine heterogeneous elements that have social, racial, and gender dynamics inscribed in them, and that are produced in a field of relations that have their own dynamics, too. These strategies of reform are plural, as Nikolas Rose and Peter Miller (1992) point out, not because they are derivations of political rationalities but rather are *translations* from the space of political rationalities to the modality of techniques and proposals used in particular locales. It is about these translations that we should worry, and not about the fidelities.

I have also been interested in how these translations have implied adopting "foreign" discourses to local arenas. In a way, my project is similar to Arjun Appadurai's when he says that, "If the genealogy of cultural forms is about their circulation across regions, the history of these forms is about their ongoing domestication into local practices" (Appadurai, 1996, p. 17). While I would speak more in terms of translation than of domestication, I find his remarks pointing simultaneously to the global and the local in the construction of cultures and knowledge.

I have claimed that the introduction of notions of "responsibility," "autonomy," "expert knowledge," which have circulated widely in Western European and North American societies, has had dissimilar effects in a field in which people are used to acting and thinking with different categories and languages. "Diversity," for example, instead of being associated with increased autonomy and respect for plural entities—whatever we may think of this respect or autonomy—has been constructed as the opposite of responsibility, and appears as the strategy design to attend those who cannot act autonomously. "Expert knowledge" has to negotiate with more encompassing traditions of the role of the intellectual, and the introduction of statistical measurements to increase accountability has been undermined by the tactics of teachers and principals.

But to reintroduce complexity and heterogeneity in the analysis of Argentinean educational policy is also to go beyond the reference to disparate mechanisms or entities, to pointing out the "mix and match" of heterogeneity. I believe that there is an important point to be made about any practice of governing: that it is never totally coherent, and that the idea of self-identity and purity are nothing but myths (cf. Valverde, 2000). To look for a "neoliberal policy," be it to support or attack it, would reproduce these myths of self-containment and coherency. Paraphrasing Nikolas

Rose (1999b, p. 97), I argue that it is through analyzing the ways in which we have become what we are that we will be able "to invent ways of becoming other than what we are." I hope that by revising our inventories, what we count and how we count it, we will be able to open up spaces for inventive politics that challenge the injustices that are causing so much suffering today.

Notes

1. A history of educational discourse in recent years is yet to be written. Southwell (2002) provides an enlightening account of the continuities and breaks between the different administrations (radicalism-peronism). Public universities have held the monopoly of critical discourse, but research about "teachers as practitioners" or the relevance of subject-matter content was initiated by their personnel. Cf. Palamidessi, in press.
2. Interestingly, in the debate around the legacy of the Normal Schools, "profession" has been pitted against "vocation." The professionalists argue that Normalists have relied on the primacy of "calling" or "vocation" over "scientific knowledge," turning teacher training into a moralizing endeavor instead of focusing on the content-knowledge that teachers have to impart. What this argument forgets is that "profession" shares the same religious roots of "calling": profession meant initially a public declaration of one's faith, an ideal of faithful service rendered to the community (La Vopa, 1988). This "forgetting" conveniently helps construct an image of the teacher as a neutral practitionner, one whose knowledge and role is prescribed by the objective sciences of teaching. No longer an "agent of the Republic," as the Normalists defined themselves, the teacher has to consider herself/himself a professional whose task is to impart content knowledge.
3. Interestingly, the experience has mobilized a strange coalition of hard-core rightist liberals and leftist libertarians. One school has been named after Eduardo Galeano, the author of an anti-imperialist best-seller in the '70s, "The Open Veins of Latin America."
4. Here Schram follows Nancy Fraser and Linda Gordon's distinctions about the idioms used to speak about dependency in their seminal article (1998).
5. Cf. Llach, Montoya and Roldán, 1999, for a pro-government account of why the Teachers' Statute should be changed. Llach was the Minister of Education in 1999–2000.
6. Homi Bhabha argues compellingly against the notion of "cultural diversity": "Cultural diversity is the recognition of pre-given cultural contents and customs; held in a time-frame of relativism it gives rise to liberal notions of multiculturalism, cultural exchange or the culture of humanity. Cultural diversity is also the representation of a radical rhetoric of the separation of totalized cultures that live unsullied by the intertextuality of their historical locations" (Bhabha, 1994:34).

7. I have developed these arguments more extensively in Dussel et al. (2000).
8. In recent events, a new collective force has emerged, whose effects are yet to be understood. Several organizations of "piqueteros," groups of unemployed, extremely poor people, politically radical and volatile, have emerged, and have been the main actors in Argentina's politics in the last months. The government has included them as central agents in the distribution of food and unemployment insurances, in what constitutes a backward move from the more individualistic policies of the 1995–2000 period.
9. There are some districts that produce such an intersection, for example when the distribution of food or unemployment insurance is tied to an expected product (be it voluntary work, a training course, or written compromise).
10. Statistics embody certain assumptions about the social world, i.e., that it can be represented "objectively" and that the numbers are neutral. Following Ian Hacking's genealogical approach, statistics appear as a technology of power that construct arbitrary classifications of people and social experiences, and configure identities that are embedded in power/knowledge relations (Hacking., 1991) As has been noted in the last 20 years, to talk about "sex" or "gender," or "race" or "ethnic origins" is not neutral or power-free. Each categorization embodies assumptions about the social world, subject positions, and the nature of experience.
11. At Flacso/Argentina, there is a study under way to analyze how statistical information is produced and used in schools. The assumption of the research team is that numbers are neither trusted nor reliable in the current culture of schools.

References

Aguerrondo, I. (2000, July 31–August 2). *Formación de Docentes para la Innovación Pedagógica* [Teacher Training for Pedagogical Innovation]. Paper presented at the conference: Los formadores de jovenes en América Latina en el siglo XXI: Desafios, experiencias y propuestas para su formación y capacitación [Youth Educators in Latin America in the 21st century: Challenges, experiences and proposals for their pre-service and in-service training]. Maldonado, Uruguay.

Appadurai, A. (1996). *Modernity at large: Cultural dimensions of globalization.* Minneapolis: University of Minnesota Press.

Auyero, J. (1997), Evita como performance: Mediacion y resolucion de problemas entre los pobres urbanos del Gran Buenos Aires [Evita as performance: Mediation and conflict resolution among urban youth in the Great Buenos Aires). In J. Auyero (Ed.), *Favores por votos? Estudios sobre el clientelismo politico contemporáneo* [Perks for votes? Studies on political clientelism] (pp. 167–233). Buenos Aires: Ed. Losada.

Bhabha, H. (1994). *The location of culture.* London & New York: Routledge.

Castel, R. (1991). From dangerousness to risk. In G. Burchell, C. Gordon, & P. Miller (Eds.), *The Foucault effect: Studies in governmentality* (pp. 281–298). Harvester Wheatsheaf: The University of Chicago Press.

Connolly, W. (1996). Pluralism, multiculturalism, and the nation-state: Rethinking the connections. *Journal of Political Ideologies, 1* (1): 53–73.

Cruikshank, B. (1999). *The will to empower: Democratic subjects and other subjects.* Ithaca, NY: Cornell University Press.

Dean, M. (1999). *Governmentality. Power and rule in modern society.* London and Thousand Oaks, CA: Sage Publications.

Dussel, I. (1997). *Curriculum, Humanismo y Democracia en la Enseñanza Media (1863–1920)* [Curriculum, humanism and democracy in secondary schooling (1863–1920)]. Buenos Aires: Oficina de Publicaciones del CBC-UBA/FLACSO.

———. (2000). What can multiculturalism tell us about difference?: The reception of multicultural discourses in France and Argentina. In C. Grant & J. Lei (Eds.), *The ideals and realities of multicultural education in global contexts* (pp. 93–114). Mahwah, NJ: Lawrence Erlbaum Associates.

Dussel, I., Tiramonti, G., & Birgin, A. (2000). Decentralization and recentralization in the Argentine educational reform: Reshaping educational policies in the '90s. In T. Popkewitz (Ed.), *Educational knowledge: Changing relationships between the state, civil society, and the educational community* (pp. 155–172). Albany: State University of New York Press.

Ehrenberg, A. (2000). *La fatiga de ser uno mismo. Depresión y sociedad* [Being tired of being oneself. Depression and society]. (R. Paredes, Trans.). Buenos Aires: Nueva Visión.

Feldman, D. (1996, July 25–27). *Quiénes son los expertos? Problemas de la reforma educativa* [Who are the experts? Problems of educational reform]. Paper presented at the Congreso Internacional de Educación: Educación, Crisis y Utopías, University of Buenos Aires, Buenos Aires, Argentina.

Foucault, M. (1980). *Power/knowledge: Selected interviews and other writings, 1972–1977* (C. Gordon, trans.). New York: Pantheon Books.

Fraser, N., & Gordon, L. (1998). Contract versus charity: Why is there no social Citizenship in the United States? In G. Shafir (Ed.), *The Citizenship Debates* (pp. 113–127). Minneapolis & London: University of Minnesota Press.

Gonzalez, H. (1993). El sujeto de la pobreza: Un problema de la teoría social [The subject of poverty: A problem for social theory]. In A. Minujin (Ed.), *Cuesta abajo: Los nuevos pobres y los efectos de la crisis en la sociedad argentina* [Downsliding: The new poor and the effects of the crisis in Argentinean society]. Buenos Aires: Losada/UNICEF.

Grant, C., & Ladson-Billings, G. (1997). *Dictionary of multicultural education.* Phoenix: Oryx Press.

Hacking, I. (1991). How should we do the history of statistics? In G. Burchell, C. Gordon, & P. Miller (Eds.), *The Foucault effect: Studies in governmentality* (pp. 181–195), Chicago: Harvester Wheatsheaf, The University of Chicago Press.

Hunt, A. (1999). *Governing morals: A social history of moral regulation.* Cambridge, UK and New York: Cambridge University Press.

Kirk, D. (1992). *Defining physical education: The social construction of a school subject in postwar Britain.* London & Washington, DC: The Falmer Press.

Kliebard, H. (1986). *The struggle for the American curriculum, 1893–1958.* New York & London: Routledge & Kegan Paul.

Larson, M. S. (1997). *The rise of professionalism: A sociological analysis.* Berkeley: University of California Press.

La Vopa, A. J. (1988). *Grace, talent, and merit. Poor students, clerical careers, and professional ideology in eighteenth-century Germany.* Cambridge, UK & New York: Cambridge University Press.

Llach, J. J., Montoya, S., & Roldán, F. (1999). *Educación para todos* [Education for All]. Córdoba, Argentina: IERAL.

McCarthy, C. (1998). *The Uses of culture: Education and the limits of ethnic affiliation.* New York: Routledge.

Palamidessi, M. (in press). *La investigación educacional en la Argentina: Una mirada al campo y algunas proposiciones para la discusión* [Educational research in Argentina: An outlook at the field and some proposals for discussion]. Buenos Aires: FLACSO.

Popkewitz, T. (Ed.). (1993). *Changing patterns of power: Social regulation and teacher education reform.* Albany: State University of New York Press.

Popkewitz, T. (1998). *Struggling for the soul: The politics of schooling and the construction of the teacher.* New York: Teachers College Press.

Popkewitz, T. (this volume). Governing the child and pedagogicalization of the parent: A history of the present.

Rama, A. (1997). *The lettered city* (J. C. Chasteen, Trans.). Durham, NC & London: Duke University Press.

Ríos, J. C., & Talak, A. M. (1999). "La niñez en los espacios urbanos (1890–1920)" [Childhood in urban spaces (1890–1920)]. In F. Devoto & M. Madero (Eds.), *Historia de la vida privada en la Argentina: La Argentina Plural, 1870–1930* [A History of Private Life in Argentina: Plural Argentina, 1870–1930] (pp. 139–161). Buenos Aires: Taurus.

Rose, N. (1989). *Governing the soul: The shaping of the private self.* London: Routledge.

———. (1999a). *Powers of freedom: Reframing political thought.* Cambridge, UK & New York: Cambridge University Press.

———. (1999b). Inventiveness in politics. *Economy and Society, 28* (3): 467–493.

Rose, N., & Miller, P. (1992). Political power beyond the state: Problematics of government. *The British Journal of Sociology, 43* (2): 173–205.

Sarti, R. (2003). *Vida en familia. Casa, comida y vestido en la Europa Moderna* [Life in the family. Household, food and dress in modern Europe] (J. Vivanco, Trans.). Barcelona, Spain: Crítica.

Schram, S. (2000). *After welfare: The culture of postindustrial social policy.* New York: New York University Press.

Schwartzmann, S. (1991). Changing roles of new knowledge: Research institutions and societal transformations in Brazil. In P. Wagner, C. H. Weiss, B. Wittrock, &

H. Wollmann (Eds.), *Social sciences and modern states: National experiences and theoretical crossroads* (pp. 230–259). Cambridge and New York: Cambridge University Press.

Southwell, M. (2002). Una aproximación al proyecto educacional de la Argentina post-dictatorial: el fin de algunos imaginarios [Educational projects in post-dictatorship Argentina: The end of some imaginaries]. *Cuadernos de Pedagogía Crítica, 10,* 53–70.

Taylor, C. (1994). The politics of recognition. In A. Gutmann (Ed.), *Multiculturalism: Examining the politics of recognition* (pp. 25–73). Princeton, NJ: Princeton University Press.

Valverde, M. (1998). *Diseases of the will: Alcohol and the dilemmas of freedom.* Cambridge, UK & New York: Cambridge University Press.

Winant, H. (1994). *Racial conditions: Politics, theory, comparisons.* Minneapolis: University of Minnesota Press.

SECTION III

THE EMBODIED SOCIAL AND WELFARE STATE

CHAPTER FIVE

CONSTRUCTING A PARENT

Ingeborg Moqvist

In the 1980s a new type of child was codified. It had been under construction for a long time; at least the better part of a century. That child is the child with human rights, internationally recognized and under the protection of the United Nations. The codification defines the status of the child as a complete human being, and might be seen as one type in a succession of child types, for example:

- The not yet fully developed little grown-up with no special sign as "child" except smallness, which Ariès (1962) allocates to the medieval times and some century thereafter.
- The innocent and vulnerable child, filled with promises and developmental possibilities that Rousseau outlined.
- The contemporaneous evil child, born in sin and loaded with a heavy burden of guilt from the original sin.
- The also contemporaneous cheap workforce child combined with the machines of industrialization, aimed to accumulate capital for others; in many respects a sibling to the medieval little grown-up, and probably yet the most common child, worldwide.
- The child in school, guarded by the state by laws to educate citizens and prevent both abuse and too much uncontrolled leisure.
- The welfare child, which is a successor to Rousseau's Emile—measured, weighed, and supported both directly and indirectly by the societies, agencies, and professions from the medical, social, educational, and psychological fields.

And now the universal child with basic human rights. It is an individualized child, considered to be an outstanding human being, not only an appendage to the parents or the family. Yet the importance of the family is stressed, and the child's right to be part of a family. Thus the necessary partner of the child is introduced, namely the parent. As Winnicott once put it, "There is no such thing as a child," meaning that a child is no child without a relating parent or at least someone else *in loci parentis* with which to establish a relationship. Emile had his devoted tutor (and Sophie her mother), the schoolchild teachers as well as parents, and the ill-fated "evil" child had a strict father administrating the repressive *schwarze pädagogik* (black pedagogy) in order to make a human out of him or her. But the laboring child? Is it not abandoned in one way or another and all alone? Could be, and its partner visualized might be the drinking, abusive, and irresponsible parent, or deceased, but the truth is it could also be an industrious low-income or in-debt parent, and either historical or out of our time.

Thus, every conception of the child has its counterpart in a conception of the grown-up. The parent is one of the significant (most would say the primary) grown-up types to consider in connection with the child (the teacher is another). But before I discuss what kind of parent would match the child with human rights, I will further outline the child itself—or rather, the conception of that child.

The conception stresses:

Individuality.
Respect.
Equality.
Competence.

Individuality does not mean that the child is seen without connections, but within the context of family and society at large the child is a unique person, to be respected for both what s/he is and might become. Thus not only the child itself should be protected and supported but also the family. But it is important not to misinterpret the concern about protecting the child's interest as an assumption of the helpless child—no, the child is seen as a competent, but vulnerable, human being. This is not only an ideological assumption; it is also what recent research in developmental psychology states (cf. Stern 2002). Equality is maybe more ambiguous than the other criteria, and should be understood relatively according to generation, while sex, class, and ethnicity ideally should be of no consequence.

In many ways, this conception of the child is similar to what parents *and* children in "ordinary" Swedish families perceive, according to several studies (Halldén, 1992; Brembeck 1992; Dahlberg, 1992; Moqvist, 1997; Karp, 2000). But the conception is by no means limited to this small and fairly well-off kingdom in Northern Europe. Since it is codified in the UN Convention of the Child's Rights, which has been accepted and ratificied by all but two governments in the world (the USA´s and Somalia´s) it must be considered as a global conception, and the child is a universal child. Ratifying the convention means that the state has an inevitable obligation to secure the rights of the children in that state, and that its efforts shall be followed up and evaluated at regular intervals.

Then, what kind of parent does the competent universal child with rights and individuality require? It is certainly not a happy-go-lucky one, with sufficient know-how only by instinct, or tradition. On the contrary, tradition has been broken and a lot of parents find themselves raising children in settings where they feel—and are—alien, brought there out of reasons they did not wish and do not control (such as political, economical, or environmental disasters). And also parents, who still live where they themselves were brought up, often find the total mental environment changed and the world different.

From my study of upbringing and parenthood in ordinary Swedish families it was apparent that parents do not take parenthood lightly (Moqvist, 1997). It was also apparent that the construction of the "normal" child requires a lot of the matching grown-ups, and of society as well. The case of Sweden shows that the welfare state has been highly supportive in several respects, for example with subscribed housing, health service, parental leave, childcare, free schooling, and subscribed leisure arrangements. But it is also well recognized that parents still find themselves somewhat at a loss in their seriously felt task to bring up and educate children in this society, changing and troublesome as it might be. And, not less important, they face difficulties to adjust their different roles in life with the role as parent. The growing demands on what skilful parenting mean have come more or less simultaneously with growing demands on the same parents as labor force.

It seems natural to find parents in need of some help and guidelines in their societal work and obligation to bring up children. And an old and trusted answer to needs like these is education. Education is a potent ideological state apparatus (Poulantzas, 1977) and seems to be the main tool or medium through which top-down and sometimes down-top changes are supposed to happen (cf. Freire, 1972). Plato made a clear connection between a political goal—the just state—and means—education—in *The Republic* (ca 300 B.C.). Rousseau wrote *Emile or On education* at the same time as *The Social Contract* (1762) and had the same purpose as Plato, though the readers of Emile through the centuries might have taken it as a handbook of child education and development. Both Plato and Rousseau can be described as moralistic as well: Human beings have an innate purpose to fulfill. Education shall promote that purpose. Less political and more Christian moralistic was the handbooks or guidelines (from the eighteenth and nineteenth centuries) that Katarina Rustschy (1977) analyzed in *Schwarze Pädagogik* (Black Pedagogy). They were meant for parents to read, learn, and practice in order to prevent the human inherited disposition to sin and thus save the child from evil, and were printed in big editions. In her international bestseller *Barnets århundrade* (The Child's Century), the Swedish author Ellen Key (Key, 1900/1909) claimed that young women should be trained in childcare, nursing, health service, and home economics, just like the young men were trained in military service. Both areas might be seen as a part of the national defense and building a good society. Key's idea remained just an idea. But when Alva Myrdal actually launched parent education classes in Sweden in the 1930s, it was a part of the social democrats´construction of a new society (Myrdal, 1941). That society was the *folkhem,* which meant that the society should be like a good home for all its subjects. By that time, new media that could be—and was—used for all kinds of educational messages had been established. Radio, and later television, popular magazines and film have all been used educationally in Sweden and elsewhere (Rose, 1999).

Education has the same advantages Donzelot (1979) recognized in the alliance between medical science and mothers´ interest in their children's health and well-being. Governing through education is generally efficient, at least if education is interpreted in its broadest sense and not as a synonym to schooling (cf. Rose 1999). Thus education in various forms have been recognized as a prominent part of governing at least since the Enlightenment (cf. Popkewitz & Bloch, 2001).

The ideas and reasons why parents need education have varied. A commonly accepted idea above those already mentioned is to prevent juvenile criminality. Another, and more direct, is to prevent the assault and battery of children and promote a good understanding of and the right support for children's development, as well as supporting the parents understanding of him/herself in connection with the child. Mixed with this reason is also how to handle the difficult life puzzle for all individuals in the family, with all their different roles and interests and with emotional intimacy preserved.

In Sweden the idea of parental courses concerning upbringing is associated with social engineering and building the welfare state (a good home for its both small and big subjects), but also of course with transition, generation gaps, and breaks in traditions. However, the first issue raised when the idea of parents' education was renewed and broadly discussed in the early 1970s, was child assault and battery and how to prevent that. At that time, a little girl was killed, and her case was well publicized. She was three years old, and had urinated in her bed. Her stepfather punished her, and at length she was beaten to death. At the trial her mother claimed that she had not realized that beating a child could be dangerous. After the girl's tragic death, it was also known that neighbors had suspected she was at risk, but had not wanted to interfere.

Though by no means unique, the fate of this girl highlighted how vulnerable children might be in the bosom of the family, with more or less able and caring parents and grown-ups avoiding to see and/or take action to support them. At this time, battering children was not exactly approved of by the authorities, but not forbidden, either.[1] The law was ambiguous. When the patriarchal law allowing the "master of the house" to punish his servants, wife, and children had been annulled in the 1890s, the children were excluded. It was allowed to beat a child with an educational aim, but not so much or in such a way that the child was hurt. This ambiguity, probably together with a great respect for parents´ but not children's rights, made both ordinary people and the authorities reluctant to interfere, and children were abused and sometimes killed without too much public notice.

However, this time people did react. The case was discussed in the Parliament, where children's vulnerability and political ambitions to support them were ventured.[2] Long-lasting opinion- and lobby groups were formed. The common consciousness about children at risk was aroused among both professionals and laymen. More specific: the connected questions of abusive parents; legal support for a certain (but small) degree of abuse; the deeply felt reluctance from both ordinary people and local authorities to interfere in what was conceived as the privacy of the family;and the ambiguity in the law about what means a parent could use in education, were all investigated and discussed in various fora, and (again) put on the political agenda. During the 1970s several parliamentarian or governmental investigations about children were done (like in several other countries), and at least two targeted children at risk.[3] One outstanding result was a law in 1979 that clearly forbid physical and psychical violence toward children, and claimed their right to caring, safety, and good education/upbringing.

It was in the context of this investigation and discussion that the idea of education for parents was placed in the 1970s, in Sweden. Better knowledge and understanding of child development and methods of how to assist the child through the various stages of development was thought to be helpful in this respect, and there was a political decision that the maternal and child health service organizations should provide classes for parental education. All parents should be invited, mothers and fathers alike. However, due to financing problems and changed priorities by health politicians, this ambitious plan was not fulfilled except regarding becoming parents and the phase of parenting that comes before the child is born, if one might say so. But, even if there was no total support for parental education in the end, lots of courses were given and attended to and became regular in some instances.[4] The curricula were mostly about stages and checklists of development; democratic or authorian and non-authoritarian methods; and, in some cases, about raising the level of consciousness of political and economical conditions of families and children.

The direct impact of parental classes compared to other influences should not be overestimated, but more and more people thought it was wrong to beat children, and that you should discuss and explain instead. Different arguments were used, including that raising children with authoritarian methods based on violence would not promote their growing up to become democratic people. Democracy was a value not to be negotiated, and discussions and explanations instead of physical violence are of course more consistent with democracy.

But it is also obvious that neither the law nor a shift in opinion on what constituted "good" methods in upbringing solved all problems. Children were (and are) still beaten, and other forms of abuse were "discovered"—above all various forms of sexual abuse. It is also obvious that the obligation of the state or government to guarantee the child's rights according to the Convention of the Child's Rights, which has been an issue on the political (and other) agenda during the 1990s, renewed the question. A subsequent Swedish parliamentarian committee investigating the subject stated in their report (SOU 1997:161) that the society has both an obligation and responsibility to support parenthood, and thus the children:

> The society has an interest in and a responsibility to give parents support in their parenthood especially due to children's need for safe and developing conditions while growing up and to the children's right to their parents´ care, concern and love. The responsibility for the best interest of the children is everybody's responsibility, and therefore it is an important urgency for the society to take charge of parents' knowledge, engagement and ability and also of their want of support and learning. (SOU 1997:161, p. 9, my translation)

It is stated in this report that the subject is to *empower* parents (the English word used in the Swedish text). It is understood that parents have ambitions and abilities, but need some help and support with a difficult task in a sometimes confusing and troublesome situation. The word "support" in exchange for education is crucial. It is understood that parents are of different kinds in various respects and that the support thus must vary with target group and stage in parenthood. If empowerment is an issue, parents must be trusted to know for themselves what advice they might need, and when. Governing this way is more indirect than through educational classes, and a good example of this type of governing is a small book produced by the Governmental Committee Against Child Abuse 1999. It is called *Föräldraboken* (Parent's book), is printed in booklet form, and is freely distributed. The text is written by a copywriter in collaboration with a psychologist and consists of short stories illuminating common problems in parenthood. There is, for example, the hardworking and tired single mother in economic stress, with her son spending long days in daycare; the worried father losing control, contact, and above all self-confidence regarding his obstinate daughter; and the father that smacks his faulting son, sadly forgetting the humiliation he felt when his father smacked him. The book ends with some short recommendations and a statement on child abuse. The text makes a good example of now common and uncontroversial opinions about upbringing in Sweden today (cf. Halldén, 1992) and can be read as an expression of mentality in a Foucauldian sense (cf. Popkewitz in this volume). And since the publication is translated into eight different languages,[5] in order to make it accessible to the current major immigrant groups in Sweden, it can also be interpreted as a lesson on what parenthood is supposed to be like, or as an outline to construct (Swedish) parents. So, how is this construction made?

First of all, parenthood is understood as a relationship, and the partners are the parent and the child. Parents can be one or two, biological or not, but the important thing is the relation between the grown-up parent and the child. However, the relation is often strained because parents are stressed. The stories exemplify stress by jobs, economic problems, or simply uncertainty of norms and which demands can rightly be put on the child, or how to handle the growing child.

It is also understood that both the child and the parent her/himself are individuals with wills, needs, and rights of their own and sometimes contending. But the child nor parent shall be diminished on behalf of the other. This states a kind of equality. The parent is not seen as supreme to the child other than in experience and wisdom, which puts a greater demand on the parent to be reasoning and patient. On the other hand, the parent is certainly not seen as a superman, but as one that sometimes does bad and stupid things to the child. When that happens, there should be some sense of humbleness toward the child that helps the parent to admit her/his fault and ask for forgiveness and/or tell the child what was the real cause behind the anger or quarrel. This humbleness should be balanced against the requirement of the parent not to abdicate and give up being a grown-up. The parent is supposed to finally set the norms, with respect and concern of the child in her/his mind, and make the child keep them.

Further, it is made very clear that a parent-child relation normally contains problems. The problems are possible to solve and are a part of the ordinary family life. It might be noted, that no uncommon or (in a Swedish setting) atypical family problems are mentioned, such as severe child abuse, juvenile delinquency, or strict demands on a daughter's behavior according to "traditional" norms in an immigrant parent's home country. This underlines the impression of a commonly shared normality; "this is about 'us.'" The concrete message is that the parent is responsible for solving the problem, and it is important that the parent keeps her/his status as a grown-up. That means that the parent shall use her/his greater wisdom and experience and be able to put up with decisions that might be uncomfortable for the time being, but good in the long run. It means also, and this is maybe the most important and consistent message, that the parent must try to see the situation from the child's point of view. This is underlined in the stories by describing the situation from both sides.

The mode to solve problems is by talking: explaining, listening, discussing, and hopefully reaching if not consensus at least a mutual understanding. Talk, or more precisely giving and taking arguments, is as I have mentioned earlier, considered an inalienable part of democracy and thus both a goal and means in a democratic education. Talk, reasoning, is also what should be used instead of punishment—it is pointed out that spanking is not allowed and that hard or scornful words can hurt just as much as blows. Furthermore, sometimes talk is needed to ask for forgiveness. It follows from both the recognition of all the "normal" problems modern life contains and the ideal of equality that a parent can do wrong against the child and thus be in the position of asking for forgiveness. In short, the construction seems to be a more humble and less self-confident variation of the authoritative parent (not to be confused with the authoritarian).

A partly different—and less official—construction of parent I want to discuss in this paper belongs to an international context, namely the European Union (EU). It was developed in a project about parents´ education with partners from six different countries.[6]

The original purpose of this project echoes well-recognized rhetoric on education and makes the Swedish statement about support to parents cited above look modest in comparision. According to the initiating seminar it was concluded that parent education programs could:

- contribute to the reduction of juvenile crime
- promote the mental well-being of families
- improve educational performance by children
- reduce the incidence of family breakdown.

These far-reaching goals connects the ambitions in this project with old, traditional ideas about governing family and children (cf. Rose, 1999). Whether they are fulfilled or not is hardly measurable. More narrowly, and possible to evalutate, it was also stressed that parent education courses should:

- target all parents, not just those perceived to be at risk
- have as leaders parents who had received training
- be offered to fathers as well as mothers.

Here I shall discuss the project as an attempt to construct a special kind of parent, partly as a match to the child with rights, partly as the *European* parent.

As it turns out, the European parent of today comes from all over the world, is of varying educational standards, and has very different traditions or mores and religious confessions (if any). Some of them live in more "modern" parts of Europe than others. Some are mothers, others fathers, some even grandparents in charge of grandchildren. Some are gay, some lesbian, and most are heterosexual. Some live in nuclear families, others are single, and some are stepparents. There is a great abundance of variation.

The agents involved in the project represented a broad range of ideologies and modes of welfare systems. Two were family centers attached to churches—the Greek Orthodox and the Catholic Church, respectively. Two were mainly educational organizations for grown-ups, one connected to the local authorities, and the other to a nongovermental organization. The three remaining were parts of the local authorities concerning social and family matters.

But, regardless of this variety concerning the administrating organization and the target groups of parents, and belonging to different policies of welfare (see chapter 1, this volume), there should be something in common to unite people as European. One aspect of that might be the (in the program) proclaimed and programmatic respect for variety both between the different countries involved and within a specific country, depending on local conditions. This declared respect might be interpreted as a caution not to impose values "from above" (and thereby eventually erase opposition opinions), or as true respect.

The basic principles that the program asked parents to adhere to were few. It should be founded on the UN Convention on the Rights of the Child, and especially article 12, on the child's right to be heard on matters of concern to it. It should be nondiscriminatory and attract different groups (regarding ethnicity, religion, occupational status, etc.) and fathers as well as mothers. Group leaders in the classes were supposed to be parents, not experts on child-rearing or education. Last, but not least, it held a positive assumption that parents are doing a good but demanding job that should be supported ("parents are doing their best and deserve support as a right").

It was also decided that the program should be a non-spanking one, though it took some discussion between the partners to reach something like consensus on that. The discussion on corporal punishment revealed different basic ideologies, which in its deepest sense was about democracy and the concept of the child.

It was decided that the written material should be kept short and promote a pedagogy based on maieutic dialogue rather than on informing. It was a major concern also that not only parents who are used to reading should feel comfortable with it, but also those less literate.

The major goals of the program (to contribute to the reduction of juvenile crime; promote the mental well-being of families; improve educational performance by children; reduce the incidence of family breakdown; target all parents, not just those perceived to be at risk; have as leaders parents who had received training; and be offered to fathers as well as mothers) directed the content. The course material thus presents topics supposed to erase discussions on:

- Family patterns
- Changes in the conditions for family life and especially being a mother and a father
- Child development and following that as a developing parent
- Communication skills that enhance self-esteem
- Emotions
- Norms, values, and setting limits
- Providing a good and safe environment for children.

The first message is that from time to time every parent faces difficult situations due to breaks in traditions or changes in the society and the conditions ruling everyday life. Some changes are outside the family, others inside. But families and parents are not supposed to handle all problems just on their own. There is help, and since parents are doing an important job they are entitled to support and help if needed. It is a right for everyone, not only those that in one way or another might be failing in this job. Besides, there is a common interest in parents doing the job and not resigning from it, even if the family itself falls to pieces and has to be rearranged.

A second message is the need to cope with the fact that family patterns are not the same as they used to be. This is so, both regarding the formal family structure and the relations within the family, and it concerns both gender and generation. For example, it follows from the Convention on the Child´s Right, that the power in the family should be shared in some way by parents and children. Especially important is the right of the child to be listened to and respected in questions of concern to it. Thus the demand is strengthened on the parent to adjust to the child according to its development. The parent must develop parallel or consequently with the child.

A third message is about communication. Raising a child in democracy takes a lot of talk and discussions. Norms and values might be constantly negotiated, especially when both the child and parent operates in a constantly changing world, and both the parent and the child are supposed to have her/his saying. But problems should be solved with verbal arguments and listening, not slaps. Communication is also an important emotional tool, and the mode of communication should promote good feelings and help the child to develop a good self-esteem.

Finally, there is a message about "what every parent ought to know" about child development and securing a safe environment for the child.

The content must be regarded as mainstream, but the pedagogy implied was not in some of the settings. The democratic study circle was the model, nonhierarchical and promoting dialogue as it is and thus was thought to enhance dialectical ways of communication—hopefully transferable into other relationships and primarily with the children and other family members. In some settings this seems to have been quite a novelty and not always well received in the beginning, when it met parents´ initial expectations on expert knowledge.

According to the evaluations the participating parents did after the courses, they were found profitable, helpful, and encouraging, and most participants were contented with both the content and outcome.[7] They appreciated the opportunity to discuss and share experiences with other parents and claimed that this had given them both new perspectives and useful ideas about several practical problems. They also said that their understanding of the process of upbringing was improved and many seemed to have gained more self-confidence as parents. The parents thought they received good and useful information and were challenged to think things over. This was done in a way they found interesting, nice, and efficient and made people feel enlightened.

The impression from the parents´ evaluations is that most of them really think they have changed and learned other and better ways to interplay with their children. And the bettering lays for a good deal of them in being more listening and reflective: "My relationship with my children has improved in that I do try to consider what is actually being said to me before I open my big mouth and say the wrong thing. Also learning to like their company more."

Changes in the way to perceive the child are also expressed. These statements are about individuality (to "recognise individual personalities of children and different needs") and to have more confidence in the children's capabilities. A remarkable change in view of her children—and of her concept of children—was expressed by a mother of four who wrote: "It learnt me to relate to my children in a different light and to look at them as human beings, not children when they are trying to speak to me." Others mention being more attentive to needs, and express positive feelings more often.

The overall and main gain seems to be on understanding, reflecting, and communicating better. Does this also indicate that children are more respected and their rights better protected? And democracy promoted? Good understanding and communication skills can of course be used in different ways, also very manipulative. Rousseau´s tutor in *Emile* is a superb example of that, as is Gordon's "transactional analysis" (represented as a handout in the manual). The only witnesses on this point are some data from the Spanish partner, who did ask children of the parents taking the course about the outcome. Fifteen children of at least school age answered a questionnaire given to them toward the end of the course.

Those children say that, as a result of the training, they expected their parents to become more aware and able to communicate and understand their needs and wishes. A little more than half of the children also declared that they had noticed a remarkable change in their parents´ behavior, for the better. However, a few children thought the change was for the worse. It may be overinterpretating, but this might suggest that the children in these cases felt that their parents were coming too close in some way and perhaps being more manipulative. Children adapt to their parents behavior, for good and for worse, and changes disturb old manners. If the child does not feel that the parent has a good will toward it, a change like that is not welcomed, and the data on child abuse and neglect does not give any support to the presumption that *all* parents are good enough parents. Most parents certainly are, and have their children's interests well in mind and as a priority, but it cannot be considered a rule without exceptions.

Then, what about the *European* parent? The data do not give much information on this, except from the discussions when the program was settled and the common guidelines finally were agreed upon.

One of the basic differences in values and social reality that was revealed was about the demand that courses should target not only mothers, but also fathers, and therefore male as well as female group leaders should be recruited. Most found this natural and good, but the Greek partner hesitated and the Spanish agency strongly objected to the realism in this as an initial demand. Men would not be interested enough, they claimed. One of the others, representing a Catholic family center in Ireland, advised that the courses directed to fathers (several thought it better to have gender-segregated groups) should be labeled other than "parents course"—his center had, he said, run that kind of stuff under the heading "how to handle stress," which certainly is an issue for many fathers (as well as mothers).

Other subjects the partners held different views on were the pedagogical style and corporal punishment, which I have mentioned before. Spanking, or slapping, caused a lively discussion. Some defended it as more or less indispensable in education; others saw it as something to be avoided but did not think it was very important; and still others saw it as a fundamental question in the context of democratic values and respect for the child.

These questions highlighted the very idea of a *common* program on parenting for different settings, both nationally and otherwise. One of the Irish talked on this, referring to the long tradition of humiliation of the Irish people and their loss of language and identity. Now, when they had gained something of that back, should they be bereft Irish family values in order to become Europeans? He argued most passionately that the program must be explicit that cultural and religious values of each country be respected. On the other hand, the social program for the delegates included a visit at an Orthodox monastery, where a bishop addressed them with a speech saying that they had come to the crib of European culture, and were welcome to take part in its wisdom, but that if they had come to impose foreign ideas or knowledge, they were not. These were somewhat extreme standpoints, but it was obvious that the idea of being European evokes hesitations, if it means having to give up some values one finds quite natural and good. The solution within the programme was, as I have mentioned, to stress flexibility, and though that was done, the evaluations from the first courses indicate that it ought to go further, and not only for reasons of nationality but also all other variables the groups contain. But, as always in educational processes, if too much respect is payed to what is already there, one might miss the object of what is supposed to come.

Conclusion

One of the children in the inquiry mentioned above pointed out that a task as dutiful as parenting surely needs training. Thus a course in parenting was good. Like many of the ideas of the welfare state, there is an alliance between personal or private interest from ordinary people and the society or state (cf. Donzelot, 1979). In this paper I have treated the child´s interest as truly stated in the UN Convention of the Rights of the Child and the child and state or "society" as allies. Following that, the parent has been regaded as an object to influence and (eventually) to change for the better. Most parents certainly want to give their children the best possible upbringing/education, and as Durkheim (1911/1975) has pointed out, no one can educate her/his children only according to her/his own wishes. The upbringing must suit the present time, societal environment, and other conditions. In education, he says, there is no such thing as *the* method or *the* goal. Every time and social setting have its ideal and appropriate means. The ideal now adopted by most governments in the world (though certainly with more or less clear ambitions to realize it) is a child with—compared to the child's status in the majority of countries—far-reaching rights and heading for democracy.[8] Since education (in all forms) is an old and trusted way of implementing reforms and new thoughts or ideologies (cf. Poulantzas, 1977), parent education thus appears to be a logical solution if the aim is to construct a partner to the newly constructed child. And in Europe, at present, this constructive work can be combined with another—the construction of the European subject.

In this chapter, I have tried to show examples of governing in the realm of family and child-rearing. It is obvious that they are tiny parts of a greater pattern of governmentality. It is also obvious that a certain kind of parent cannot be constructed by texts or courses alone. What the two cases reveal are two slightly different but very firm conceptions of what children are entitled to and what parents are supposed to provide, and two possible ways to reach that. One is somewhat subtler than the other but both aims at governing through mentality.

Notes

1. However, corporal punishment was forbidden in the Swedish schools as of 1958. It was then considered incompatible with the school´s common goal to foster democratic citizens.

2. The report of proceedings of the Swedish Parliament, March 25, 1971.
3. *Barn som far illa/Children at risk* by Socialstyrelsen/National Board of Health and Social Service, och Allmänna barnhuset/The Common Orphanage 1971–1975 and *Barnets Rätt/The Child's Rights,* by a parliamentarian committee (SOU) 1977–1987. Other investigations concerned children's conditions in the contemporary society in general, schooling, daycare, and family support. See for example Rose (1999) for an overview of similar investigations in the UK. The international interest in questions on children at the time was manifested by the UN committee on children's rights. It was established in 1979 and presented to the UN Convention on the Child´s Rights in 1989.
4. Different agents have hosted these kind of classes, mostly attached to common health service or to educational organizations, and then mostly non-governmental.
5. Arabic, English, Kurdish, Persian, Somali, Sourani, Spanish, and Turkish.
6. The origin of this particular program was a transnational seminar on Parent Education held in Waltham Forest (a suburb of London) in June 1995. The seminar was funded by the European Initiative for the Prevention of Urban Delinquency and Reintegration of Young Offenders. It was a Socrates Dialogue project, and dialogue was stressed in different ways. My part has been the evaluator´s, together with Lydia Sapouna, Department of Social Work, University College of Cork in Ireland. The evaluation report is unpublished.
7. When all six agencies that took part in the program had run their first courses (which are those the evaluation concern), a total of 127 parents with almost 300 children, from newborn babies to middleagers, had taken part. Of these parents 10 were fathers, and of the related children a majority was boys. Most participants were native in their respective country, but differed in occupational and educational status; mainly between groups but also within.
8. As Woodhead (1999) has pointed out, the declaration is founded on values and conditions for children in the Western world.

References

Abrahamsen, G. (1997). *Det nödvändiga samspelet.* [The necessary interplay]. Lund, Sweden: Studentlitteratur.

Ariès, P. (1962). *Centuries of childhood: A social history of family life.* New York: Vintage Books.

Brembeck, H. (1992). *Efter Spock: Uppfostringsmönster idag* [After spock: Childrearing patterns today]. Doctoral dissertation. Gothenburg, Sweden: Skrifter från Etnologiska föreningen i Västsverige. 15.

Dahlberg, G. (1992). The parent-child relationship and socialization in the context of modern childhood: The case of Sweden. *Annual Advances in Applied Developmental Psychology,* 5: 121–137.

Donzelot, J. (1979). *The Policing of Families.* New York: Pantheon Books.

Durkheim, E. (1911/1975). *Opdragelse, uddannelse og sociology* [Education and sociology]. Copenhagen, Denmark: 11x18 Samfund.

Freire, P. (1972). *Pedagogy of the Oppressed.* Harmondsworth, UK: Penguin Books.

Halldén, G. (1992). *Föräldrars tankar om barn* [Parent's thoughts about children]. Stockholm, Sweden: Carlssons.

Karp, S. (2000). *Barn, föräldrar och idrott: En intervjustudie om fostran inom fotboll och golf* [Children, Parents, and Sports]. Doctoral dissertation, Umeå, Sweden: Department of Education, Umeå University.

Key, E. (1900/1909). *The century of the child.* New York: G. P. Putnam's Sons.

Moqvist, I. (1997). *Den kompletterade familjen. Föräldraskap, forstran och förändring i en svensk småstad* [The Augmented family]. Doctoral dissertation. Umeå, Sweden: Department of Education, Umeå University.

Myrdal, A. (1941). *Nation and family: the Swedish experiment in democratic family and population policy.* New York: Harper.

Plato. (2000) *The Republic.* Edited by G. R. F. Ferrari. Cambridge, UK: Cambridge University Press.

Popkewitz, T., & Bloch, M. N. (2001). Administering freedom: A history of the present. Rescuing the parent to rescue the child for society. In K. Hultqvist & G. Dahlberg (Eds.), *Governing the child in the new millennium.* New York and London: Routledge Falmer.

Poulantzas, N. (1977). *Fascism and dictatorship: The third international and the problem of fascism.* London: NLB.

Rose, N. (1999). *Governing the soul: The shaping of the private self* (2nd ed.) London and New York: Free Association Books.

Rousseau, J. J. (1762/1988). *Émile or On Education.* New York: Basic Books.

Rutschky, Katarina. 1977. *Schwartze Pädagogik.* [Black Pedagogy] Berlin: Ullstein.

Statens Offentliga Utredningar (SOU). (1997). *Stöd i föräldraskapet:* Betänkande av Utredningen om föräldrautbildning. [Support in parenthood: Report from the Commission about parents´ education] (No. 161). Stockholm, Sweden: Fritzes.

Stern, D. N. (2002). *The first relationship: Infant and mother.* Cambridge, MA: Harvard University Press.

Woodhead, M. (1999). Combating child labour: Listen to what the children say. *Childhood, 6* (1), 27–49.

CHAPTER SIX

EARLY CHILDHOOD EDUCATION

THE DUTY OF FAMILY OR INSTITUTIONS?

Loïc Chalmel

When one compares the organization of the various school systems in Europe at the dawn of the twenty-first century, it has to be said that the often laborious construction of a unified economic community is certainly far from having encompassed in its wake the development of a set of coherent teaching curricula; thus today it is impossible for the advised observer to draw the outline of a single European educational model (Vaniscotte, 1996). This diversity in the methods of reception of pupils at the various levels of their schooling as well as the diversity to be found in the aims of teaching, seems to become even more complex concerning education in early childhood, which is often excluded from the time of compulsory schooling.

The Contemporary Approach

A disparate network of structures

The institutional forms assumed today by preschool structures throughout Europe are multiple: nurseries, crèches, kindergartens, nursery schools, etc. The modes of public or private financing characterize the political choices of the countries concerned regarding prescolarization, and on the whole determine the conditions of access the various social classes have to a particular category of institution. Moreover, families often have the choice, within the same geographical context, between several places of reception to anticipate the compulsory schooling of their child, which makes analysis of the situation even more complex. However, it is rare to find states that have developed institutions with an educational purpose that are not integrated into the global school system (EURYDICE 94[1] & OMEP 98[2]). For example, the coexistence of structures that have resulted from different historical social actions constitutes an alternative to the family choice. Also, the attempt to diversify what is offered to families corresponds with the will displayed by European authorities to propose major centers that have a more flexible operation than within the school framework, often better adapted to the needs of families: "to encourage the flexibility and the diversity of the children's child care services within the framework of a strategy having as its aim to increase choices and to answer the preferences, needs and specific circumstances of the children and their parents, while preserving a coherence between the various services" (CCE, 1992[3]).

The age of access to compulsory schooling differs appreciably from one country to another: from 4 years old (Northern Ireland and Luxembourg) to 7 years old (Denmark). Thus, to the late nature of entry into "school" institutions in Scandinavian countries, is opposed the postulate of the school that is considered the universal reception structure for children after three years in the Latin countries (France, Italy, but also Belgium) (Vaniscotte, 1996, p. 34). Once again, this great internal diversity in Europe concerning the earliness of compulsory schooling is related to the family and educational policies specific to each country, and illustrates the social effort that is authorized regarding early childhood. If the Eastern European countries of the old Eastern Bloc—strongly influenced by Marxist ideology—and those in Northern Europe—of social democratic sensitivity—develop the idea of a collective responsibility for the reception, education, and health of all children whatever their age, the development of prescolarization within the community of free enterprise countries is entrusted to specialized staff, who have undertaken further training in conformity with the requirements of the global education system. In this case, the financial investments relate to the institutional framework of the school: the public money invested in educational measures is thus clearly separated from the budget devoted to social measures. Consequently we shouldn't be astonished by the massive and generalized regression of prescolarization in the Eastern European countries as a consequence of their passage into the bosom of a "free exchange" economy and cultural system (OMEP, 1997). This apparent dichotomy must nevertheless be moderated by a de facto situation: certain countries, whatever their ideological membership, have never really developed a coherent preschool system. It is generally those countries that develop individual responsibility as opposed to a collective assumption of responsibility (United Kingdom). The type of assumption of responsibility for early childhood—between collective or individual education is related to the family model, the educational or social purpose of the preceived of responsibility and political and ideological choices. It also depends on specific traditional cultural representations related to childcare and education within each community.

However, contemporary research in preschool education, with its various approaches—psychological, sociological, didactic—attests in a relatively homogeneous way to the extreme importance of initial learning for the future psychological balance of a child. Therefore can we legitimately ask ourselves why certain countries still refrain from creating nursery schools, when others are convinced of their importance and their role in the development of a child? We can also ask ourselves, Why do many states see in prescolarization a key place for the future structuring of learning and personality, whereas others only conceive it as a nursery or a place of socialization?

Misleading convergences

Beyond undeniable intra- and international divergences previously described, a certain number of recent changes nevertheless strive to bring institutional operations closer together, aiming at reducing the effects induced by cultural ideology or practices. In particular, beyond local singularities, the adoption by the majority of the European states of an apparently more consensual position, on a certain number of common ideas—the determination of the age of compulsory schooling to six years old or the systematized reception of all three-year-old children—should be related to the results of research in education since the 1970s that has focused on the genesis and the psychological development of young children. It is only Scandinavian countries that are still unsure about bringing into alignment their legislation with other European states. Some of their neighbors, among whom the possibilities of reception for early childhood were traditionally limited, have for their part lowered the age of obligatory prescolarization to at least five years old (United Kingdom, Netherlands) and, for some (e.g., Luxembourg), even to four years old (EURYDICE, 1995).

However, the idea according to which institutional attempts at harmonization significantly modify practices in the field hardly stands up to analysis. First of all, the methods of application of reforms vary according to the framework of ideological reference that is appropriate for the implementation of the various family policies. Then, a progressive modification of routine behavior cannot occur without interfering with the values and the cultural practices that underlie the educational practices in a given context. Thus, although the interest in the education of a young child develops generally, as a result of the conclusions of research, undertaken in the framework of objective scientific knowledge, and despite the fact that the large majority of young Europeans are sent to nursery school at least by the age of five, differences in opinion on the aims of preschool education remain.

Latin countries and those from the BeneLux countries are concentrating all their efforts on the development of a modern nursery school, par excellence tool of socialization and integration. A large consensus on its privileged role in the reduction of handicaps of a social origin make it possible to ensure everyone that, from a point of view that is often idealized in terms of equal opportunity, a cognitive development in conformity with a national standard exists in these countries. This policy of compensation for sociocultural handicaps is coupled with an objective regarding preparation for fundamental learning: a nursery school is certainly first and foremost a school, an integral part of the education system in spite of the characteristics related to the age of its pupils who confer upon it its relative identity:

> The nursery school is thus certainly a school because we start to structure learning there. Teachers are very close to pupils. They give them support and stimulate them while trying to understand their difficulties. A nursery school is a privileged place for detecting problems, and even handicaps and for acting in order to prevent them as early on as possible. It is also the first place where children understand the meaning of writing and language codes. They handle books, locate titles, establish a difference between the letters and the pictures and cultivate their memory. At nursery school, we use the rough data of experience and little by little we help children to structure this data. (Roussel, 1995, p. 2875)

In contrast to countries in the South, for whom the concern of curing social inequality with early schooling is of primary importance, the states of Northern Europe set up less school-like institutions, and instead create reception facilities that are more ludic (playful) and more open toward the world of the family. The process of socialization within a kindergarten, for example, is carried out within a more individualized framework, purporting to respect the rhythm and the needs of each individual. The normative nature of early learning that is omnipresent at the nursery schools in the Latin countries and in Benelux countries is here considerably undervalued, focusing instead on the general development of personality. Moreover, if the rates of prescolarization can appear close on consulting statistical data specific to each institution, these rates must be balanced according to the relative duration of the time during which pupils are taken in charge: in conformity with the school rhythm of the country of origin at nursery school, children may be in the different institutions only a few hours in the day or even in the week for many kindergartens. This very short time regarding assumption of responsibility in the institution compels mothers to have greater availability, which isn't without problems concerning adapting to modern life (OMEP, 1997). On the other hand, these mothers retain a privileged status as early childhood teachers, which contributes to the implementation of a double process of socialization for their children, simultaneously internal and external to the family world. The family policies of the various states arrange, with more or less pleasure, the entry of women into the labor market by supporting the choice of working or taking care of their children, while making it possible for mothers to resume employment after parental leave, and/or by financing minding the child at home by the mother at least until the age of three. This implicit status of privileged teacher, complementary to the institution, recognized in a singular way by mothers and more generally parents, has induced legitimate concerns about training for "the job of being a parent":

> Our contemporary society hardly provides any training for the job of being a parent. We do not suddenly put a secretary in an office without having taught him or her how to typewrite . . . simply saying to him or her : "you are a secretary, go ahead!" But when a woman becomes a mother, it is as if society abruptly says to her : "You are 'a mother', we have not taught you a lot about the job, but go ahead, do your best!" . . . There is no job in our society that is more difficult than being this one of a kind all-rounder, working twenty-four hours a day, both a child psychologist and a teacher whom we name a parent. (Dodson, 1972, pp. 23–24)

Historical sources

Thus the educational purposes of nursery or preschools and those of kindergartens or childcare programs are not identical, in both policies and practices. The former concentrate the essence of their efforts on preparation for elementary school, whereas the latter, who do not necessarily depend on the ministry of education, focus more attention on the intellectual, physical, and moral development of a personality in the making. The underlying ideological designs, heirs of traditions, history, and culture that are specific to the various countries, diverge on principles as essential as the role of play in learning, a socialization accentuating the primacy of the collective in relation to a more individualized structuring of identity, the openness to families, and the singular place granted to mothers in initial education. The adaptation of preschool education to modern living conditions also resists the weight of history, and the reception of children under three years in Germany as in Switzerland, despite the insistent requests of families, is not a priority for the political authorities. The Fröbelian tradition permeates kindergartens and a recent law confirms the traditional role granted to the family by German society (Kinder und Jungendhilfegesetz, 1991). Therefore we should hardly be surprised if certain states affirm that a child has sufficient cognitive maturity to approach certain learning criteria at the age of four, while others take several years to access knowledge of a comparable nature.

Beyond impassioned teaching debates to determine the most judicious mode of prescholarization for the *future* Europe, none of the currently privileged options is the fruit of chance: it therefore seems legitimate to wonder about the historical origins of these divergences, by supposing that a common starting point paradoxically corresponds to various orientations that are sometimes radically opposed. The examination of the teaching steps suggested by the precursors of education for very young children confirms the assumption of a common source of inspiration evolving in a subtle dialectical debate, with Christian tradition as a reference, an essential context for any pedagogic prepositivist reflection. The differentiated developments of preschool education in countries of Protestant tradition and the French-speaking or German-speaking communities of federal states like Switzerland or Belgium indisputably present the marks of this contrasting evolution. To look for possible coherence between the origin and the development of the idea of prescolarization in Europe, it is necessary for us to temporarily abandon the contemporary educational field in order to explore the theological and historical context that generates it, without losing sight of the essential focus of this chapter: What is the import of this original message? In what condition does it reach us and in what way does it still influence modern practices?

Historical Approach

The highlighting of significant meeting points between the evolution of the innovative spirit of Protestant pietism and the teaching steps suggested by the precursors of preschool education, from Jean-Amos Komensky

(Comenius) (1592–1670) to Jean-Frederic Oberlin (1740–1826), Jean-Henri Pestalozzi (1746–1827), and Friedrich Fröbel (1782–1852) to Pauline Kergomard (1838–1925), invites us to trace the ways in which the values embraced by pietists influenced the development of two of the original institutions for preschool education: the German kindergarten and the French nursery school. However, these institutes, nourished by the same wish on the part of pietists to organize in an early way the young child's sensitive link to the world and his creator, correspond to two dissimilar contemporary teaching models, at least in appearance. It is the genesis and the evolution, in a determined economic and social context, of this surprising contradiction that I now wish to analyze.

Comenius' alternative

The Piétistic-Moravian ideal is forged in the sphere of influence of an original attempt to reform the Church, and through it, all feudal society in Bohemia (Chalmel, 1996, pp. 7–18). With regard to pre-school education, the famous educationalist Jean-Amos Komensky (known as Comenius), heir to the educational Moravian tradition, wrote the first significant treaty in connection with the education of very young children. In this referential text, he locates the development of early childhood between two universes, the two terms of its alternative being irreducible to each other. He recognizes the determining importance of family education and singularly that of mothers, but at the same time affirms that a child never develops any better intellectually, physically, and morally than in the presence of his peers (Comenius, 1992). Thus he adopts a middle position, voluntarily ambiguous. On the one hand, the mother appears irreplaceable, whatever the socioeconomic conditions of the moment in question: "My opinion is not to remove children from their mothers to entrust them to teachers before the age of six. Here are my reasons: initially, young children require more care than a teacher who is responsible for a crowd of children can give; it is thus preferable that children remain surrounded by the care of their mothers . . ." (Prévot, 1981, pp. 169–170).

On the other hand, the collective nature of teaching overrides parental or maternal tutorship: "parents and tutors are very useful to children, but the company of other children even more so. They tell each other stories and play amongst themselves. Equal children by age, knowledge and courtesy, mutually sharpen the spirit. Nobody will doubt that a child contributes better than anyone to sharpening the spirit of another child" (Prévot, 1981, p. 157).

The Comenien speech thus remains resolutely ambiguous and is confined to a compromise with vague limits between a natural maternal education and the recognized superiority of community teaching over tutorship. In the line of Comenius, the educationalists of the eighteenth century recognized the essential role of the mother for education in early childhood; neither science nor reasoning are useful where instinct, guided by blood ties, is enough: "It is in the child's first years, when we rely on the development of his faculties, without any help from art, with the great impulse of nature, there lacks generally and above all a mother who, favoured with the advanced training of her own faculties, has at the same time, in a conscious way, the sharp, balanced and matured stimulant of what she should naturally be for this child" (Pestalozzi, 1826/1947, p. 282).

Mothers or Institutes?

Pestalozzi adopts the first term of Comenius' alternative as his own, regarding the mother as the single and essential mediator in the sensory apprehension of the world by her child. If the mother is absent or if her personality is unimportant, the educational relationship is null or ineffective and the child is left to his own devices; if the mother is possessive or authoritative, the place left to the child hardly enables him to be the initiator of his learning and makes this learning inoperative or ineffective. From this point of view, Pestalozzi constantly insists on the concept of a "happy medium" in his educational speech. Conscious that any idea of balance is essentially fugitive and precarious and does not seem very compatible with the concepts of progression or continuity, he is not deceived by the ability of mothers to achieve the noble task that he sees as theirs by nature.

We then witness a change in the pedagogic question of the young child toward his mother: the education of the latter then becomes a great issue for the Swiss educationalist. For this purpose, a treaty published in 1803 is entitled "The book of mothers," and unambiguously announces its finality: It is a question of making available to them "A whole series of processes which will enable them to prolong the relationship between their heart and their child until the day when the means likely to facilitate virtue, associated with the means likely to facilitate the intelligence of things, will allow the child, matured by the exercise, to adopt an autonomous position in all that relates to right and duty" (Pestalozzi, 1801, p. 220).

Mothers' responsibilities, for Pestalozzi, are considerable, weighed down with consequences as they condition the later success of the methodical process. Deeply convinced of the fact that the bonds linking women and children are the paradigm of what the later union between man and his Creator should be, Pestalozzi grants mothers, in the first years of human life, a decisive and exclusive role: "Let elementary teaching never be the business of reason, let it remain the business of women for a long time before becoming the business of men" (Pestalozzi, 1801, p. 216). Thus, this state of relative dependence, during which the child thinks, discovers, feels his familiar surroundings through the eyes of his mother, decreases gradually with the years, in a way inversely proportional to the development of his directions and reason according to Pestalozzi. It then enters into a transitional period during which the "home" universe (*Wohnstube*) becomes gradually too narrow, where interest is focused on objects that are outside the familiar environment, where the need for autonomy starts to be faced with the protective maternal will: "The feelings of the very small child disappear and make way for the first impressions that are independent from the mother and caused by the charm of the world" (Pestalozzi, 1801, p. 213).

It is at the conclusion of this transitional period that Pestalozzi locates the essential crossroads of any successful education. It is truly a new birth; the child leaves the protective "egg" of the *Wohnstube,* in which he perfected a sensitive and moral structure that makes him ready to face the world. It is at this crucial time that the school itself must be able to ensure the relay for the educational mother by deploying all its art to preserve intact in the pupils' hearts the feelings of love, gratitude, confidence and obedience. Pestalozzi makes out of this stage of development a major phase, where all is played irremediably without the possibility of going back on oneself. Either the child, guided by an enlightened educationalist, will make a success of the shift in natural feelings binding him to his mother toward a creative God, or, indeed delivered to the exclusive material pleasures of the world that surrounds him, he will be irremediably corrupted and will become the toy of his own significant impressions: "All the natural teaching which will be given to him in this new career can be found in narrow and stimulating cohesion with what he endorsed under the gaze of his parents. The family direction thus closely joins the school and what is taught there" (Pestalozzi, 1826/1947, p. 291).

The Pestalozzian process subjects maternal education to the risks of a balanced mother /child relationship, which again imperceptibly induces the idea of "lack." The introduction of a third initiator would have the advantage of placing the educational act outside this dualistic relationship, thus saving the mother from an unrealistic image of exemplification fraught with consequences, and enabling the child to establish a link with the multifaceted world. This idea of "lack," understood and analyzed by Friedrich Fröbel, leads to the introduction of a mediator referee in the bivalent relationship, the kindergarten mistress, in the genesis of the end product of the "German" approach to a preschool concept: the kindergarten.

Jean-Frederic Oberlin builds the institution of the "knitting rooms" *(poêles à tricoter)* around the second axis of Comenius' proposals, namely the idea of early collective education. The Pastor of "Ban de la Roche" by no means disavows the irreplaceable role of the mother in the first years of life, but this militant of the evangelical faith cannot be satisfied with an idealistic position completely removed from the socioeconomic reality of his time. If the mother/child relationship can be privileged in the most favorable cases, extending this type of tutorship to the entirety of a population inevitably leads to the conservation or even the progressive aggravation of social differences. However, a more fraternal society, where every man is able to enter into spiritual research, cannot accept ignorance and financial and intellectual precariousness. The "knitting rooms" constitute the driving force of the intellectual and social transformation in "Ban de la Roche" and allows young women, "the leaders of tender youth" *(conductrices de la tendre jeunesse),* to be recognized in their role as teachers. For Oberlin, female sensitivity, which is irreplaceable in leading a small child toward his future, does not necessarily imply the presence of a mother. A girl, recognized for her moral qualities, can favorably substitute in a number of cases for the defective mother, thus allowing women to pass from the implicit recognition of an active role in the education of their children to the status of teachers as a whole, paid and socially recognized for their work. The families excluded from the educational process as such are nevertheless regularly associated with the examination of its products and informed of pupils' progress, due to the system of public "recitations," developing the "school" nature of the "knitting rooms" within the population.

We find in the directing principles of the teaching practices developed by "the leaders of tender youth," the essence of the Comenius approach, summarized in the formula: "from the bottom upwards." Any teaching is concrete initially, the progressive distance in relation to the object of study makes it possible to build, by successive reformulations, the concept: the child learns within an ascending spiral in which he remains the center. The body constitutes from this point of view the privileged physical and sensorial medium to build initial knowledge and it is solicited under multiple aspects (physical, hygienic, etc.). In the same way, if the heat of a stove is welcome in winter, when deep snow covers the cold ground, the return of spring incites walking, with active discovery in symbiosis with the medium of life. It is a true awakening pedagogy with which the "leaders" try to associate their pupils. Another founding principle of the pedagogy of the "knitting rooms" relies on the fact that any learning necessarily generates production. If knitting constitutes the paradigm of it, this precept also applies to the constitution of herbaria, the drawing or the impression of plants. Lastly, the education of tender youth is synonymous with joy and a good mood, with respect to the rhythm of the child. For this purpose, a large place is left to repetition so that the new facts have many occasions to be memorized in spirits: "Here, song accompanies work; sometimes we spell by heart; we tell instructive stories within the reach of childhood; in the summer we pick plants, from which we learn the names, the distinctive aspects, virtues. Drawing, the illuminator of geographical maps of a small size which Mr. Oberlin engraves in wood and prints, in a sufficient number, maps of Ban de la Roche, France, Europe, the Planisphere, have passed from these schools to families, and still provide today the recreation of Sundays" (Neufchateau, 1818, p. 20).

Play quite naturally finds its place as a privileged didactic means in the daily practice of "leaders." The need to play, characteristic of early childhood is present in learning. Several "tools" therefore come to the rescue of teachers to kindle the desire for learning: teaching toys imagined by the Pastor, collective or what are called "society" games, and objects that come from the natural history collection.

A resolutely modern debate

The dialectical and mutual questioning that is remotely established between the Master of Yverdon and the Pastor of "Ban de la Roche," outlines the contours of the various current options concerning the education of very young children. Starting from a corpus of values and common purposes, their fundamental divergences appear to be centred on the conditions of its development. In fact, the finalities of a reasoned maternal education that Pestalozzi wishes to promote with his "Book of mothers" and those of Pastor Oberlin's small knitting schools are the same: it is primarily about moral training, which makes it possible to weave insoluble bonds between the small child and his Creator, in order to build a fairer and more fraternal society in the future. However, the means to achieve this are appreciably different. The installation of the "knitting rooms," when it exists in a specific way, only differs from that of *Wohnstube* (the home/hearth) in its higher dimensions allowing the reception of about 30 children; the usual surroundings of the life of small schoolchildren is perfectly taken into account in the institution. But teaching does not remain confined to between the walls of the small school; in the beautiful season it is in the open book of surrounding nature that pupils and "leaders" gather the matter that is necessary for the learning process. In this sense, the universe thought out by Oberlin seems an opening onto others and the world. It is opposed to the protective environment of *Wohnstube,* which is presented in the form of a closed space, a laboratory for the development of the human spirit in which one period of intellectual maturation complementary to the biological time of maturation is achieved. Oberlin, in the spirit of Comenius' *Orbis pictus,* causes a triangular interaction in the center of which he places one by one his small pupils. The three interacting poles of this triangulation are represented by the "leaders," substitutes for the Pestalozzian mother (by nature, in her physical and human dimensions) and by the other children whose similar but different behavior interferes with the "leaders" of interaction with the children. The mediation, the numerous connections between the various participants in the educational relationship are made possible by manual work, the active principle from which intellectual learning is built. The relationship with the other makes it possible for the young child to be alternatively co-actor of his learning, in an identification process, but also an observer of the steps and behavior of the alter ego, which allows the progressive structuring of an original personality within the framework of a dialectical identification/differentiation. Thus, the child can alternatively be perceived either in the center of the educational relationship or in its periphery.

The *Wohnstube* schoolchild for his part is included in a bi-directional relationship that only functions in conformity with Pestalozzian principles when the two poles (mother/child) are partners in a participating dualistic relationship. But if the later methodical process is based on taking into account intuitive knowledge acquired within the family, *Wohnstube* is in fact excluded from the school system thought out by Pestalozzi. With the development of his autonomy, the child progressively needs his mother less and less. It is with the generalized development of the methodical process in elementary schools that Pestalozzi hopes to ensure perenniality and to *reinforce* the effectiveness of a maternal and family education:

> But if the Method could perhaps one day be applied to primary schools, then there is reason to hope that the girls who have studied it and who will become wives and mothers will, if they want to, be able to carry out the wishes of its author, while becoming themselves the first teachers of their children. These exercises, as we have already been able to see, are only a conversation which she can have with her child at any hour of the day, I will even say that in the middle of any occupation, she can start at the house and begin again in the countryside; in other words all this asks so little from the

teacher, the steps of the latter are so fully traced, that in very little time the mother of a large family is able to make her eldest children the tutors of her juniors. (Chavannes, 1805)

Contrary to this, Oberlin's "small school within school" becomes the paradigm of all the other institutes, an essential element for the generalized diffusion of the French language and the development of evangelical values: "From the bottom to the top." We find similar contemporary divergences when we observe the objectives posted respectively by the educational structures of a school or nonschool type. E.g., the knowledge acquired by the child at nursery school (the ecole maternelle of France, for example), appears essential to his later success, whereas at kindergarten (the *jardin d'enfants* or "child's garden"), it is first and foremost the development of his personality that will make it possible to establish, in a natural way, and at the appropriate time, useful knowledge without true preparatory work.

The idea according to which any learning generates production remains a driving principle of the activity developed by the "leaders" in the "knitting rooms." The practice of knitting know-how with economic and commercial aims, constitutes the general policy developed by Oberlin to improve the precarious economic status of his parishioners. This concept of utility corresponds to the spirit of the Pestalozzian experiment of "Neuhof," in which the Swiss educationalist addresses older pupils who come from a working-class background. Paradoxically, his speech on early childhood proposes a very spiritualized option, which is much more distant from the economic constraints that incontestably differentiate dialectical confrontation from the intelligence of the child with the reality of the surrounding world. Thus, Pestalozzi's project seems better adapted to well-off backgrounds than to socially underpriviledged populations, which is not the least of paradoxes for someone who is also defined as the teacher of the people. In addition to the moral teaching of their young pupils, "the leaders of tender youth" are concerned with their intellectual development and anticipate their socioprofessional future. The institution of "knitting rooms" is not an additional part, but one of the components of a broader, global, and coherent educational system.

In the nineteenth century Friedrich Fröbel and Pauline Kergomard, heirs of the pietistic Moravian thought, attempted the gamble of a difficult closer connection between the two terms of Comenius' alternative (Chalmel, 1996/2000). Between the school and the family, kindergarten mistresses, mediators of the educational relationship between mothers and their children, as well as teachers in nursery schools, responsible for replacing in the best way possible the workers occupied ten hours a day in manufacturing, confirm the specific recognized role of women within the framework of two different educational systems. The teaching models destined for early childhood, inspired by Moravian pietism, thus moved in time to mainly give rise to two different institutions, the French nursery school and the German kindergarten, whose finality actually remains of the same order.

Socialization or identity building? Enclosing or opening? Mothers or teachers? These questions, the fundamental heritage of the Comenius alternative, define the terms of the debate that has been incessantly renewed since the eighteenth century, on the methods of care and education of early childhood in Europe (and in many other countries where these ideas were imported). They forecast the difficulty of engaging in a constructive dialogue between European educators who are in charge, without regarding the historical heritage of the existing systems as a prerequisite for exchanges. None of the options that are currently being developed are free from this historical and cultural marking. To recognize such a dimension contributes to a better understanding of the historical and cultural contexts for the divergences, and may aid in the difficult construction of a more consensual idea of preschool education in a unified Europe.

Notes

1. EURYDICE is the information network on education in Europe.
2. Organisation Mondiale pour l'Education Préscolaire (World Organization for Early Childhood Education).
3. Centre Coordonné de l'Enfance.

References

Centre Coordonné de l'Enfance (1992). Unpublished manuscript.

Chalmel, L. (1996/2000). *La petite école dans l'école. Origine piétiste-morave de l'école maternelle française* [The small school within the school. The Pietist-Moravian origin of the French Maternal school]. Paris, Berne: Peter Lang.

Chavannes, A. (1805). *Exposé de la méthode élémentaire de Henri Pestalozzi.* [An expose of the basic method of Henri Pestalozzi]. Paris: Levrault et Schoell.

Comenius, J. A. (1632/1992). *La grande didactique* [The great didactic: Setting forth the whole art of teaching all things to all men]. Paris: Klincksiek.

———(1658/1727/1887). *The orbis pictus.* Syracuse, NY: W. C. Bardeen.

Dodson, J. (1972). *Osez discipliner* [To dare to discipline]. Kehl-am Rhein: Trobisch.

EURYDICE. (1994). *L'enseignement préscolaire et primaire dans l'Union Européenne* [Preschool and primary school teaching in the European Union]. Bruxelles.

———(1995). *Les systèmes d'enseignement et de formation initiale* [Systems of teaching and pedagogy in initial training]. Bruxelles.

Kinder und Jungendhilfegesetz (1991). Law published by German government.

Neufchateau, F. D. (1818). *Rapport fait à la société Royale et Centrale d'Agriculture sur l'agriculture et la civilisation du Ban de la Roche* [Report made to the Royal Society and Center of Agriculture on Agriculture and Civilization in Ban de la Roche]. Paris: Huzard.

Organisation Mondiale pour l'Education Préscolaire. (OMEP). (1997). *Éduquer le jeune enfant en Europe* [To educate the young child in Europe]. Paris: United Nations Educational Scientific and Cultural Organization.

Pestalozzi, J. H. (1801). *Comment Gertrude instruit ses enfants* [How Gertrude teaches her children]. Berne et Zürich, Sweden: Gessner.

———(1826/1947). *Le chant du cygne* [Swansong]. Neuchâtel, Switzerland: La Baconnière.

Prévot, J. (1981). *L'utopie éducative de Comenius* [The utopian education of Comenius]. Paris: Belin.

Roussel (1995) *Ecole maternelle* [The maternal school]. Paris: Bulletin Officiel. (No. 95–836)

Vaniscotte, F. (1996). *Les écoles de l'Europe* [The schools of Europe]. Paris: Institut National de Recherche Pédagogique.

World Organization for Early Childhood Education (OMEP). (1998). *Our World?* Östervåla, Sweden: Elanders Tofters.

CHAPTER SEVEN

TEENAGE PARENTHOOD IS BAD FOR PARENTS AND CHILDREN

A FEMINIST CRITIQUE OF FAMILY, EDUCATION, AND SOCIAL WELFARE POLICIES AND PRACTICES[1]

Miriam David

Introduction

"Teenage parenthood is bad for parents and children" (Social Exclusion Unit, 1999a, p. 90) was a statement from an official British document prescribing poverty and education policies to tackle social exclusion arising from teenage pregnancy. The U.S. government made similar statements in developing "programs for abstinence education" for teens to teach "that bearing children out-of-wedlock is likely to have harmful consequences for the child, the child's parents, and society" (Social Security Act #510 1996). Most recently, President Bush refused to endorse a UN declaration on children's rights unless UN plans for sexuality and health education taught only abstinence before marriage (*New Statesman,* May 17, 2002).

Teenage sexuality as an educational and social problem is high on global public agendas. The "risks" are seen as consequential not only for teenage parents, but for future generations of children and for "society." Teenage parenthood is identified as "bad" or "harmful" for the young women involved who are deemed to be too young, not married, and assumed not to have planned their pregnancies. But their feelings and desires are not addressed, and neither are issues of sex and gender, directly. Policies to prevent these problems and ensure personal responsibility and social inclusion through education for employment and citizenship have become major public debates under regimes of economic and neo-liberalism. Both British and U.S. governments, whether ostensibly conservative (Thatcherite or George W. Bush) or democratic (New Labour under Blair or New Democrat under Clinton) in the last decade, have justified their approach to sex and relationships by identifying international transformations in family life, the economy, and social welfare. Young parenthood, constructed in gender-neutral language, is seen as a moral and economic issue. Education of traditional family values or of "abstinence-only until marriage" is proposed to ensure personal responsibilities rather than reliance on the state and social welfare. Previous liberal solutions to pregnancy, particularly among young women, and birth "out-of-wedlock" (and the latter term itself engenders a notion of imprisonment in marriage) were centered on private approaches such as adoption or abortion. The focus of these neoliberal regimes is on the reassertion of traditional family values as a public solution to personal problems.

The aim of this chapter is to analyze official discourses about teenage sexuality, and the public rather than private approaches to personal intimate issues through welfare policies. The debates on teenage parenthood are emblematic of changing welfare discourses about families and the balances between children and parents through global transformations. I explore the complexities and contradictions in these shifting discourses as proposed solutions to socioeconomic and familial transformations and contrast the neoliberal approaches with those of more social democratic regimes. Gender, race, and social class matters have also become interwoven in the fabric of policy debates about the complexity of sex and family, but not with any fully explicit recognition. The education policies for personal and economic responsibilities are double-edged solutions to complex sex/gender issues since little attention is paid to traditional gender-based approaches to family life, which contrasted the responsibilities of motherhood with those of fatherhood. I conclude by considering the reasons for these cultural shifts and their implications for changing family lives, and those of teenagers especially, where gender has been neutralized.

The New Debate about Teenage Sexuality as a Personal and Social Risk

The debate about young people's sexuality has been raised dramatically by neo-liberal regimes, but the public policy question of teenage sex is not new. What is relatively distinct in the neoliberal discourses is the emphasis on personal issues on a public educational agenda (Burdell, 1995/6; Lees, 1997; Kelly 2001; Zellman, 1981). These discourses have addressed a number of socioeconomic changes entailing the risks and safety of young people. Certain groups of children have been constructed as "at-risk," in terms of health, such as HIV/AIDs and drugs, and lack of families' support, social care, and long-term employment. Social and economic changes have indeed transformed family lives, from moves away from the legal institution of marriage and divorce to either single parenthood or consensual and cohabiting unions, and the employment of parents of young and dependent children (Haskey, 1998; Duncan and Edwards 1999; David, 2001a). More generally, transformations in family lives and education have affected social class composition (Ball, Maguire and Macrae, 2000; Bernstein, 1996; Ehrenreich 1989).

The shifts are part of a far more complex pattern of restructuring of social welfare, education, and employment (Clarke and Newman, 1998). Traditional liberal approaches were seen as incompatible with socioeconomic transformations, and neoliberal strategies have centered upon policies about social exclusion (Bullen, Kenway and Hey, 2000; Kelly, 2001). The traditional social democratic emphasis on collective, social responsibility has been transformed under economic and neoliberal regimes toward more individual personal responsibility and, at the same time, more diversity and individuality, making for more question of risks from the wider society (Scott, Jackson & Backett-Milburn, 1998; Jones 2001). Social welfare changes toward individualization from social protection have emphasized more personal responsibilities, and the need for more constant assessment and reflection on our lives. Indeed, as I will show, the term "personal responsibility" is now entrenched in legislation in the United States (Schram, 2000 p. 27–58).

These policy discourses also raise the question of addressing these shifting agendas through education for employment rather than through the more traditional channels of social welfare. New Labour in Britain have, for instance, developed a whole raft of policies reinvigorating education with this personal and social agenda; shifts in personal, social, and health education, along with citizenship education (Arnot and Dillabough, eds. 2001). Yet these education policies, deriving from policies of social exclusion, focus mainly on those young people deemed "at-risk"; vulnerable groups within the population, rather than as general policies for all young people within school. The risks for such young people are presumed to be manifold, but center on their lack of opportunities for employment. In this respect traditional gender roles within the family have been discounted; in the drive to place employment center stage, although in contradictory fashion, traditional family values of marriage underpin the revisions to educational approaches to sex and relationships. Despite acknowledgement of family and social transformations, traditional family values are still articulated, in the context of religious values about marriage and family life (Monk, 2001; 2000). Moreover, the emphasis lacks an explicit concern with gender; forms of exclusion are assumed to be about all socially disadvantaged young people who are deemed "at-risk." This is despite the acknowledgement that educational, social, and economic changes have affected boys and girls differentially, and especially how social networks and social capital have affected new generations of men and women differently (Franklin, 2000; Walkerdine, Lucey and Melody, 2001). Bullen, Kenway, and Hey (2000) have argued that "redefined notions of equity and social justice inform New Labour's understanding of the risks of teenage pregnancy and motherhood and how its teenage mother policy functions as risk management."

Social Liberalism and the problem of teenage sexuality and motherhood

Despite official claims about changing sexual and social conditions (SEU, 1999a), there is evidence to suggest that such questions of teenage sexuality and parenthood have been of concern to international policymakers over the last 50 years (Allen, 1987, Burdell, 1995/6, Lees, 1997, Middleton, 1998, Phoenix, 1991, Hudson and Ineichen, 1991, Kelly, 2000, Lesko 1990, Zellman 1981). The tenor of the current debates, entailing public matters as personal rather private, is what Driver and Martell (1997) refer to as the "cultural turn." The debates have developed on an international scale, but there is variation among European countries such as Holland (Lewis and Knijn, 2002), and in Canada (Kelly 2000), Australia (Kenway and Bullen 2001), and New Zealand (Middleton 1998 and Allen, 2001).

In the postwar era of social democracy in Britain, education to prevent teenage pregnancy and parenthood was not addressed, but the social problems of lone parent families (of whom many were young mothers) became public (Kiernan, Land and Lewis, 1999). Pressure groups such as the National Council for One Parent families focused on the question of girls who were "pregnant at school" (1976). They showed that there was an issue of a small number of such girls, less than 5,000 per annum, and proposed alternative educational solutions, such as special educational units for teenage mothers (Ineichen and Hudson 1991; Kelly 2000). The terms of debate about the problem of young and teenage women becoming pregnant have changed (Allen, 1987).

In the United States the question of teen pregnancy and parenthood was also a long-standing public policy issue, although framed differently as a problem of "urban poverty." Social welfare debates throughout the 1960s and 1970s era of social liberalism centered on aid to families with dependent children (AFDC), crucially what was termed welfare in the United States. This focused mainly upon large metropolitan areas, although there was also a growing need in some rural areas (Piven and Cloward, 1976). The rising rolls constituted a constant and significant problem, seen as a fiscal rather than purely moral concern. Most targeted for moral censure however were those young and usually teenage women in receipt of welfare, nicknamed "welfare queens or mothers" (Schram, 2000 p. 27).

Moreover, in most Western countries, the ages of marriage and/or motherhood have dramatically changed over the past two centuries. It was both more acceptable to be a young mother in the nineteenth century and not seen as related to employment or education, and yet the issue was also hidden and covered up (Lees, 1997; Kelly, 2000). Thus even in the early postwar period, whatever social welfare solutions were available, such as private adoptions or even "shotgun marriages," were relatively private. The availability of abortions in Britain for instance remained private until the late 1960s and early 1970s, and even after the passing of the 1967 Abortion Act, its availability for young women was difficult (Young and Greenwood, 1975). Similarly in the United States, after the Supreme Court decision of *Roe v Wade,* in 1973, the availability of abortions for young and single mothers continued to be contested (Petchesky, 1980).

The particular solutions to teenage sexuality, pregnancy, and parenthood contrasted with approaches adopted in other European countries such as the Netherlands (Ingham and van Zessen 1998; Kelly, 2001; Ketting & Visser, 1994; Phillips, 2001). In the Netherlands, a strongly social democratic regime, sex education constituted a component of public education throughout the postwar period, and part of emotional, health, sexual, and psychological programs (Lewis and Knijn, 2002). In Australia, New Zealand, and Canada, educational programs were also less punitive (Middleton, 1998; Allen, 2001; Kelly, 2000). There was less emphasis on marriage, sex within marriage, or abstinence outside of a traditional marital framework, or on the problems of "at-risk children." By contrast, there has been more emphasis on normalcy and necessity of teaching not only about sexual bodies but also about social and sexual relationships as psychologically healthy. Thus the constructions of sex and sexuality have focused upon how integral a part of normal biographies and family lives they are (Allen, 2001; Walkerdine, Lucey and Melody, 2001).

"Personal Responsibility" as a public discourse for educational policy

In the 1990s in both the United States and Britain, policy discourses on sex and sexuality education were transformed, in association with social welfare shifts. I argue that Britain "borrowed" (to use Silver's term, 1991) quite unashamedly from the United States in reconstructing its social welfare policies, particularly those for young people in "public" or state education. Together these policy developments around sex and sexuality education, in association with the personal responsibility elements of social welfare, constitute the emblematic shifts in public policy. They combined a traditional focus on fiscal conservatism with a moral agenda on family values and a cultural shift in our understandings and perspectives on young people and young women as mothers in particular.

During the Thatcherite era in Britain there was a re-invocation of Victorian family values, especially through education (Arnot, David and Weiner, 1999). This marked a dramatic transformation in public policy perspectives on sex and sexuality, heralding the first public debate on sex. It was the occasion for the inauguration of a preventative strategy through sex education, and the emergence of public debates about sexuality and sex as subjects for schools. But such strategies were part of a broader approach to reinvigorating traditional "family values" (David, 1993). Parents were to be involved in decisions about the teaching of sex education and allowed a choice in whether their children should attend or be withdrawn (Epstein and Johnson, 1999). Marriage was re-invoked as a necessary component of such teaching. While this remains central to developments in the curriculum, a shift in focus toward wider debates about values and relationships has been raised as necessary for such education, representing an aspect of the "cultural turn" (Driver and Martell, 1998; 1999).

In the 1980s in the United States, also a period of neo-liberalism influenced by Reaganism, educational programs began to emerge as potential solutions to such problems of young people in school (Burdell, 1995/6; Lesko 1990; Zellman 1981; Schram, 2000). Here too there was a growing focus upon "family values" and a re-invocation of marriage. Emergent programs of "abstinence-only-until-marriage" education were developed initially to deal with the economic and fiscal problems of poverty in relation to welfare rolls. However, it was only closely linked to funding in the 1990s, as part of a much wider compromise about social welfare supports in the contexts of wider social and economic transformations (Schram, 2000). As Fine and Weis (1999) have shown, these strategies did not address the specific problems of teenage mothers in many urban settings.

The discourse of personal responsibility and social inclusion through education

The neoliberal discourses emerged out of economic liberalism of Thatcherism and Reaganism in the late twentieth century, but New Labour and New Democrats created their own distinctive approach to social and public policies (Giddens, 2000; Driver and Martell, 1998). Their focus was more on individualism and personal responsibilities for changing social and sexual identities, since gender and sexual identities were seen as changing in response to wider social and familial transformations. Indeed, part of the process has been toward new kinds of critical reflection, reflexivity in constructing and reconstructing social identities and gendered subjectivities (Scott, Jackson and Backett-Milburn, 1998; Jones, 2001). New ways to tackle social change and the effects of globalization, the changing political and socioeconomic context, and the need to address social, family, and cultural diversity have emerged as critical to public discourses (Oakley, 2000). The shifts remain highly complex and contradictory in relation to public discourses of education and employment and how they draw selectively on evidence, ignoring aspects of research about global family life changes toward social, cultural, racial/ethnic, and sexual diversity and their relation to educational changes (David, 2001a). The interconnections between education and employment have become critical to the proposals in the neoliberal policy discourses among social democrats, such as New Labour in Britain and New Democrats in the United States.

This neoliberal policy discourse became dominant in the United States in the early 1990s as part of welfare reforms, drawing initially on economic experiments and experiences in Wisconsin (Bloch, 1999; Glennerster, 2000; Schram, 2000). It also drew on a moral debate about the growing underclass (Murray, et al., 1996). New Labour in Britain adopted it wholesale, restructuring its practices accordingly. Canada and Australia, however, developed new, contrasting, and less punitive public policies for children. (Joseph Rountree Foundation, 2000).

Neoliberal public discourses put the emphasis on paid employment and the necessity of paid work as a solution to poverty, reconstructed as social exclusion or inclusion in Britain (Levitas, 1998). There have been a plethora of British initiatives in which all expect and require more individual responsibility, including parental involvement that has become entirely normative (David, 2001b; David, 2002a and b). Paid employment has become the global mantra for this policy discourse, which is about training and skills for paid employment for young people—boys and girls alike—as specific measures (O'Connor, Orloff, and Shaver, 1999). The plethora of initiatives disregards the gender of parenthood. It thus includes providing services to enable mothers of young children to seek and find paid employment as well as young people who are single and do not have dependent children. It aims to ensure paid work variously through policies borrowed from the United States (the New Deal) and support for children through taxation, working families tax credits, child credits, childcare services, and early education initiatives (David, 1999; David, 2001a).

Teenage Sexuality and Education for Social Inclusion

An expression of the public policy discourse in Britain was found in a policy review document, from the Labour Government's Social Exclusion Unit set up by New Labour specifically to develop new strategies across a range of social welfare issues. It was to reconstruct approaches to children and young people from socially and economically disadvantaged families (Tomlinson, 2001). In their report on *"The Changing Welfare State: Opportunity for All: Tackling Poverty and Social Exclusion,"* a clear approach using educational and training strategies was initiated *de novo:*

> The Government believes that everyone should have the opportunity to achieve their potential. . . . That means making sure that all children, whatever their background and wherever they live, get a first class education, giving them the tools they will need to succeed in the adult worlds . . . our goal is to end child poverty in 20 years. . . . In order to eradicate child poverty we need to provide opportunities for their parents to work. For most people of working age, the best way to avoid poverty and social exclusion is to be in paid work. But the enormous economic and social changes of the past 50 years . . . the changing nature of employment patterns and family structure—have left key groups . . . unable to compete in the labour market. (SEU first annual report, 1999b, pp. 3, 9)

There were specific strategies targeted on groups of children and young people in schools deemed to be in danger of social "exclusion." Social inclusion was defined in terms of access to work or employment, rather than in relation to the traditional notions of social class, and for men and women alike, regardless of family responsibilities such as parenthood. Education and training were identified as necessary to enable young people to learn how to find work rather than rely on social welfare to deal with potential poverty.

The Social Exclusion Unit (cm 4342, 1999a) also produced a report entitled *Teenage Pregnancy.* Its goal was "to develop an integrated strategy to cut rates of teenage parenthood, particularly under-age parenthood, towards the European average, and propose better solutions to combat the risk of social exclusion for vulnerable teenage parents and their children" (1999a, p. 2). It addressed the question starkly, using "evidence" from international studies (excluding, however, quite curiously the United States), to substantiate the claim about the extent of the problem. However, using American language, it drew on studies about "what works" and how to deal with reducing the scale, focusing on the importance of education and employment. The prime minister, Tony Blair, in his foreword to the report, was typically melodramatic:

> Britain has the worst record on teenage pregnancies in Europe. . . . Teenage mothers are less likely to finish their education, less likely to find a good job, and more likely to end up both as single parents and bringing up their children in poverty. . . . The report . . . shows how we . . . must improve education on relationships and sex for teenagers . . . and the real responsibilities of being a parent, including the financial responsibilities of being a father . . . if they do have a child they should be strongly encouraged to complete their education and keep in touch with the jobs market. . . . (SEU, 1999, pp. 4–5)

The report itself offered slightly more measured terms and yet still emphasized employment and education, even for young teenage mothers where it was also somewhat ambiguous. It also identified figures for the "problem" that were not particularly grave, in terms of most social problems:

> In England, there are nearly 90,000 conceptions a year to teenagers; around 7,700 to girls under 16 and 2,200 to girls aged 14 or under. Roughly three-fifths of conceptions—56,000—result in live births . . . the UK has teenage birth rates which are twice as high as Germany, three times as high as in France and six times as high as in the Netherlands. *Some other countries—notably the US—have rates even higher than the UK.* [sic] . . . the specific issue of teenage pregnancy needs a specific remedy as well . . . reducing the rate of teenage conceptions . . . getting more teenage parents into education, training or employment, to reduce their risk of long term social exclusion. . . . (SEU, 1999a, pp. 6–8)

It set out a specific strategy for education, as well as health, which entailed developing detailed curricular guidance and extracurricular activities.

There were moves to ensure individual and familial responsibilities for this essentially "personal" matter, part of the wider process of restructuring public and private individual responsibilities for welfare. It also redrew boundaries between personal and professional educational matters. Moreover, with developments in the so-called knowledge economy, through ICT, what counts as education as an individual personal matter changed (Blackmore, 1999; Peters, 2002). Changes in information technology also affect what education professionals do. It also has implications for families and schools and for home-school relations that are barely recognized in these debates.

Second, it also set employment opportunities for all as the overarching goal of its policy initiative for reducing teenage pregnancy and parenthood. Here the route out of poverty was seen as through job opportunities, copying the United States. Yet again the process of individualizing responsibilities was entailed in this approach, and no account taken of the responsibilities of parenthood, and the historical emphasis on motherhood as an occupation or preoccupation for mothers of very young children (David, 1993). In this respect, the spectre of the nineteenth-century or Victorian approach to social welfare of children in families in poverty was raised (Piven and Cloward, 1966). The approach was to ensure individual responsibilities coupled with moral censure and limited financial support. There was veiled reliance on a version of the language of the Elizabethan poor law and its revisions in the nineteenth century (Piven and Cloward, 1966). This "language" was revisited under neoliberalism of the late twentieth century (Arnot, David and Weiner, 1999, chapters 3 & 6).

The main strategy was that of education, developed in a policy and curriculum document containing guidance for sex and relationship education. It had multiple, competing aims and diverse views on the challenging nature of the issues to be addressed. However, a specifically religious and moral dimension of family life remained uppermost in the document. Yet this had been widely contested in the consultation process about its content, as it had in relation to other New Labour policy consultation documents on families (David, 1999).

The guidance specifically set out:

> This is the first time that schools have had a national framework . . . as part of sex and relationship education, pupils should be taught about *the nature and importance of marriage for family life* [sic] and bringing up children . . . pupils should learn the significance of marriage and stable relationships as key building blocks of community and society. Care needs to be taken to ensure that there is no stigmatisation of children based upon their home cir-

> cumstances. . . . SRE . . . is lifelong learning about physical, moral and emotional development. It is about the importance of marriage for family life, stable and loving relationships, respect, love and care. It is also about the teaching of sex, sexuality, and sexual health. It is not about the promotion of sexual orientation or sexual activity—this would be inappropriate teaching. It has three main elements: attitudes and values . . . personal and social skills . . . knowledge and understanding. . . . (SRE Guidance, DfEE 0116/2000, pp. 4–5)

The language of the national guidance revealed both the economic and moral ideological underpinnings of the approach. On the one hand, policies for teenage pregnancies were justified in terms of the economic necessity for the well-being of the nation, in respect to international economic competitiveness. In particular, the UK's declining position in relation to other European countries provided a key rationale for this development. On the other hand, complex moral arguments and justifications for such social intervention peppered the guidance document. It emphasized the centrality of and gave explicit support for marriage, despite widespread international evidence to demonstrate the absolute decline in marriage as a key to family life over the last two to three decades. In England, for instance, two out of five children are now born to mothers who are not married and may be living alone (Allan, 2002, Haskey, 1998; O'Connor, Orloff and Shaver, 1999; Kiernan, Land and Lewis, 1999). Moreover, debates about diverse forms of sexuality were heavily contested in the consultation, especially in relation to political debates, at the height ofThatcherism, about the promotion of homosexuality (Epstein and Johnson, 1998). Nevertheless, these issues were proscribed.

Three aspects to the curriculum were suggested as how to teach these topics, namely *values, skills, and knowledge.* The emphasis was pedagogical, with an assumption that information is the way to deal with what is deemed to be an essentially rational problem. Despite the rhetorical shift to include relationships in the curriculum, the emphasis remained—as in previous neoliberal guidance about sex education, emanating from a Conservative government in the mid-1980s—on information about conception and contraception. First it hoped to develop a pedagogical practice to prevent teenage pregnancies and to ensure that schoolgirls and teenage girls avoid immature parenthood by committing them to study. Second, it aimed to ensure that all young people, including boys, learn about the responsibilities of parenthood.

These aims were to be ensured through teaching about skills, knowledge, and understanding and attitudes across curricular subjects, such as national curriculum science, and within a personal, social, and health education framework, linked with new developments in citizenship education. Other subjects such as English, drama and religious education were also expected to contribute to the development of these skills and understandings. Moreover, there was also an expectation of crosscurricular presentations, and the use of individual tutor strategies. Working with parents and local communities was also seen as an essential element to the approach.

The document set strong parameters for developing sex and relationship education, such as support for particular sexual relationships, and especially marriage, and in developing attitudes toward responsibilities including involving parents and the local community. Yet at the same time it was ambiguous and equivocal about how these issues should be developed in the curriculum. It thus provided a space and opportunity for an innovative and nonpunitive pedagogy for teaching about relationships and sex. It also allowed for a different pedagogy on gender and sexual identities, in association with reflective accounts of these social processes (David with Smith and Alldred, 2001b). In particular, while it was emphatic about marriage, there was also scope for developing refreshing innovative approaches to masculinities and femininities, in relation to jobs, careers and citizenship education. These may also be considered in relation to private and personal matters. It thus illustrates the ways in which reflexivity has begun to permeate policy documentation and social practices more generally.

Discourses on sexuality education and personal responsibility in the United States

"Abstinence education" started within the era of economic liberalism in the 1980s (Burdell, 1995/6; Kelly, 2001). However, it became of far more critical importance during the 1990s as part of the restructuring of social welfare and a wider settlement of welfare for single mothers (Gordon and Fraser, 1997). The values behind this were the assumption of a moral framework for sexual activity. A rationalist notion that all sexual activity that resulted in pregnancy was "unplanned," "unwanted," and "unintended" disregarded young people's own views of their reasons for such activity. Another assumption was that "self-sufficiency" is about the refusal to become a parent outside of a particular moral framework. Thus "abstinence-only-before-marriage" education became tied to major elements of federal funding for education in the United States. Individual states had to commit to providing such "morally correct family values" within programs of sex education in order to qualify for federal funding for wider educational programs such as for educationally disadvantaged children, charter schools, and other choice mechanisms. Its aims were, however, argued as being to prevent teen pregnancies and parenthood for children "at-risk" who were in compulsory education.

As stated within the legislation, the "purpose was to enable the State to provide abstinence education, and at the option of the State, where appropriate, mentoring, counselling, and adult supervision to promote abstinence from sexual activity, with a focus on those groups which are most likely to bear children out-of-wedlock" (SSA #510 1996). Thus the question of risk became a central feature, and the term "abstinence education" was defined as "an educational or motivational program" for teaching about various aspects of the harm of sexual activity before and outside marriage. Eight separate features were itemized, and the term "harm" as a simile for "risk" was peppered throughout the list. It was linked to both health and psychological problems for parent and child. These were mentioned as "out-of-wedlock pregnancy, sexually transmitted diseases, and other associated health problems"; "the likelihood of harmful psychological and physical effects," "harmful consequences for the child, the child's parents and society" (as cited above); and "how alcohol and drug use increases vulnerability to sexual advances." However, there were also particular positive but individual aspects to the program. These were stated as: social, psychological, and health gains; the normal and "expected standard for all school age children"; that a "mutually faithful monogamous relationship within the context of marriage is the expected standard of human sexual activity"; "the importance of attaining self-sufficiency before engaging in sexual activity."

This was legislated during Clinton's administration in 1996, and all states had to commit to abstinence education to receive federal funding in 1998 and beyond. This process has been tightened and reinforced under the Bush administration (as noted in the introduction). Initially 49 states did make the commitment, the exception being California. Interestingly, in all the commentaries there was an absence of gender in discourses about teens, adolescents, and parenthood.

However, the focus of attention was, and is, on young women, as can be seen by the parallel developments in social welfare legislation. The shifts in welfare discourses and social legislation started under the Clinton administration but have continued under Bush. Several commentators have linked these two major policy shifts, especially in their implicit if not explicit focus on young women. Throughout the 1980s there had been a growing concern about the welfare rolls, that is, the increases in the number of single- and female-headed households in need of social and fiscal support (Piven & Cloward, 1986). The problems with the Aid to Families with Dependent Children (AFDC) reached a crescendo in the early 1990s and Congress finally legislated for its abolition (Schram, 2000). Clinton reluctantly had to accept and sign the legislation as a compromise for support for other social programs. While AFDC had been criticized by liberals and other social commentators for its relatively punitive aspects and how it created "dependency," especially for single women/mothers, it was seen as more generous than its successor in its commitment to categorical support (Gordon & Fraser, 1997).

In its place, a program of temporary aid for needy families (TANF) was proposed, based on a program developed to deal with a fiscal crisis within Wisconsin. Here the central element in the revised policy was the new emphasis on "personal responsibility," particularly for mothers, and through work, not dependency on the social state (Schram, 2000).

The shift in public support for single mothers with dependent children living below the poverty line was toward temporary financial aid rather than regular and dependable social and financial support. Moreover, the emphasis also became that of forcing such women into forms of low-paid employment rather than reliance on the state, even when they had very young children or babies. Thus teenage parents/mothers would also be affected by this fiscal shift, with its attendant cultural aspects, especially the emphasis on "family values of marriage" for sexual activities. In all of these policy shifts there are also clear cultural changes; moves from popular fears and anxieties about dependency of women on the state to revised focus on ill-defined concepts of "needy families" (Gordon and Fraser, 1997). Moreover, the risks for young and teenage parents become not only about physical and psychological health but also about lack of employment and financial support. Thus economic factors are also used to influence young people's sexual activities and behaviors, returning to nineteenth-century and Victorian family values.

Similar issues were raised in Britain under New Labour, building as the United States did on Wisconsin's poverty programs, named "Wisconsin works." Sex education, as noted, has been legislated for in Britain, modifying the previous sections of the 1986 Education Act, through the Learning and Skills Act, 2000 (Monk, 2000; Thompson, 2000). This built upon a moral "family values" framework about the centrality of marriage, which has been specified in various policy documents, such as Supporting Families (David, 1999).

A particular feature of the policy discourses in Britain was the focus on teenage parenthood, as a key factor leading to social exclusion, defined in terms of poverty. New Labour identified this as a matter of serious import for young people—both men and women. It led to the setting up of a special government unit on teenage pregnancy in 2000. The first annual report of the Independent Advisory Group on Teenage Pregnancy repeated its concern that teenage pregnancy should remain high on the government's agenda (Independent Advisory Group on Teenage Pregnancy [IAGTP], 2001). The argument was based on "evidence" of the increasing scale of the problem by comparison with other countries, with which Britain is deemed to be in competition. Teenage pregnancy assumed the status of a "moral panic" despite the extremely modest statistical scale compared to other social welfare problems, demonstrated by the official evidence offered in the various policy documents (SEU, 1999a; IAGTP, 2001). The educational approach, while addressing personal issues in public, has only achieved more rhetoric.

Contradictory facets of the public educational discourses on teenage parenthood

In both the United States and in Britain, the revised welfare discourses for young/teenage, single, and poor mothers and the attendant policies for sex and sexuality education raise contradictory issues about the overarching shifts in public discourses and the personal and lived experiences of such women and their families.

Much of the focus of attention is on young people's—boys' as well as girls'—*personal* responsibilities and need for education. Ambiguously, yet another aspect of the identification of the problem is that of young mothers, seeing them and their children as denied educational and social opportunities. Thus a solution for future generations of children is to emphasize their need to be in education, employment, or training. This focus on young women and young mothers as a key part of the problem flies in the face of other evidence about the changing balance of educational achievements between boys and girls (Arnot, David and Weiner, 1999). Thus the arguments about how to address the problems appeared as a veiled attack on young women and their emerging social and sexual identities, at a time when other young people are being positioned to take their own responsibilities and individual risks. This illustrates the contradictory rhetorical approach to the problem and the solutions.

The social research evidence about changing gender and education is indeed complex. On the one hand, there is overwhelming statistical evidence about how girls from all social classes and locations now "outperform" boys at all levels of education, such that there has been "a closing of the gender gap" (Arnot, David & Weiner, 1999). Walkerdine et al. (2001)

address differences between girls in terms of social class; Frosh et al. (2001) look at young black men and these have to some extent been addressed in education policy debates. On the other hand, there is also a question about "closing the achievement gap," addressed in the United States as social and racial diversity rather than gender differences (Moe, 2001). There is also research evidence that suggests that these discourses lead to policies that may have contradictory implications for families, especially young mothers, with a growing expectation of involvement in paid employment and gender-neutral parenthood (Bullen, Kenway and Hey, 2000; Kelly, 2001; Lewis and Knijn, 2002).

Moreover, the shifts in social welfare are part of much broader cultural, economic, and social transformations, in the contexts of globalization of family changes, economics, and political values. Thus there is a new emphasis on what has variously been called individualization or forms of self-reflection and the taking of personal responsibility for one's life, especially in economic and financial terms (Beck, 1992; Beck, Giddens and Lash, 1996; Bauman, 1998). The traditional gender codes about employment and family values have been replaced by the relatively gender-neutral values of work and responsibility. So girls are no longer to be taught the values of marriage, maternity, and full-time motherhood but rather those of personal responsibility and education, therefore, as a route to employment. This is curious, and contrary to the invocation of marriage as fundamental to normal and healthy sexual behaviour. Yet, in some complex and contradictory way, parental involvement in their children's education still remains, and indeed has become a more important facet, of public education (David, 2001a). Thus for the majority of parents, parental involvement in education is as naturalized and normalized mothers; fathers remain distant from the regular and normal involvement in their children's education.

Nevertheless, the key issue of "risk" of "harm" for young and teenage parents/mothers is of social exclusion; particularly from employment, since educational achievement is suggested as the key to employment opportunities. These educational and employment opportunities then give rise to "personal responsibility." In the United States, a more recent general educational problem concerns how "to close the achievement gap" between social class groups, especially racial and ethnic minorities, who might be deemed to be in poverty or in "needy families" (Moe, 2001; Stambach & David, 2002). While this specification does not raise issues of gender, all the evidence and rhetoric in the abstinence education programs is of teenage girls and teenage mothers "at-risk" of social and psychological disadvantages and unhealthy development. In Britain, by contrast, the specification has been about how "to close the gender gap in achievement," identifying now young men, from poor/working-class and socially disadvantaged backgrounds, as the more problematic. However, when it comes to the specification of policies around social exclusion and teenage pregnancy and parenthood, the issue of gender becomes much more muted. Thus girls who are "at-risk" of teenage pregnancy and parenthood become those also identified as being socially and educationally disadvantaged and on whom such programs of sex education should be targeted.

Thus the shifts in these public policy discourses identify particular social and minority ethnic or racial groups of women as in need of special and particular help. In this respect, the shifting discourses have come to target particular women for special help for their perceived sexually inappropriate and personally irresponsible behavior. The discourse of social welfare compared to "public" education divides and differentiates on class, racial, and gender. Girls "at-risk" become the subject of special help and provision. This reveals one of the major contradictions at the heart of this neoliberal approach to public policy and restructuring patterns for parents and children. Emphasizing individual responsibilities may allow for more opportunities and also for diversity, but it also creates obstacles to wider social and collective responsibilities of government and communities for their economic and social futures.

For children and young people to learn about educational and employment opportunities may be essential to tackling the question, but it cannot provide the whole solution. On the one hand, educational strategies cannot hope to alter deeply held cultural beliefs about sex and relationships, and although they may raise issues, they may not deal with attitudes and values that are not susceptible to rational argument but are about emotions. Views about masculinities, femininities, sexual relations, and parenting are all part of a much broader canvas.

Conclusions

This critical analysis of policy discourses about teenage sexuality, pregnancy, and parenthood has illustrated the challenges, complexities, and contradictions in them. It has pinpointed the ways changing balances between parents and children have been restructured within welfare policies and practices with respect to families. A focus on teenage pregnancy and parenthood reveals the tensions in policy discourses between children and adults, especially around sexual and gender identities, while these may be concealed in policy rhetoric. While neoliberal discourses as applied by New Labour in Britain and New Democrats in the United

States have as an overarching aim educational and employment opportunities for all, it can be concluded from this analysis that such language is mere rhetoric. It is not related to the possibility of achieving such aims for all young people. The policy discourse of employment opportunities for teenage parents contradicts the discourse of parental involvement in children's education (David, 2001a). No evidence has been provided to demonstrate the necessity of such involvement in paid employment except the assertion of the importance of paid employment as a route out of poverty. At the same time, as the title of this chapter illustrates, such parenting would be "bad" for children as well as their parents. There seems to be no recognition of the dilemmas that may be created for teenage and/or lone mothers with respect to paid work, personal responsibility, and the necessity of parental involvement in their children's education.

Policymakers have not attended to the "evidence" of the voices of mothers nor the narratives and accounts of young people. The perspective has ignored notions about emotions, feelings, and sexual desire. Yet as we have noted there is research evidence that highlights the dilemmas for mothers—especially young, single, and teenage—living in a variety of constrained circumstances and situations of poverty. More complex questions about changing gender and sexual identities in relation to social and educational transformations are barely alluded to, yet the research literature is replete with fascinating examples, as researchers have convincingly demonstrated (Kenway, 2001; Walkerdine, Lucey & Melody, 2001; McInGaill, 1998; Mikel Brown, 1998). In particular, there is evidence from the perspectives, understandings, and narratives of new generations of boys and girls (Arnot et al., 1999). These give voice to how young people develop new and more complex gender and sexual identities, and in relation to work and family, than previous generations (McGrellis et al., 2000; Ball, Maguire & Macrae, 2000; Kenway, Willis, Blackmore & Rennie, 1997; Lingard & Douglas, 1999). Some of this has to do with changing social contexts, and some to do with changing policy, toward making individuals cope with their own emerging identities of masculinities and femininities in emerging new contexts of risk, where reflexivity and self-assessment become more important (Walkerdine et al., 2001; Frosh et al., 2001). Drawing on this kind of research evidence would begin the process of contributing to understandings of reflexivity, risk, and individualization in relation to global and social transformations. It would also aid understandings of transformative social and sexual identities, not only for teenagers but also the balances between children and adults, personal and political, education and family.

Note

1. This chapter derives from a paper originally prepared for the Umea University, Sweden and UW–Madison joint invitational conference in Sweden held on May 18–20, 2001, on *Restructuring the Governing Patterns of the Child, Education and the Welfare State,* and the invitational Keele University Department of Education conference, June 26–27, 2001, on *Travelling Policy.* It has also been presented at two seminars in New Zealand at the Universities of Auckland (August 14, 2001) and Waikato (August 17, 2001) and used as a basis for a summer school course on *Families, gender and education: Issues of policy and practice* in the Department of Education Policy Studies and Curriculum and Instruction, University of Wisconsin–Madison (July 14–August 9, 2001). I am most grateful to the participants at these various seminars for help with revisions to this paper and its arguments. I am especially grateful to Erin Haley, a graduate student at UW–Madison, for her work on the teen pregnancy and prevention programs in the United States and to Dr. Pamela Alldred and Ms. Pat Smith, my colleagues on the SREPAR project at Keele for help in preparing this chapter. It is a version of a work in progress on understanding global, national, and local discourses about education, particularly for young people, in relation to aspects of social inclusion and exclusion. It considers how such official discourses have been borrowed and transported between various countries, especially the United States and UK.

References

Allan, G. (May 1, 2002). *Friendships and intimacies,* inaugural professorial lecture. University of Keele.

Allen, I. (1987). *Personal, social and sex education.* London: Policy Studies Institute.

Allen, L.(2001). Closing sex education's knowledge/practice gap: The reconceptualization of young people's sexual knowledge. *Sex Education, 1* (2), 109–123.

Arnot, M., David, M., & Weiner, G. (1999). *Closing the gender gap: Post war education and social change.* Cambridge: Polity Press.

Arnot, M., & Dillabough, J. (2000). (Eds.), *Challenging democracy and citizenship.* Cambridge. Polity Press. See especially the chapter by Sue Lees, "Sexuality and Citizenship Education."

Ball, S., Maguire, M., & Macrae, S. (2000). *Choice, pathways and transitions post-16; new youth, new economies in the global city.* London: Routledge Falmer.

Bauman, Z. (1998). *Post-modern ethics.* Cambridge: Polity Press.

Beck, U. (1992). *The risk society.* Cambridge: Polity Press.

Beck, U., Giddens, A., & Lash, S. (1996). *Reflexive modernisation.* Cambridge: Polity Press.

Bernstein, B. (1996). *Class codes and control.* London: Routledge.

Blackmore, J. (1999). *Troubling women: Feminism, leadership and educational change.* Buckingham: Open University Press.

Bloch, M. N. (July, 1999). *Welfare state reform and child care: A critical review of recent U.S. and international trends.* Presented at the Child and the State: Restructuring the governing discourses of the child, family, and education. Conference at University of Wisconsin–Madison.

Bullen, E., Kenway, J., & Hey, V. (2000). New Labour, Social Exclusion and Educational Risk Management: the case of "gymslip mums." *British Educational Research Journal, 2* (4), 441–456.

Burdell, P. (1995–1996). Teen Mothers in High School: Tracking Their Curriculum. In M. Apple (Ed.), *Review of Research in Education, 21.* Washington, DC: American Educational Research Association.

Clarke, J. & Newman, M. (1998). *The managerial state.* London: Macmillan.

David, M. E. (1993). *Parents, gender and education reform.* Cambridge: Polity Press.

———(1999). Home, work, families & children: New labour, new directions and new dilemmas. *International Studies in the Sociology of Education, 9* (3), 209–229.

———(September, 2001a). Gender equity issues in educational effectiveness in the context of global, social and family life changes and public policy discourses on social inclusion and exclusion. *Australian Educational Researcher, 28* (2), 99–125.

———(2001b). Family, Gender and Education: Issues of Policy and Practice. In F. Smit, K. van der Wolf, & P. Sleegers (Eds.), *A bridge to the future: Collaboration between parents, schools and communities.* Nijmegen, Holland: Institute for Applied Social Sciences, University of Nijmegen.

———(2002a). A feminist critique of gender equity issues in public discourses. In S. Benjamin, S. Ali, & M. Mauthner (Eds.), *Gender and education.* London: Palgrave.

———(2002b). Gender equity and public policy discourses. In A. Griffith & C. Reynolds (Eds.), *Equity and Globalisation in Education.* London and New York: Palgrave.

Department for Education and Employment (DfEE). (2000). *Sex and relationship education guidance.* July 0116/2000.

Driver, L. & Martell, S. (1998). *New labour: Politics after Thatcherism.* Cambridge: Polity Press.

———(1999). New labour: Culture and economy. In L. Ray and A. Sayer (Eds.), *Culture and economy after the cultural turn.* London: Sage.

Duncan, S. & Edwards, R. (1999). *Lone mothers, paid work and gendered moral rationalities.* London: Macmillan.

Ehrenreich, B. (1989). *Fear of falling: The new middle classes in the USA.* New York: Basic Books.

Epstein, D., & Johnson, R. (1999). *Schooling sexualities.* Buckingham: Open University Press.

Fine, M. & Weis, L. (1999). *The unknown city: The lives of poor and working class young adults.* Boston: Beacon Press.

Franklin, J. (2000). What's wrong with new labour politics? *Feminist Review, 66,* 138–142.

Frosh, S., Phoenix, A., & Pattman, R. (2001). *Young masculinities: Understanding boys in contemporary society.* London: Palgrave.

Giddens, A. (1999). *The third way.* Cambridge: Polity Press.

Glennerster, H. (2000). *The welfare state.* London: London School of Economics and Suntory Toyota Research Centre pamphlet.

Gordon, L. & Fraser, N. (1997). A genealogy of "dependency" tracing a keyword of the US Welfare State in N. Fraser 1997. *Justice Interruptus: Critical reflections on the post socialist condition.* New York and London: Routledge.

Haskey, J. (1998). Family and marriage in late twentieth century Britain. In M. David (Ed.), *The Fragmenting Family.* London: IEA Health and Welfare Unit. Choice in Welfare. No. 44.

Hudson, F., & Ineichen, B. (1991). *Taking it lying down: Sexuality and teenage motherhood.* London: Macmillan.

Independent Advisory Group on Teenage Pregnancy (IAGTP). (2001). *First Annual Report.* London: HMSO.

Ingham, R. & van Zessen, G. (1998). From cultural contexts to international competencies: A European Comparative Study, paper presented at AIDS in Europe, Social and Behavioural Dimensions cited in IAGTP.

Jones, A. (Ed.). (2001). *Touchy subject; teachers touching children.* Dunedin, New Zealand: University of Otago Press.

Joseph Rowntree Foundation. (2000). *Reforming children's benefits: International comparisons:* http://www.jrf.org.uk/knowledge/findings/socialpolicy/090.htm).

Kelly, D. (2000). *Pregnant with meaning: Teen mothers and the politics of inclusive schooling.* New York: Peter Lang.

Kenway, J. & Bullen, E. (2001). *Consuming children: Education-entertainment-advertising.* Buckingham: Open University Press.

Kenway, J., Willis, S., Blackmore, J. & Rennie, S. (1998). *Answering back: girls, boys and feminism in schools.* London and Sydney: Allen and Unwin.

Ketting E., & Visser, A. P. (1994). Contraception in the Netherlands: The low abortion rate explained. *Patient Education and Counselling, 23* (3), 161–171.

Kiernan, K., Land, H., & Lewis, J. (1998). *Lone motherhood in 20th century Britain: From footnote to front page.* Oxford: The Clarendon Press.

Lees, S. (1997). *Ruling passions: sexual violence, reputation and the law.* Buckingham: Open University Press.

Lesko, N. (1995). The leaky needs of school-aged mothers: An examination of US programs and policies. *Curriculum Inquiry, 25* (2), 177–205.

Levitas, R. (1998). *The inclusive society?* London: Macmillan.

Lewis, J. and Knijn, T. (2002). Comparative perspectives on sex education in GB and Holland. *Journal of Social Policy, 29* (4), 669–695.

Lingard, B. & Douglas, P. (1999). *Men engaging feminisms pro-feminism, backlashes and schooling.* Buckingham: Open University Press.

Mac An Ghaill, M. (1999). New cultures of training: Emerging male (hetero)sexual identities. *British Educational Research Journal, 25* (4), 427–444.

McGrellis, S., Henderson, S., Holland, J., Sharpe, S., & Thomson, R. (2000). *Through the moral maze: Young people's value.* London: The Tufnell Press.

Middleton, S. (1998). *Disciplining sexuality: Foucault, life histories and education.* New York: Teachers College Press.

Mikel Brown, L. (1998). *Raising their voices: The politics of girls' anger.* Cambridge, MA: Harvard University Press.

Moe, T. (2001). *Schools, vouchers and the American public.* Washington, DC: Brookings Institution.

Monk, D. (2001). New guidance/old problems: Recent developments in sex education. *Journal of Social Welfare and Family Law, 23* (2), 271–291.

———(2000). Theorising education law and childhood: Constructing the ideal pupil. *British Journal of Sociology of Education, 12* (3), 355–370.

Murray, C., et al., (1996). *Charles Murray and the underclass: The developing debate.* Choice in Welfare No 33 London: Institute of Economic Affairs and The Sunday Times.

National Council for One Child Families. London: National Council for One Parent Families.

New Statesman (May 17, 2002). Published weekly. London.

Oakley, A. (2000). *Experiments in knowing.* Cambridge: Polity Press.

O'Connor, J., Orloff, A., & Shaver, S. (1999). *States, markets and families: Gender, liberalism and social policy in Australia, Canada, GB and USA.* Cambridge: Cambridge University Press.

Petchesky, R. (1987). *Abortion: Women's choice.* New York: Basic Books.

Peters, M. (2001). National Education Policy Constructions of the "Knowledge Economy": Towards a Critique. *Journal of Educational Inquiry, 2* (1).

Phillips, A. (2001). A dash of moral outrage. *The Guardian November 29th.*

Phoenix, A. (1991). *Young mothers?* Cambridge: Polity Press.

Piven, F. F., & Cloward, R. (1966). *Regulating the poor.* New York: Basic Books.

———(1986). *The new class war.* New York: Basic Books.

Joseph Rountree Foundation. (2000). *Reforming children's benefits: international comparisons.* http://www.jrf.org.uk/knowledge/findings/socialpolicy/090.htm.

Scott, S., Jackson, S. & Backett-Milburn, K. (1998). Swings and roundabouts: Risk anxiety and the everyday worlds of children. *Sociology, 32* (4), 689–750.

Schram, S. (2000). *The new welfare.* New York: Basic Books.

Silver, H. (1991). *Educational policy analysis.* London: Macmillan.

Social Exclusion Unit (June, 1999a). *Teenage pregnancy.* London: The Stationery Office Cm 4342.

———(1999b). *The changing welfare state: Opportunity for all: Tackling poverty and social exclusion.* First Annual Report Sept Cm 4445. London: The Stationery Office

Stambach, A. & David, M. E. (In press). Feminist theory and educational policy: How gender has been involved in the home-school debates about school choice. *Signs: Journal of Women and Culture.*

Thompson, R. (1994). Moral rhetoric and public health pragmatism: The recent politics of sex education. *Feminist Review, 48,* 40–60.

Tomlinson, S. (2001). *Education in a post-welfare society.* Buckingham: Open University Press.

Walkerdine, V., Lucey, H. & Melody, M. (2001). *Growing up girl: Psychosocial explorations in gender and class.* London: Macmillan.

Young, J. & Greenwood, V. (1977). *Abortion in demand.* London: Pluto Press.

Zellman, G. (1981). *The Response of the schools to teenage pregnancy and parenthood.* Santa Monica, CA: RAND Corporation.

CHAPTER EIGHT

CHILD WELFARE IN THE UNITED STATES

THE CONSTRUCTION OF GENDERED, OPPOSITIONAL DISCOURSE(S)

Gaile S. Cannella

> Reagan's positions on public issues, which he had been espousing for some time, were very clear. He disliked welfare recipients, whom he believed to be "cheats" and "free loaders" who should be forced back into the labor market. (Trattner, 1999, p. 363)

> Welfare is bad for you—on that proposition liberals and conservatives now seem to agree. (Whitman, 1987, p. 1)

> The presidential campaign of Bill Clinton clearly put welfare reform on the national political agenda in the early 1990's. Under the catchy slogans of "ending welfare as we know it" and "making work pay," welfare reform and EITC (earned income tax credit) expansion became central planks of Clinton's presidential campaign. (Weaver, 2000, p. 127)

> The absence of married parents is related to poor academic performance. . . . The absence of married parents leads to intergenerational illegitimacy. . . . Lack of married parents, rather than race or poverty, is the principal factor in the crime rate. (Rector, 1996, The Heritage Foundation)

> There are several reasons why so many children are growing up in poor, single parent families. The first is the continuing high rates of nonmarital births . . . marriage has declined precipituously, especially in poor communities. The 107th Congress will almost certainly return to legislation designed to promote marriage among poor families. . . . Are there policies that

can promote marriage and reduce teen and out-of-wedlock child-bearing? (Haskins, Sawhill, & Weaver, 2001, pp. 3, 4, 8)

The quotes above represent the discourses that have since the 1960s gained increased hegemony over public conversation(s) in the United States regarding welfare, specifically challenging and removing child and family rights to and resources for public assistance. These discourses are of blame and catastrophe, and have become entirely gendered by creating women as immoral beings who require either state or spousal corporeal, economic, and domestic control. In this context, the welfare of children both constitutes and is constituted by these gendered public discourses. Women (and especially poor women) have literally become the instruments for delivery of children to the patriarchy (Grumet, 1988).

From a global, crossnational perspective, some would assume that the equitable welfare of citizens can be promoted by state provisions. The purpose of the "welfare state" can be viewed as modifying social and market forces to achieve greater equity for citizens (Ruggie, 1984). Modernist theoretical work attempting to explain the constructions of the "welfare state" has included this notion that the state can generate possibilities and equality for citizens by altering state and market relations, social rights, and decommodification of labor. State differences in the ways such factors are handled are illustrated in Esping-Andersen's (1990) discussions of the liberal, conservative-corporatist, or social-democratic dimensions of welfare. Feminists have included attempts to achieve equity but proposed definitions of the welfare state that are much more extensive; these proposals include the wide range of services that provide social support to all citizens—housing, medical services, education, childcare, and special supports for the variety of human conditions. Unfortunately, this broad definition has not been included in the construction or reconstructions of the welfare state in the United States. Recent welfare reform in the United States has not only reproduced a historical belief that assistance can be used as a means of social control (Trattner, 1999), but has imposed on women a "survival of the fittest" market theology that is judgmental and that demonizes females as immoral and lazy.

This chapter will examine the construction of contemporary, oppositional (contemptuous and even punitive) welfare discourse(s) beginning in the 1960s, the historical time period in the United States in which the civil rights and other social movements resulted in progress toward equity for racially and sexually marginalized groups. Orloff (1996) posits that state social provision or interventions into civil society may or may not produce greater equality among citizens, but that most often reconstitution of the welfare state has been related to gender relations. Therefore, studies of the welfare state cannot be gender blind. For this reason, the method of analysis that I have chosen is a bricolage (Kincheloe, 2001; Denzin & Lincoln, 1994), a methodology that is emergent and involves a variety of techniques. Specifically used in this analysis are: (1) the problematization of child welfare as constructed within a gendered context; (2) the multiple lenses provided by feminist thought as juxtaposed upon conservative activities and language over the past 30 years; and (3) analyses of public data sources. These methods reveal power orientations that further oppress, marginalize, and reproduce inequity.

The Gendered Constitution of Child Welfare

Child welfare in the United States has always to some extent been tied to agendas that were not, or were minimally, concerned for those who are younger. Viewed as the key to social control, children have been the objects of "pity" and "compassion" (Trattner, 1999, p. 109), in the name of "the man of the future" or the "destiny of America" (p. 110). For example, the history of Euro-American education as both construct and practice is filled with a focus on social and/or religious control, especially of the poor (Rose, 1990; Walderdine, 1984). More specifically tied to the generic concern for the life conditions and physical welfare of those who are younger, multiple examples abound. To illustrate, although not entirely insensitive to children identified as in need, Charles Loring Brace used the New York Children's Aid Society (founded by him in 1853) to remove from society (by placing them as workers on farms in the west) the homeless, vagrant children that he referred to as the "dangerous classes" (Trattner, 1999, p.115). This early foster care was to be a "moral and physical disinfectant" that would provide protection for the property of the well-to-do, reduce the evils of the city, and provide a hard-working life with the most solid of all citizens, the farmer. No recognition was given to what children or their parents, who lived in the city, wanted.

Further, the public dialogue concerning beliefs about and welfare of those who are younger has not only been tied to socioeconomics, but to women as mothers and consequently to the creation of an "appropriate" behavior for females. This gendering of child welfare is grounded in Euro-American Enlightenment psychology and medicine and the use of science, economics, and religion in the United States to legitimate the control of women. For example, Foucault (1978) has demonstrated the presence of the dominant discourses of Victorian Europe in Freud's construction of female experience. The twentieth-century obsession with controlling the sexuality of women and children is revealed—the construction of motherhood as the natural state for females creates a discourse in which women and children are chained to each other. Further, although Freud's specific theories of sexuality have been questioned (and, some believe, totally rejected), the perception that early experience with one's mother results in either healthy or maladaptive societal adjustment abounds (both in more recent psychological and medical theories and, perhaps more importantly, in public discourse).

The medical field in the United States continues to be influenced by conceptualizations of French physicians in the eighteenth century who were concerned about the protection of children raised by servants of the rich, by domestic nurses, and in foundlings homes (Donzelot, 1979). Wet nurses and servants were described as self-interested, hateful to those who are younger, incompetent individuals practicing unhealthy habits, and natural enemies of the rich man's child; these wet nurses and servants were, of course, women. At this historic time, when medicine was emerging as an "expert" position of power, male physicians constructed "household medicine" (p. 16), methods of care and child-rearing that could be prescribed medically (by male physicians) and administered by women. A system of surveillance, scientifically determined hygiene, and practice was established over women as they were to protect, preserve, and control children.

Specific to circumstances in the United States, in the early 1900s public health reformers focused on economic roots of poverty and child well-being, rather than engaging in the earlier moralist discourse. Many of them believed that illness is the leading cause of dependency. Based on the realization that thousands of children were being committed to public institutions either because their widowed mothers became ill from worry and overwork or because parents were not able to support them, widows' pension laws were passed all over the country. Clearly, the dialogue of the day was that children would be better served to be home with their mothers, who should be serving society by remaining at home to care for their children (Trattner, 1999). Court judges even declared that the family should be kept together by paying the mother some of the funds that would go to asylums or foster homes. Gordon (1994) clearly demonstrates, however, that this focus on a particular type of benefit for "widows" led to a 1930s gendered welfare state that constructed women (and minorities) as receiving low-status, dependency support while white men received high-status supports like unemployment compensation and old-age pensions. Additionally, Coontz (1997) has illustrated the ways in which women's behavior was controlled (e.g., living in "appropriate" neighborhoods, dressing in the morally correct way) through pension requirements. The dialogue was child protection (Finklestein, Mourad, & Doner, 1998)—the result was a method of controlling women.

During World War II, women themselves instituted a massive program of support through the construction of 24-hour childcare for the multitude of women who flooded the work force as a duty for the country and the fighting men who were overseas, challenging not only the polarization of male/female roles but generating a socialist childcare methodology that provided support for all children and families. However, social scientists (through the growing construction of psychological theories) were generating a hegemonic knowledge base that would blame women for everything, from colic to delinquency (Spitz, 1945; Levy, 1944). True to the notions espoused by psychoanalysis and medicine, mothers (and by implication, all women) were considered the roots of pathology, responsible for all problems, and ruthless and power hungry (Wylie, 1942; Eyer, 1992). When the war ended and men returned home wanting jobs, this social science was used to disqualify both women's successes in the labor force and their design for social provision. Magazines and newspapers were used to extoll home and the woman's domestic role, identifying those who continued to work as wicked child abusers whose children would be placed in institutions and foster care.

This gendered focus on women and children continued in psychology (and was reinforced by American Protestant constructions of Christianity) and influenced scholarly, political, and other public discourses of family and child welfare. From the late 1940s through the later part of the twentieth century, American women (through scholarship in both the United States and Europe) have been bombarded with psychological constructs that label them as ultimately tied to successes and problems (especially failures and abnormalities) of children.

These misogynous constructs have: likened women and children to some type of generic animals placed in torturous conditions by male researchers (as in the work of Harlow, 1959); indicted women as murderers of their children by depriving them of "appropriate" social security (as in the "death sentence" for children with working mothers suggested by Bowlby, 1951); required women to provide both "feminine attachment" and "masculine detachment" in the achievement of independence for their children (Ainsworth, 1967; Eyer, 1992); and placed the fate of women and children in the hands of medical "experts" who would constitute their identities through hormones, chemistry, and crucial developmental periods

(Klaus & Kennell, 1976; Eyer, 1992). The scholarship and resultant public discourses (whether intentional or not) have focused on keeping women at home with children (with the exception of poor women, who are constructed as not knowing how to provide the best—read as middle-class—environment for their children), and on controlling the behavior of both. These perspectives are clearly implicated in the dominant construction of child welfare as a gendered-female responsibility.

The practice of education is the child welfare discourse that has actually gained the greatest attention in the United States. Child development and pedagogical knowledge bases (Bloch, 1987; Silin, 1987; Cannella, 1997) have been imposed on teachers in the name of the child's best interests. Predominantly female teachers have been taught to desire white middle-class language and reasoned behaviors (reflectivity, inquiry, discovery, logic) in themselves and the younger human beings with whom they work. Further, expectations for a truth-oriented professionalism that would appear apolitical project a view of the "good teacher" as grounded in the science of child development. This professionalism (Ginsburg, 1987; Popkewitz, 1994) positions teachers as responsible for all educational problems. If teachers (again predominantly females) would just teach the "right way," children would all be successful. Again, the welfare of the child is tied to women.

Although various provisions of child welfare have improved life for children and included such issues as increased opportunities through education, healthcare (of the entire community), child labor regulations, and the improvement of settlement house conditions, one must question whether it is possible to separate the state (and the construction of the "welfare state") from the complexities of society that are integrated with constructions of power and its perpetuation. While not all child welfare issues have tied women and children together, most have either reinforced the chain or demonstrated some form of "patriarchal will to power." Again to illustrate: child labor regulations were not specifically tied to women in the United States, but were at least partially gendered because of underlying attempts to generate more work for adult male labor; further, recent American and European attempts to globally eliminate child labor (however well-intended and probably needed when involving human exploitation by multinational corporations) have employed so-called Western constructions of the child, functioned with the imperialist belief that cultures of color must be guided by the more "advanced" perspectives of white cultures, and have ignored diverse human conditions, family cultural beliefs, and economic circumstances—an agenda that is at best intellectually and economically colonialist, and is most likely also patriarchal. Child welfare in the United States has been clearly tied to multiple sites of power. Much of this power is located within patriarchy and most often a burden is placed directly on the backs of women.

Feminist Analyses of Societal Welfare

Placing child welfare within the lenses of gender relations requires analyses of at least some of the various theories of the welfare state that take gender into account as both constituted by and constituting cultural, social, political, and economic processes. Increasingly, social analysts express the belief that because gender is implicated in all social relations, the state cannot be considered without the recognition of gender and its embeddedness within politics and policy. These perspectives, which would be considered feminist (Orloff, 1996), reveal gender inequalities, differentiations, and hierarchies that are historically critical in a given society at a particular moment in time. The author reminds the reader that these theories are presented as sites from which contemporary actions can be viewed, not as universals explaining particular truths of the welfare state.

Feminist analyses of the welfare state have emerged that provide the double-voiced possibility that gender relations are constitutive of the welfare state and are constituted by the welfare state. As examples, feminist critiques of neo-Marxist theories have pointed to the role of the state in the reproduction of patriarchy as well as capitalism (Abramovitz, 1988). Others have interrogated the ways in which liberalism and democratic perspectives construct and are constructed by gender relations (Vogel, 1991; Lister, 1990). These lenses position the analysis of the welfare state from within a broader construction of power and control (Orloff, 1996; Shaver, 1990).

Gender, Inequity, and Patriarchy

Issues of gender equity/inequity as integrated within constructions of the welfare state have been viewed from at least two perspectives. The first examines the state as implicated in the reproduction of patriarchy. The second views the state as the vehicle for the elimination of social inequities, most often through the transfer of income or through buffering of women's poverty.

The emergence of the modern welfare state is viewed by feminists who would examine the reproduction of patriarchy as imposing a structural change from private or personal individual male dominance to the substitution of the state, as the public structure for male dominance (Ungerson,

1985; Gordon, 1990). Gender hierarchy is believed to be maintained through: (1) the division of labor that places women in caregiver positions; (2) a family wage system that results in higher wages for males legitimated through the belief in support for dependent women and children; and (3) marriage, which includes a traditional family ethic and gendered labor. Scholars have proposed that these mechanisms may exist and contribute to overall patriarchal relations, even though a system may provide support for family failures (Lewis, 1993; Lister, 1990). The reinforcement of traditional gender roles and the construction of gendered citizenship (as independent and male) has been examined. Some scholars discuss a two-tiered system in which "masculine" and "feminine" programs have emerged, with masculine programs providing support for market failures and feminine programs targeting women as related to family problems (Bryson, 1992; Fraser, 1989). The second tier is systematically restrictive and regulatory. Patriarchal mechanisms are thus associated with the exclusion of women from political power (Nelson, 1984). However, Orloff (1996) points out that while exhibiting a sophisticated understanding of gender relations, there may be some state social provisions that do advance gender equity. This possibility leads to discussion of the second perspective that would address equity.

The belief that welfare states can actually eliminate gender inequities leads supporters to call for income-transfer programs that, while not removing labor vulnerabilities in which women have been placed, at least buffer women's poverty (Orloff, 1996). These feminist perspectives have discussed crossnational variations that demonstrate better positions for women in countries in which benefits and other social programs are relatively generous (Mitchell, 1993; Norris, 1984), and have also labeled some welfare states as low spenders and as providing little support. The complexity of increasing equity financially has, however, been recognized as more challenging than simply understanding the amount of state spending and benefits offered—but is also tied to gender equity in the market as well as female participation in the labor force. Additionally, in the United States the push for increased benefits may actually create more difficulties for the construction of improved and equitable life conditions because of the racial, socioeconomic, and gendered stereotypes that dominate (Roberts, 1995; Fraser & Gordon, 1994; Quadagno, 1994).

Although examination of these perspectives reveals differing positions of power and a diversity in the provision of benefits needed for immediate assistance, expectations are fairly uniform and to a large extent exhibit a linear, simplistic focus on the state as agent of financial equity. Further, women's roles in the actual creation of policies are not examined through these lenses, nor are the interactions among class, gender, and other powered social relations.

Maternalism as Shaping Public Policy

Recent studies of women reformers have revealed that most early activists entered the political world on the basis of "difference," expressing the belief that women should stand for, protect, and construct policy regarding women and children because of women's own unique role as mothers. Most all supported a division of labor based on gender, and legislation that was specific to gender. These women (and their male counterparts) generally spoke the language of maternalism, believing that women's roles involved morality and nurturance because of the "capacity to mother" (Orloff, 1996, p.13). Groups that supported this perspective were as diverse as those concerned with population control, believers in family wage that supported males as breadwinners, and feminists who called for the "endowment of motherhood" (Koven & Michel, 1993, p.4), a concept that is not consistent with the second-wave of American feminist thought that espoused equality in difference. However, much of the public in the United States accepted some form of cult of the true woman (Cannella, 1997; Grumet, 1988). Both working-class groups and white middle-class women played a role in the creation of benefit policies based on women's role in reproduction (Skocpol, 1992).

Crossnational examinations of women's movements have revealed that women's political actions have not, however, necessarily resulted in either gendered protection or equity for women and children. Public discourses and historical circumstance play important roles. As examples, Hobson (1993) showed how a public concern about population decline led to the development of new protections for women workers in 1930s Sweden; no similar concern was expressed during the New Deal era in the United States as a gendered system emerged. In France and Britain, Pedersen (1993) demonstrated that the economic balance of power may be an issue; while women wanted recognition as both mother and qualified worker, employers wanted cheap, subordinate labor, and workers wanted women to be constructed in the role of wife. Further, religion and conservative agendas may also play a role in social policy that is maternalist in nature.

The United States is a country identified as having a strong women's movement, but weak public social provision. Although women's movements in the United States have had a strong equal rights character and focused on a universal-breadwinner strategy, equal pay, employment, and compensation have not yet been attained (Sainsbury, 1999). Further, Sklar

(1993) has argued that in most countries the organizing principal for early welfare politics was class, with organized labor and working-class parties playing significant roles as champions for social provision. However, in the United States, gender as constituted by middle-class women substituted for class. Voluntary organizations created and controlled by middle-class married women played a major role in the construction of policy. Today social programs for women are inferior to those for men, yet women designed most of the provisions (e.g., AFDC) (Gordon, 1994). Finally, maternalist provisions were not equally designed for or accessible to women of color, demonstrating that concerns for motherhood and children were bounded by racist forces and assimilationist purposes (Boris, 1995; Mink, 1994; Gordon, 1994).

This involvement of women in the construction of gendered, inequitable, oppressive social programs reveals the complexities of social policy. Besides the perspectives of the women and men involved in the conceptualization and development of social provision, feminist analysts have pointed to historical circumstance, strength of working-class groups, and the actual political mobilization of women, which may be profoundly influenced by the ways that traditional gender relations continue to exist (Orloff, 1996). From attempts to address this complexity (and most likely with a concern for power as a factor), the notion of social policy regime has emerged.

Gender and Regimes of Social Provision

First, policy regimes are defined as "a complex of rules and norms that create established expectations" (Sainsbury, 1999, p. 5). Gender regimes are the rules applied to gender relations. Social policy regimes are patterns of functioning that have been institutionalized within the welfare state, creating systematic relations of accommodation, conflict, and domination (Orloff, 1996; Shaver, 1990). Policy regimes may contain political, economic, legal, and even discursive elements.

Although believing that the perspective is gender blind, feminists have taken Esping-Anderson's (1990) work, which discusses three regime types in capitalism—liberal, conservative-corporatist, and social-democratic—and expanded the view. This very capitalist (which is what it claims to be) and class analysis orientation includes state/market relationships, social rights, and decommodification of labor as characteristics that differentiate regime types. Feminist scholars have expanded the perspective in a variety of ways. One example is the introduction of the issue of "care." Taylor-Gooby (1991) suggests that traditional capitalist regime types vary as to expectations regarding unpaid care work and dependent populations; gender relations and views of gender equality constitute and are constituted by perspectives on unpaid work and how this work is shared by men and women (Lewis, 1997). Even the most social-democratic regime can be limited by an inequitable regime of care. McLanahan, Casper, and Sorensen (1995) have proposed that in any regime type, the only ways that women have an equal chance for reduced poverty is to either be linked to a man through marriage (as illustrated in the Italian reinforcement of marriage) or be like a man in working for pay (as illustrated in Sweden's promotion of female employment). The unpaid care work, which has been so much a part of women's lives, continues to make them vulnerable to poverty.

Feminist Discourses and New Possibilities

Feminist analysts continue to introduce lenses of possibility that can influence the construction of the welfare state. As examples, Orloff (1996) points out the ways in which women's interests are conflictual and contradictory. The interests of children, women, and men may compete. Women do not all share the same values, interests, or privileges. Access to political power (and the maintenance of that power) is an issue for particular groups of women (just as with particular groups of men). Conflicts of interest are to be expected in heterogeneous societies. Further, Orloff (1993) and O'Connor (1993) have attempted to gender the lenses within traditional examinations of the welfare state. To illustrate, both have argued that while decommodification (potential for greater independence from the labor market for survival) is important for everyone, not all social groups have access to jobs that allow for benefits and independence. Further, Orloff (1996) argues for examination of supports that lead to the "capacity to form and maintain an autonomous household" (p. 38). Embedded in the expectation that women should provide care work, this capacity has not been fostered "for" women but would give them the opportunity to adequately support family through a breadwinner-level income without feeling the necessity to marry (Sainsbury, 1999).

Sainsbury (1999) has generated strategies toward the elimination of gender inequities in the welfare state based on Fraser's gender-equity models (1994, 1997). The first strategy, the universal-breadwinner, has as the goal that all women and men would be engaged in paid equitable work that provides social benefits tied to that work. The second strategy, the caregiver-parity perspective, would equalize the roles of caregiver and earner, making the rewards between careers and those in traditional work roles comparable by creating equitable social entitlements based on both principles of care and work. Finally, the universal-caregiver strategy would call for the redesign of policies (not simply ways that an individual family might choose to function) in ways that would systematically break down traditional divisions of labor between males and females and construct equitable sharing of care and public work.

Additionally, feminist analysts have demonstrated that political parties and both the historical and immediate complexion of governments make a difference in social provision. Politically left parties have tended to be more open to social provision, while those to the right may even be hostile (Sainsbury, 1999). O'Connor (1999) provides an excellent example (obvious to those of us in the United States). When democrats have controlled particular state governments within the United States, comparable-worth work programs have been adopted; at the national level, all recent Republican administrations (Nixon, Reagan, and Bush, Sr. and Bush, Jr.) have refused to support legislation focused on equality in employment.

Finally, poststructural feminists have argued for pulling away from attempts to determine how women's welfare interests can be best represented, to examining ways that these interests (and other structures) are constituted (Pringle & Watson, 1992). Power is considered key to this examination of the conditions through which particular forms of reason are/have been inscribed (Foucault, 1978). Questions might be asked like: What are the apparatuses or technologies that have constructed a specific complex of governmental power (e.g., patterns of the welfare state)? Are there technologies outside the state that produce a state context of domination and subordination (Popkewitz & Brennen, 1998)? Can understanding the production of particular welfare discourses lead to possibilities for social change? It is from this poststructural position using the juxtaposition of feminist theories of the welfare state that I move to examine oppositional child welfare discourses that have emerged in the United States.

A Conservative Agenda and the Reconstitution of the State: Backlash and (Old or New) Forms of Power

During the 1960s, citizens' activist movements changed the nature of government in the United States. Boycotts, sit-ins, marches, and demonstrations (often reported in the media), and other nonviolent citizen grassroots group activities created a national awareness of discrimination. Although Supreme Court decisions and civil rights legislation were necessary to enforce particular, just, and equitable behaviors among some individual citizens, grassroots groups were successful in developing public support for these civil rights court decisions and legislation (Berry, 1997; Garrow,

1978). During this period, the anti–Vietnam War movement was fueled by citizens groups. Although not always popular, the groups kept the problems with the war at the forefront, resulting in pressure on both the Johnson and Nixon administrations. Hispanic farm workers organized in California to achieve more equitable treatment. Most conspicuously, women became involved in their own citizens' movement, using the model provided by civil rights groups to address their own issues of subjugation, injustice, domination, and power. "To adult professional women in the early 1960's the growth of civil rights insurgency provided a model of legal activism and imaginative minority group lobbying" (Evans, 1979, p. 25).

Grass roots citizens were successful in promoting liberal public interest agendas, and were successful in constituting views of the state and their power within it. Power blocks to individual freedom (especially for women and people of color) were at least legislatively removed. Those who had been traditionally excluded from power were now part of governmental influence and decision making. A grassroots, nonviolent activist model was generated that would be used by student groups, environmentalists, American Indians, and citizens groups around the world (Morris, 1984). A perspective was generated that supported the continued involvement of diverse citizens in policy making and in the valuing of diversity.

These successes did not set well with conservatives—those who wanted small government, the maintenance of power found in a patriarchal status quo that controls diversity (often imposed through religion), and believe that the free market is the avenue for solving problems. Conservatives were disturbed by the successes of the citizens movements of the 1960s, especially the civil rights, women's, and environmental movements (Berry, 1997). A new right emerged in the 1970s that believed that the American way of life (or at least their own way of life) was threatened by the "liberal" imbalance created by the demand for equity by diverse groups of peoples. A politics of resentment emerged that used the language of "moral decay" (p. 35), previously used against the poor, to a level that implicated anyone who did not espouse traditional family values (including woman's place as mother in the home). Feminism, which can generally be defined as concern for both sexism and women's lives (hooks, 2000), was presented by conservatives as an evil concept that did not fit with American values, with Christianity, or with any moral form of being. Feminism was offered as the root cause of divorce, out-of-wedlock births, growing welfare cases, and a multitude of other societal situations described by conservatives as societal failures.

Certainly this response to the successes of marginalized groups fits within the feminist lens that would examine the reproduction of patriarchy. At the point of "backlash," conservatives were personally and individually angry. However, this anger was turned into an organized effort to inscribe (and in some cases re-inscribe) public discourses that, rather than focusing on oppression, discrimination, equity, and social justice, were patriarchal, narrow, and imposed limitations on all those who would not accept the discourse. Through a reproduction of patriarchy, family and marriage became signifiers of morality, values, "truth, justice, and the American way."

Political "Right" Reactions

When Ronald Reagan became president, he and other conservatives interpreted the election as a mandate for redirection (rather than understanding that the results may have had much more to do with economic concerns and personality dissatisfaction) and moved toward the directions of many of his early twentieth-century predecessors (like Hoover). Reagan disapproved of any kind of welfare and followed the ideas of conservative scholars like Anderson (1978), who declared that the war on poverty was won in his book *Welfare: The Political Economy of Welfare Reform in the United States,* and Glider (1981), *Wealth and Poverty,* who espoused the belief that the poor needed the motivation of poverty to succeed (Trattner, 1999). Further, Reagan's sweeping tax cut generated a loss of $1 trillion, creating a large federal deficit; this loss was then used to create a direct assault on the welfare state.

The assault was successful. Even political liberals declared war on the poor in the name of reform and accepted a conservative fiscal model that ignored systemic inequality supported by institutions and a discriminatory hidden unemployment. Poverty began to be discussed as relative, and as caused by the moral failure of the poor combined with government entitlement actions that perpetuated laziness. During this time of fiscally capitalist conservativism and blaming the poor, liberal women's groups (those who had been profoundly active in constructions of the welfare state and in the 1960s liberation movements) lost power; their perspectives were ignored and their influence decreased (Costain, 1992).

Constructing Apparatuses of Power

The construction and inscription of oppositional public discourses involved an organized plan and mobilization of resources. Following the 1960s liberal successes, conservatives joined together to create new sites of power and new communications technologies, and to undermine and reconstitute American institutions—major activities involved the construction of conservative foundations with the purposes of redirecting such

American institutions as academia, media, religious institutions, Congress, and the judiciary (Covington, 1998). These institution include: "the Lynde and Harry Bradley Foundation, the Carthage Foundation, the Earhart Foundation, the Charles G. Koch, David H. Koch and Claude R. Lamb Charitable Foundation, the John M. Olin Foundation, the Henry Salvatori Foundation, the Sarah Scaife Foundation, the Smith Richardson Foundation," and the Heritage Foundation (p. 2).

Over the past 20 years, the foundations have granted a total of $89 million to fund conservative scholarship in university programs, $27 million to train the next generation of conservative journalists, and awarded $41.5 million to build conservative media and pro-market legal organizations. The conservative-capitalist state orientations of these foundations are gender blind and construct citizenship as within a narrow parameter. The following discussion of "program interests" from the Lynde and Harry Bradley Foundation in reference to "service-providing institutions" illustrates: "citizenship is being challenged today, however, by contemporary forces and ideas that regard individuals more as passive and helpless victims. . . . This systematic disenfranchisement of the citizen and the consequent erosion of citizenly mediating structures pose grave threats to the free society . . ." (The Bradley Foundation, p. 3).

Gendered, Oppositional Discourse

Conservative political rhetoric and organized, conservative-capitalist efforts to change the direction of major American institutions have also resulted in discourses of the welfare state that focus on females; this use of females to express a supposedly gender neutral view of citizenship evidences a blatant public patriarchal will to power. A punitive discourse that would blame women for having children without resources, yet provides no equalizing mechanisms for labor or care, has been accepted and is used by both political parties, the media, and many in the general public. A maternalist perspective has been "turned up-side-down" to create not a privilege but a method of demonizing women who are caught without a male—who were not either married or dependent on the paid work of a male. Examples of the discourse abound: In an analysis of documents from a variety of groups, females as mothers are the continued objects of discussion—women are inscribed as not married, low-income, single parents, teen parents, low skilled, out-of-wedlock mothers, incompetent, and lacking responsibility. Males are barely mentioned, while women are indicted as placing their children at risk for abuse and neglect. Marriage is privileged and women are demonized as creating all of society's ills through teen and out-of wedlock births (Marshall, 2001; Haskins, Sawhill, & Weaver, 2001; Rector, 2000).

Even reports from organizations like the Brookings Institute (which has historically made a concerted effort to report multiple perspectives on an issue), use language that is negative toward females in demonstrating the focus that was included in the 1996 welfare legislation and is expected in the 2002 reauthorization (Bush, 2002). The language assumes an ideal women—one who is married. As examples: "Too many children are being reared by single mothers" (Haskins, Sawhill, & Weaver, 2001, p. 1). "Are there policies that can promote marriage and reduce teen out-of-wedlock child-bearing" (p. 8)? This language would obviously disqualify single parents, gay/lesbian families, and any form of diversity that is not grounded in a male/female legal (and likely religious) nuclear relationship.

Women and children (especially those who are poor and have less resources) have been reconstituted as the morally deficient who corrupt society. Further, they have been newly constituted as the human subject of backlash against the equity, diversity, and social justice movements in the 1960s—movements that challenged traditional sites of power, patriarchy, and capitalist truths.

Conclusions

Those of us who would support child and overall human welfare as part of a socially just system of human equity and respect did not realize the resentment and fear of loss of power that would result from gains made through the 1960s acceptance of diversity (whether based on gender, race, or socioeconomic class). As feminist analysts (Roberts, 1995; Fraser & Gordon, 1994; Quadagno, 1994) have explained, attempts to equalize rights for marginalized groups actually exacerbated the equity problem; poor women became the targets of unending assault and the subjects of the new discourse.

Further, we did not realize the organized, divisive, and funded efforts in which so many would engage; we did not understand the strong will to power or the resentment toward the historically marginalized who would stand for equal treatment. Perhaps these conservative successes embody the dangers of capitalism that so many have espoused. If so, should our efforts to construct a welfare state that equalizes opportunity and provides just and equitable support for everyone use this gendered capitalist tool, attempt to dismantle it, or work toward a kind of reconceptualization and reconstitution of the capitalist state? What roles would children, women, and other marginalized groups play in such a reconstitution? What should be our strategies?

Finally, conservative rhetoric calls for traditional values. Perhaps we should play a much more public and vocal role in this discussion. Can we generate a critical public discourse that is respectful but moves toward equity and resources for everyone? Have our own perceptions, beliefs, and desires as a public actually been newly constituted? Do we believe that others are evil beings always ready to take advantage of us (and our money, especially if they are different from us)? Or have busy, caring individuals been overwhelmed by redirected conservative messages without analyzing underlying beliefs and obtaining information about people? How do we choose to individually govern ourselves? Ultimately, do we publicly "care" about each other? Are we as a society (locally and globally) willing to engage in the discourses of care that are embodied in notions like the universal caregiver (Sainsbury, 1999)?

Child welfare may never escape the regimes of power within which it is embedded. To fundamentally address the welfare of children (and for that matter all people in a society), gendered, oppositional discourses cannot be tolerated. Power agendas within regimes of social provision must be revealed and directly addressed. We must engage the diverse forms of power that we create and that constitute us.

References

Abramovitz, M. (1988). *Regulating the lives of women: Social welfare policy from colonial times to the present.* Boston: South End Press.

Ainsworth, M. D. S. (1967). *Infancy in Uganda: Infant care and the growth of attachment.* Baltimore: Johns Hopkins University Press.

Anderson, M. (1978). *Welfare: The political economy of welfare reform in the United States.* Stanford, CA: Hoover Institution.

Berry, J. (1997). *The interest group society.* New York: Longman.

Bloch, M. N. (1987). Becoming scientific and professional: An historical perspective on the aims and effects of early education. In T. S. Popkewitz (Ed.), *The formation of school subjects* (pp. 25–62). Basingstoke, England: Falmer.

Boris, E. (1995). The racialized gendered state: Constructions of citizenship in the United States. *Social Politics, 2,* 160–180.

Bowlby, J. (1951). *Maternal care and mental health.* Geneva: World Health Organization.

The Bradley Foundation. Current program interests. Milwaukee: Retrieved April 19, 2001, from http://www.bradleyfdn.org/program.html.

Bryson, L. (1992). *Welfare and the state.* London: Macmillan.

Bush, G. (February 2002). *Working toward independence.* West Wing Connections. www.whitehouse.gov/new/released/2002/02/welfare-reform-announcement-book-all.

Cannella, G. S. (1997). *Deconstructing early childhood education: Social justice & revolution.* New York: Peter Lang.

Coontz, S. (1997). *The way we really are: Coming to terms with America's changing families.* New York: Basic Books.

Costain, A. (1992). *Inviting women's rebellion.* Baltimore: Johns Hopkins University Press.

Covington, S. (1998, Winter). How conservative philanthropies and think tanks transform U.S. policy. *Covert Action Quarterly, 63,* 1–8. Retrieved January 24, 2001, from http://mediafilter.org/CAQ/caq63thinktank.html.

Denzin, N. K., & Lincoln, Y. S. (1994). (Eds.). *Handbook of qualitative research.* Thousand Oaks, CA: Sage.

Donzelot, J. (1979). *The policing of families.* New York: Pantheon.

Esping-Andersen, G. (1990). *The three worlds of welfare capitalism.* Princeton, NJ: Princeton University Press.

Evans, S. *Personal politics.* New York: Knopf.

Eyer, D. E. (1992). *Mother-infant bonding: A scientific fiction.* New Haven, CT: Yale University Press.

Finkelstein, B., Mourand, R., & Doner, E. (1998). Where have all the children gone? The transformation of children into dollars in Public Law 104–193. In S. Books (Ed.), *Invisible children in the society and its schools* (pp. 169–182). Mahwah, NJ: Lawrence Erlbaum.

Foucault, M. (1978). *The history of sexuality.* New York: Pantheon.

Fraser, N. (1997). *Justice interruptus: Critical reflections on the postsocial condition.* New York: Routledge.

———. (1994). After the family wage: Gender equity and the welfare state. *Political Theory, 22,* 591–618.

———. (1989). *Unruly practices: Power, discourse, and gender in contemporary social theory.* Minneapolis: University of Minnesota Press.

Fraser, N., & Gordon, L. (1994). A genealogy of "dependency": Tracing a keyword of the U.S. welfare state. *Signs, 19,* 309–336.

Garrow, D. (1978). *Protest at Selma.* New Haven, CT: Yale University Press.

Ginsburg, M. (1987). Reproduction, contradiction and conceptions of professionalism: The case of pre-service teachers. In T. Popkewitz (Ed.), *Critical studies in teacher education: Its folklore, theory, and practice* (pp. 86–129). New York: Falmer Press.

Glider, G. (1981). *Wealth and poverty.* New York: Basic Books.

Gordon, L. (1994). *Pitied but not entitled: Single mothers and the history of welfare.* New York: Free Press.

———. (1990). (Ed.) *Women, the state, and welfare.* Madison: University of Wisconsin Press.

Grumet, M. R. (1988). *Bitter milk: women and teaching.* Amherst: University of Massachusetts Press.

Harlow, H. (1959; 1973). Love in infant monkeys. In T. Greenough (Ed.), *The nature and nurture of behavior: Developmental psychobiology* (pp. 94–100). San Francisco: Freeman. (Original work published in *Scientific American* in July, 1959).

Haskins, R., Sawhill, I., & Weaver, K. (January 2001). *Welfare reform reauthorization: An overview of problems and issues.* Washington, D.C.: The Brookings Institute. www.brokings.edu

Hobson, B. (1993). Feminist strategies and gendered discourses in welfare states: Married women's right to work in the United States and Sweden. In S. Koven & S. Michel (Eds.), *Mothers of a new world: Maternalist politics and the origins of welfare states* (pp. 396–430). New York: Routledge.

hooks, b. (2000). *Feminism is for everybody: Passionate politics.* Cambridge, MA: South End Press.

Kincheloe, J. L. (2001). *Describing the bricolage: Conceptualizing a new rigor in qualitative research.* American Educational Research Association, Seattle.

Klaus, M., & Kennell, J. (1976). *Mother-infant bonding.* New York: Wiley.

Koven, S., & Michel, S. (1993). *Mothers of the new world: Maternalist politics and the origins of the welfare state.* New York: Routledge.

Levy, D. (1944). *Maternal overprotection.* New York: Norton.

Lewis, J. (1997). Gender and welfare regimes: Further thoughts. *Social Politics: International studies in gender, state, and society, 4* (2), 160–177.

———. (1993). Introduction: Women, work, family, and social policies in Europe. In J. Lewis (Ed.), *Women and social policies in Europe: Work, family, and the state* (pp. 1–24). Aldershot, UK: Edward Elgar.

Lister, R. (1990). Women, economic dependency, and citizenship. *Journal for Social Policy, 19,* 445–467.

Marshall, W. (February, 2001). *Welfare reform: A progress report.* Democratic Leadership Online Community. www.ndol.org

McLanahan, S., Casper, L., & Sorensen, A. (1995). Women's roles and women's poverty in eight industrialized countries. In K. Mason & A. M. Jensen (Eds.), *Gender and family change in industrialized countries* (pp. 258–278). New York: Oxford University Press.

Mink, G. (1994). *Wages of motherhood: Inequality in the welfare state, 1917–1942.* Ithaca, NY: Cornell University Press.

Mitchell, D. (1993). Sole parents, work, and welfare: Evidence from the Luxembourg income study. In S. Shaver (Ed.). *Comparative perspectives on sole parents' policy: Work and welfare* (pp. 53–89). University of New South Wales Social Policy Research Centre Reports and Proceedings no. 106.

Morris, A. (1984). *The origins of the civil rights movement: Black communities organizing for change.* New York: The Free Press.

Nelson, B. (1984). Women's poverty and women's citizenship: Some political consequences of economic marginality. *Signs, 10,* 209–232.

Norris, P. (1984). Women in poverty: Britain and America. *Social Policy, 14,* 41–63.

O'Connor, J. (1993). Gender, class, and citizenship in the comparative analysis of welfare state regimes: Theoretical and methodological issues. *British Journal of Sociology, 44,* 501–518.

———(1999). Employment equality strategies in liberal welfare states. In D. Sainsbury (Ed.), *Gender and welfare state regimes* (pp. 48–74). New York: Oxford University Press.

Orloff, A. S. (1993). Gender and the social rights of citizenship: The comparative analysis of gender relations and welfare states. *American Sociological Review, 58,* 303–328.

———(1996). *Gender and the welfare state.* Institute for Research on Poverty Discussion paper no. 1082–96. University of Wisconsin–Madison.

Pedersen, S. (1993). *Family, dependence, and the origins of the welfare state: Britain and France, 1914–1945.* New York: Cambridge University Press.

Popkewitz, T. (1994). Professionalization in teaching and teacher education: Some notes on its history, ideology, and potential. *Teaching and Teacher Education, 10* (1), 1–14.

Popkewitz, T. S., & Brennan, M. (1998). Restructuring of social and political theory in education: Foucault and a social epistemology of school practices. In Popkewitz, T. S. & M. Brennan (Eds.), *Foucault's challenge: Discourse, knowledge, and power in education* (pp. 3–38). New York: Teachers College Press.

Pringle, R., & Watson, S. (1992). Women's interests and the post-structuralist state. In M. Barret and A. Phillips (Eds.), *Destablizing theory* (pp. 53–73). Stanford, CA: Stanford University Press.

Quadagno, J. (1994). *The color of welfare: How racism undermined the war on poverty.* New York: Oxford University Press.

Rector, R. (2000). How congress reformed the welfare system. In *Reviewing the revolution: Conservative success in the 104th congress.* Washington, DC: The Heritage Foundation.

Rector, R., & Fagan, P. (1996). How welfare harms kids. Backgrounder 1084. Washington, DC: The Heritage Foundation.

Roberts, D. (1995). Race, gender, and the value of mothers' work. *Social Politics, 2,* 195–207.

Rose, N. (1990). *Governing the soul: The shaping of the private self.* London: Routledge.

Ruggie, M. (1984). *The state and working women.* Princeton, NJ: Princeton University Press.

Sainsbury, D. (1999). Gender, policy regimes, and politics. In D. Sainsbury (Ed.), *Gender and welfare state regimes* (pp. 246–275). New York: Oxford University Press.

Silin, J. G. (1987). The early childhood educator's knowledge base: A reconsideration. In L. G. Katz (Ed.), *Current topics in early childhood education* (pp. 17–31). Norwood, NJ: Ablex.

Shaver, S. (1990). *Gender, social policy regimes, and the welfare state.* Paper presented at the annual meeting of the American Sociological Association, Washington, DC.

Sklar, K. (1993). The historical foundations of women's power in the creation of the American welfare state, 1830–1930. In S. Koven and S. Michel (Eds.), *Mothers of a new world: maternalist politics and the origins of welfare states* (pp. 43–93). New York: Routledge.

Skocpol, T. (1992). *Protecting soldiers and mothers.* Cambridge, MA: Harvard University Press.

Spitz, R. (1945). Hospitalism: An inquiry into the genesis of psychiatric conditions in early childhood. *Psychoanalytic Study of the Child, 1,* 53–75.

Taylor-Gooby, P. (1991). Welfare state regimes and welfare citizenship. *Journal of European Social Policy, 1,* 93–105.

Trattner, W. I. (1999). *From poor law to welfare state: A history of social welfare in America.* New York: The Free Press.

Ungerson, C. (1985). *Women and social policy: A reader.* London: Macmillan.

Vogel, U. (1991). Is citizenship gender-specific? In U. Vogel and M. Moran (Eds.), *The frontiers of citizenship* (pp. 58–85). New York: St. Martin's Press.

Walkerdine, V. (1984). Developmental psychology and the child-centered pedagogy. In J. Henriques, W. Hollway, C. Urwin, C. Venn, & V. Walderdine (Eds.), *Changing the subject: Psychology, social regulation and subjectivity* (pp. 153–202). London: Methuen.

Weaver, R. K. (2000). *Ending welfare as we know it.* Washington, DC: Brookings Institution Press.

Weir, M., Orloff, A. S., & Skocpol, T. (1988). (Eds.). *The politics of social policy in the United States.* Princeton, NJ: Princeton University Press.

Whitman, D. (1987, June). The key to welfare reform. *The Atlantic online.* Retrieved from wysiwg://105/http://www.theatlantic.com/politics/poverty/whitmanf.html.

Wylie, P. (1942). *Generation of vipers.* New York: Rinehart.

CHAPTER NINE

GLOBAL/LOCAL ANALYSES OF THE CONSTRUCTION OF "FAMILY-CHILD WELFARE"

Marianne N. Bloch[1]

The Reason of Welfare about Families and Children

In this analysis, I focus on cultural reasoning about "Family-Child Welfare" at the beginning of the twenty-first century in relation to reasoning about "Family-Child Welfare" in the past. Specifically, I draw on Foucault's notions of sovereign, disciplinary, and bio-power (Foucault, 1977; 1980, 1988) to examine new patterns of regulation, new "governmentalities" (Foucault, 1979/1991) regarding the child, family, welfare, care and well-being. I examine "global" and "local" changes in governing strategies at the end of the nineteenth century and the rise of what some have characterized as the "social state" at the beginning of the twentieth century (Donzelot, 1979/1997; Rose, 1999). The global changes in power/knowledge relations and discourses governing "the subject" (the child, childhood, family, welfare, education, and childcare) are related to a local hybridization of these discursive languages and practices.[2] I illustrate these ideas by discussing broad shifts in governing discourses, and, drawing on O'Malley's (1998) notion of "traveling" and "indigenous governance," I examine specific ways in which "global" and "local" discourses travel and get translated within three specific contexts: Senegal, Hungary, and the United States.[3]

In the final sections of the chapter, I examine what Rose (1996) and others have described as the demise of the "social" during a period some call postmodern, postindustrial, postsocialist, and/or postcolonial. Specific examples of global discourses and their "translation" within Senegal, Hungary, and the United States will be used to illustrate how discourses "travel" into different geopolitical-historical contexts (see Bhabha, 1994; O'Malley, 1998).

Reason of Welfare about Families, Childhood, and Nation-State

In the introduction to this volume, illustrations of the perceived differences between "weak" liberal welfare states, corporate conservative welfare models, and "strong state" centralized social democratic and communist welfare state models were drawn. While these pictures are useful in describing different national policies related to the distribution of social resources and "benefits" or "provisions" (in health, education, and childcare, for example), they ignore cultural systems of reasoning embedded within notions of welfare, childhood, family, and the nation-state oriented toward care for the well-being of its "population." In this analysis, I focus on cultural and historically contingent ways in which cultural systems of reasoning have combined to govern and constrain the "reason of state" practice.

While welfare regimes, as discussed by Esping-Anderson (1996), Sainsbury (1999), Orloff (1996), and others use notions of "sovereign power" and the nation-state as fixed, known geographical entities, I use the idea of "nation-state" as an imaginary and imagined historical space, constructed by cultural, historical, and geographical mappings related to power/knowledge relationships (Anderson, 1991; Bhabha, 1990; Escobar, 1995; Gupta & Ferguson, 1997; Said, 1978). In this conceptual framework of "nation-state," the state is no longer an essentialized actor that can hold power or empower others. I approach the "state" as a construct related to power and knowledge relations that govern conduct, action, and subjective identities. The cultural systems of knowledge and power, rather than being centrally located in "the state" as an entity or institution, circulate across different institutions and individuals. They are embedded within reasoning systems of schools, family, or government in different laws and policies. According to this reasoning, the "state" is also embedded within our own "private" understandings about childhood, good or poor parenting, and desire for certain types of "well-being." The power/knowledge relations help to construct "reason" that fabricates the traditional boundaries among "state," civil society, family, and individual that are embedded in past and current "welfare state" philosophies.

If the social state involved new patterns of governing family, nation, and well-being during the nineteenth and twentieth centuries, the "demise of the social state" (Rose, 1996) signals a new rupture in patterns of regulation at the end of the twentieth century. The new discourses are associated with new patterns of globalization, postindustrial, and postsocialist shifts in economic and political relations related to increases in global poverty, and new cultural systems of communication and knowledge. New discursive patterns of regulation are associated with what some call a postmodern shift in cultural anxieties about self and others, increased insecurity and uncertainty about the future, and uncertainty about the promises of progress and rationality associated with the Enlightenment and modernity. The concept of dependence on the state, as one depends upon a mother/father or family, to care for its children (future citizens) has been broken apart by these uncertainties, with new calls for local communities and private alliances to take the place of "nation" in ensuring the welfare of self, individual, and community.

New Governmentalities and the Rise of the Social State: Global and Local Discourses in the Changing Patterns of Governing

In Foucault's different volumes, *Madness & Civilization* (1965), *Discipline & Punish* (1977), *and the History of Sexuality* (e.g., 1988), he described the shift from sovereign rulers or monarchs holding "power" to new forms of governing subjects of emerging nation-states that required different state reasoning (raison d'etat). These included diverse technologies that could be used to form the reasoning that would govern childhood, family identity and conduct, as well as the self. The new technologies embedded different forms of surveillance, regulation, discipline, and policing across social institutions (Donzelot, 1979/1997). The new technologies also involved new forms of expertise and experts, the development of new ways to categorize, differentiate, and normalize populations, and strategies that could be used to govern everyday moralities, and truths to facilitate self-governing. There were also new ways to administer reason and conduct, new desires and pleasures, and new understandings of self as actor and citizen.

In this chapter, I examine the rise of the social state across different geographical contexts and conceptual spaces (e.g., childhood, the family, welfare, care, national identity). In this section, I look at the movement from

European and Western sovereign states to the idea of nation-state, with its changing patterns of governing and regulation of childhood, family, and national identity.

Constructing the sovereign state

Citing Foucault, Gordon (1991) describes changes in religious and secular state governing practices in the sixteenth and seventeenth centuries that are important in the eventual rise of the social state in the late nineteenth and early twentieth century.

> Foucault singles out the emergence of doctrines of reason of state in sixteenth-century Europe as the starting point of modern governmentalities, as an *autonomous* rationality. The principles of government are no longer part of and subordinate to the divine, cosmo-theological order of the world. The principles of state are immanent, precisely, in the state itself. . . . As Etienne Thuau has written: "the notion of state ceases to be derived from the divine order of the universe. The point of departure for political speculation is no longer the Creation in its entirety, but the sovereign state. Reason of state seems to have perverted the old order of values. . . . Born of the calculation and ruse of men, a knowing machine, a work of reason, the state encompasses a whole heretical substrate. . . . Set above human and religious considerations, the state is thus subject to a particular necessity. . . . Obeying its own laws, *raison d'etat* appears as a scandalous and all-powerful reality . . ." (Gordon, 1991, pp. 8–10)

In Gordon's quotation, there are at least four important discursive shifts signaled. First, there was a break between the secular and the religious state in governing. Second, there were new philosophical beliefs about the ways in which rationality and logic of the mind could be divorced from the body, emotions, and the spiritual world. Third, there was a new importance attached to the need for autonomous rationality in *modern* governing. Fourth, there was a movement to a sovereign state, "set above human and religious considerations," governed by men (not God or the church) and by reason.

The shift toward rationality over the spiritual and toward the sovereign state from the church were important ruptures in discourses of governing because this appeared to centralize power in monarchs' hands. But the sovereign ruler, according to Donzelot (1979/1997) and Foucault (1977), had to develop laws and police the security of its populations within distinct territories. This development of technologies for policing populations developed into multiple technologies of governing, the notion of welfare as policing security and the economy for families, and the population of sovereign territories.

The rise of the nation-state, and a disciplinary rather than a sovereign power

As sovereign rulers were gradually replaced by democracies in late-eighteenth-century Europe and in the United States, disciplinary forms of surveillance were developed that inscribed self-governance in individual action. For new democracies, it was important to govern as part of a social contract with individuals, families, and entire populations and to govern in ways that enabled individuals to see themselves as self-problem-solving, autonomous participants in the newly developing democracies (Rousseau, 1762/1979). New laws and regulations and the growth of new organizations of social administration and police in the nineteenth and twentieth centuries were instrumental in the social administration of these ideas, desires, and freedoms (Foucault, 1977; Popkewitz & Bloch, 2001; Rose, 1999).

The new strategies of governing were intertwined with the rise of new cultural systems of reasoning that incorporated the enlightenment faith in autonomous rationality that differentiated and prioritized logic over emotion and that privileged reason and science as progress and truth. The rise in an evolutionary notion of progress, the growth of technology, and the discovery of natural laws of science led to a belief that science could discover universal laws of social and cultural life that would define progress and an evolutionary development. The new reasoning examined the proposition that the development of individuals as well as nations progressed from uncivilized to civilized, immature to mature, and undeveloped to developed. These discourses became embedded gradually within the reasoning of care and child-rearing, schooling for modern childhood, and welfare policies that were to define and govern the modern nation-state (Baker, 2001; Cannella, 1997).

However, the reasoning about the child, family, and welfare had embedded within them critical conceptions that governed subjectivities about self and others who were different. The new ways of reasoning were related to the rise of multiple strategies for policing the family and population, including a variety of interventions into families *within,* and *across,* nations to guarantee the welfare of the child, the family, the populations of individual nations, and of nations that were under colonial care.

Modernity as an era of technological invention, and industrialization

The growth of natural and social sciences, laced with Romanticism and Enlightenment beliefs, resulted in new discourses, in which religious ideas of salvation after death shifted to secular notions of progress, modernization, and ameliorating life on earth through science and reason. The nineteenth century saw shifts in Europe and the United States from family farms and economies toward new forms of technologies and industries, including a large increase in the rise of small mercantile shops and factories in urban areas. The new type of work often required work outside the home and new gendered divisions in domestic and childcare labor at home. Men were seen as the natural workers outside the home, while women's *natural* work was discursively associated with maternity, and the reproduction of future citizens within the home.

Constructing new "empires"

As European countries and the United States carved themselves into new nation-states and territories in the eighteenth and nineteenth centuries (and then again throughout the twentieth century), the West (Eastern and Western Europe, Britain, the United States) also saw new territories as potential economic markets and as ways to enhance their territories or empires. The expansion of economies and of political power depended upon the expansion of a European and Western conception of modernity and civilization, or at least the idea that to be modern, Western, or European was good and to be desired. This required the construction of new systems of reason and the translation of new governing discourses that were designed to reconstruct identity and conduct, and to be inclusive of those capable of becoming modern, and Westernized, while excluding other ways of acting or thinking from what was regarded as reasonable and civilized.

Entire regions of the world were reconstructed; new countries and populations were defined and categorized and distinguished from each other. In addition, mass immigration resulted in extended families being separated, with a frequent resulting romanticized nostalgia for the lost family, nation, or culture. This resultant "memory and forgetting" resulted in a hybridized notion of identity that drew from historical memory and a nostalgia for a remembered, national past. The new countries often mixed similar ethnic and linguistic groups, placed them into new categories with new national names that homogenized and erased differences where they existed, and made new differences where previously there had been none.

In the United States, expansion of territories included the post–Civil War colonization of parts of Mexico, many that had been previously colonized by Spain, and the remapping of most indigenous Indian land into territories owned by European-American citizens or the government of the United States. In Eastern Europe in the nineteenth century, different wars forced renegotiation of territories and a constant refabrication of citizenship for different ethnic groups. One result was the creation of the Austro-Hungarian empire, which was populated by diverse ethnic and linguistic groups that had not been homogenized as citizens of the empire.

After centuries of a European and American slave trade that exported Africans (many from Senegal and the west coast) to other countries, postwar Europe used the 1884–85 Treaty of Berlin to expand territories into the constructed continent of "Africa." The territory previously named Senegambia, as one example, was divided into Senegal (a French colony) and Gambia (a British colony)—with members of several ethnic groups living near the River Gambia being divided into two different countries, colonized by two different European nations with two different languages and cultural, colonial customs.

The fabrication of citizenship involved new governmentalities and technologies. These included mapping citizenship geographically, the importation of languages and customs to homogenize populations, and the use of statistical and populational reasoning to differentiate them from each other (e.g., see Anderson, 1991). But the technologies of governing were not the same in all places; nor were the new mentalities of modernity, or a Western or European civilization identical, as these developed in different places, and at different times.

The translation of ideas of what the normal and civilized child, family, parent, and country were to be was what made the new governing strategies within and across nations so important. While missionaries, and other travelers, were to begin to spread the ideas of what civilized and uncivilized or primitive behavior and conduct were like in Africa, it was the disciplinary practices from newly emerging scientific fields, such as anthropology and the natural sciences, and the modern Western school, among others, that were to identify how a civilized family or educated child was to act, how one identified oneself subjectively. It was the textbook in European schools and colonies that fabricated literacy as the capacity to write and read, and language as formally French, German, Portuguese, Spanish, or English. But it was the specific ways by which these discourses were introduced into countries and were hybridized, or translated, that were critical.

O'Malley (1998) suggests one way in which this process might take place. She suggests that translation implies a process that highlights the way indigenous governance made sense of "governing from a distance," a concept she draws from Latour (1986). Indigenous governance ignores "aspects which are 'incomprehensible,' thinking of practices as if they were situated within a familiar rather than an alien culture, 'correcting' obvious 'errors', assigning significance according to familiar rather than to alien priorities, and so on" (O'Malley, 1998, p. 162). The significance of translation is "in producing unanticipated shifts in the operation of rule between the intentions of the initiating programmers and the practices which are put into effect at relationally distant points" (O'Malley, 1998, pp. 162–163). Translation, then, is multidirectional and complex, developing a hybrid pattern of governance that travels from local systems of reasoning to the global, and back again into local concepts and specific, contingent spaces. In the chapter, I examine the notion of translation, and some of the possibilities of hybrid governing systems as I illustrate constructions of global and local reasoning about childhood, modernity, citizenship, welfare, and the well-being of families and nations.

Circulating discourses of the modern citizen

The modern citizen, child, family, or nation was defined in part by seventeenth- and eighteenth-century European philosophers (for example, Locke's 1693 *Treatises on Government* or Rousseau's *The Social Contract* [1762/1993] or *Emile* [1762/1979]). The reasoning embedded in the new governing defined a nature of childhood (natural-romantic, blank slate) rather than of the child born of sin. The good parent would guide the reason and conduct of the child through education (rather than punishment), whether by tutor (Rousseau, 1762), parent, or school. A need for modern schooling, a common written and spoken language, or the wisdom of childcare or early education inside or outside the home, were all related to constructions of how to normalize the family, children, and childhood in the image of good, moral (male, white) citizens. This imaginary of the good family and the male child-citizen was one that included the idea of responsible, autonomous families, citizens, and children who would learn self-governance, and extend this to governing the modern democratic nation. Females' natural role, in contrast, was to rear children to be good future citizens of the state (Martin, 1986). While the rationalities of governing and imaginaries of good citizenship encompassed elite/middle-class/working-class/poor, male/female populations within and outside national borders, the reasoning included only some identities, beliefs, and behavior as normal, while excluding others as abnormal.

Childhood as a stage of development

The global discourses on childhood included the differentiation of childhood into a distinct stage, different from adolescence and adulthood. The idea drew from the theories of Comenius related to his *School of Infancy* (originally part of *The Great Didactic,* 1632, see Lascarides & Hinitz, 2000, p. 65, note 42) and Rousseau (1762/1979), and Locke (1693). In their philosophies, *childhood* was a separate stage; it became reasonable to see the child as dependent, less powerful, less developed, more emotional and irrational and in need of maturing through specialized care and early education.

Although in many locations around the world, the child is *still* an intricate part of family economic and cultural systems, including engagement in labor for the family, by the eighteenth century in Europe and in the United States, the child in bourgeois and elite families was distinguished from the adult, with separate activities, methods of instruction, and disciplinary strategies. By the nineteenth and early twentieth century, this vision of the autonomous, but still developing child, in need of a special stage-related education, which was originally constructed for the child of a particular class and cultural group, was discursively translated as an imaginary of a child and childhood that embodied the rationalities of civilization and progress. With biological maturity, and the right child-rearing methods at home and scientific pedagogies at school, the child would become more reasonable, less emotional, and more adult-like. Childhood became an imagined space where all children were in need of care, as dependent and primitive, as an immature form of the civilized adult. This image, circulated throughout Europe in the early twentieth century, was transferred through a variety of technologies, including the spread of modern Western schooling, the use of statistics to categorize and differentiate those who were civilized from those less civilized, and through the spread of literacy and literature that transferred images that defined modern, Western/Northern civilization.

Scientific reasoning and the globalization of laws about the universal child/family and his/her development

Darwin's *On the Origin of the Species by Means of Natural Selection* (1859), and the growth of natural sciences, fostered a belief in science as the instrument that would discover universal truths about childhood and family. Status was attached to new scientific methods, technological innovations associated with new industries, and new beliefs about the relationships between spirituality and the natural world that, in combination, could create modern man. Spencer's evolutionary biological beliefs about the cultural survival of the fittest merged with Darwin's research and with missionary and anthropologists' scientific descriptions of native "others" to reinforce political/cultural and religious philosophies of evolutionary superiority, as well as oinferiority—those who were different, such as the primitive/uncivilized/less developed.

Laced with the ideas that science and industry were new engines of truth and progress, research institutes, universities, and other centers for development were formed in many nations. In the development of new ideas, new university disciplines and departments (e.g., economics, sociology, social work, psychology, anthropology, educational psychology, child development), and new experts, new truths and new identities for whole peoples and geographical spaces were imagined; but the process was not unidirectional, since these ideas were translated and transformed within different colonial contexts into internalized desires, conduct, and local hybrid reasoning systems (see Bhabha, 1994; Chakrabarty, 2000; Escobar, 1995; O'Malley, 1998; Said, 1978). While the patterns were not identical, the governmentalities related to a mapping of the natural sciences into a making of universal laws and truths about humans (man/woman/child), their development, learning, social behavior, and progress crossed the borders of nations, whether from Europe to Japan, Korea, or Taiwan, or from the United States to Latin America.

The circulation of discourses about the modern child and family passed quickly through the travels of new educational and scientific experts (e.g., Horace Mann in the nineteenth century) to other places, the joining of progressive ideas in published texts (e.g., Darwin's *Origin of the Species,* or Froebel's *The Education of Man),* and the idea of the kindergarten. They also traveled through reports of missionaries, traders, and other voyagers to other lands and cultures. In the early twentieth century, they included the works of John Dewey, Vygotsky, Montessori, G. S. Hall, Piaget, Freud, and Boas. As students studied their texts, received education in American or European universities, or heard lectures in Japan or China, the circulation of ideas were related to notions of progress through scientific discoveries, the ability to discover universal laws of child development, and the ability to assess normality/abnormality.

The early-twentieth-century work by Franz Boas and his students, on American Indians, and British anthropologists' reports on natives in India,

China, and Africa were particularly influential texts that helped form conceptions about civilization and development, and what and who were not civilized. The descriptions of others were opposed to the conception of universal development based on what was known as "hard scientific evidence" in the West; this, then, helped in the construction of imaginaries of the universally developing child, man, and nation that would be scientific, progressive, reasonable, and modern, as distinct from others who needed civilizing, intervention for remediation of deficiencies, or salvation of other kinds.

Technologies of social administration: Psychology and education

During the late nineteenth and early twentieth centuries, a dispersed form of governmentalities and strategies that ranged from schooling, to laws, to doctors' and educators' advice to mothers increasingly governed childhood. A new class of experts developed—child psychologists, social settlement workers, sociologists, and social welfare experts. With the help of philanthropists, they designed and paid for kindergartens, and built schools and playgrounds to instill proper conduct and moral behavior. While not *all* children fit into the image of the universal good, dependent child, modeled on the imaginaries formed for children from western and European, bourgeois, elite, and largely white backgrounds, the increasing push for a law of universal development, verified by science, reinforced the discursive construction of a universal sameness or difference in childhood. These conceptions formed the foundation for new approaches in social, family, and educational welfare interventions (Popkewitz & Bloch, 2001; Cannella, 1997).

Socially administering the "normal" family

The construction of the normal family as a nuclear two-parent (father/mother) Western family, where the father was the autonomous provider, and the mother the nurturer, gradually also took on a naturalness and a normality that was originally based on the conception of the Western elite family as the model for normality, maturity, civilization, and the economy. This conception of the family was gendered and sexual; it was also based upon class-related and racial discourses of difference (see Stacey, 1996). In the normal family, biological and other legal ties existed between members; the abnormal family had no legal ties and often involved multiple caretakers, not just the natural mother. The *normal* family was European or European-American in cultural background, typically white, and from a bourgeois or upper-class family background; this normal family usually owned property and was economically autonomous, whether that meant the provider was employed in what became known as a white-collar job, or lived off the proceeds of an estate or from inherited money. The universal mother became associated with natural mothering, emotion, and caregiving. The abnormal mother (whether a widowed, single, or divorced mother) and the abnormal family from a different cultural/racial/linguistic background were considered immature, racially inferior, uncivilized, and uneducated.

The biological, legal, philosophical, social/cultural, political, and economic discourses that formed the reason of the normal, good, and civilized family constructed and constrained identities and opportunities, subjective experience, and conduct. As with the notion of a universal child and normal childhood, they assumed a sameness that was also based on difference and abnormality. The normal family/abnormal family, good parent/poor parent (or nonparent) conceptions that emerged as universal images of the family were governing discourses that laid the foundation for Froebelian U.S. kindergarten programs that were used to intervene into the lives of poor families and young children in the United States. These governing discourses were circulated to other countries in Europe, Asia, Latin America, and Africa (often as part of church and missionary programs) (see Wollons, 2000). Early efforts at parent education in the nineteenth and twentieth centuries and other medical, social, and economic interventions also embodied normalizing discourses that combined images of religious and secular cultural salvation (e.g., see Popkewitz & Bloch, 2001; Popkewitz, this volume).

Research and population and statistical reasoning

The technologies of governing families and children and the reason of welfare appeared global, although they were constructed differently in different places. Scientific research was a critical strategy used to construct truth about who was normal and which children or families were perceived as abnormal and in need of different social interventions. Statistical categories for the normal family were constructed from narrow samples and contrasted with demographic population facts about different families and different cultural/social/religious/economic organizations.

In the United States, for example, categories about unmarried mothers or single-parent families were used to develop policies that provided different forms of schooling, child care/early education (nursery schools), and health care for populations that were or were not "of the norm." In Hungary, those who were considered nationals were contrasted with Jewish people, Roma (gypsies), and other non-nationals, with different critically important policy interventions related to both the normalization and the exclusion of these children and their families from schooling, jobs, and the conceptions of what it meant to be a good national citizen. In Senegal, those who were elite Senegalese (noble families, government civil servants who had become rich, or those educated in France) became model colonial, cosmopolitan citizens, appropriate to the imaginary of the good French citizen; others were considered primitive, culturally/racially inferior, and unworthy of being made a legal citizen of France, or a cosmopolitan member of Europe, or the modern West.

Modern schooling as a technology of administering normality

The modern school developed in response to the new needs for homogenizing children and their families. European and colonial languages, customs, morals, and aesthetics were taught as though these represented civilization, or the highest forms of human cultural development. In the United States, schools, and urban social welfare settlement-house kindergartens and day nurseries were used to regulate and govern the morality and conduct of Eastern, Southern, and Irish European immigrants' behavior and cultural morality. Teachers taught immigrant children English and appropriate conduct to help in the Americanization of young citizens; parent education extended the ideas of a certain fabricated American normality to mothers. African Americans in cities such as Minneapolis as late as 1920 set up their own kindergartens—in which they (African Americans) used the term "missions"—since they were shut out of the white settlement houses.[4]

In Hungary, at the turn of the century, in similar ways, though in a considerably different geographical and political/economic context, modern schools taught European conduct and Hungarian languages and culture, in a way that assimilated other nationals into a hybrid form of what was thought of as "high culture" from Western Europe and the "high Hungarian national (pure) culture," defined as being of Magyar ethnic descent. In Senegal, the French used an assimilationist policy to teach "capable" Senegalese European (or French) habits of mind as quickly as possible; those who were considered uneducable, or naturally primitive, were left without Western schooling or language instruction. While Koranic schooling, other educational practices, and Arabic and African language instruction were still the dominant forms of education, these were not modern, civilized, or developed systems of education and, according to the French, required different interventions.

The reasoning of the welfare state as part of a social administration of the family, child, and nation

The social state that emerged during the first third of the twentieth century embedded liberal philosophies of freedom, rationality, science, and progress into notions of what it meant to be a citizen of a modern, democratic nation. While the traditions of a welfare state differentiate between those states that provide more freedom and autonomy for individual action and choice and those that centralize control of welfare in the state, the distinctions between the public and the private embedded in these traditions were effects of the new governmentalities related to the need to care for populations, the rise of science, statistics, new groups of experts, and different technologies to socially administer families and children. While the welfare state traditions that are labeled liberal or social democratic have different ways of thinking about social rights and responsibilities of the state and individual, the rationalities of governing were directed at public and private reasoning, subjective identities, and public and personal conduct.

In the United States, Great Britain, Western and Eastern Europe, and in the colonial education systems in Africa, the inculcation of personal ideas of agency, the notion of an autonomous rationality, and an actor who could participate actively within the different democratic nations were critically important concepts. Part of the new governmentalities was the construction of a sense that one behaved in a certain way (e.g., in order to become "modern"), that there was freedom to participate or to make choices. Rose (1989) referred to this as "governing the soul." Governance encompassed a subjective sense of self, how one acted toward, thought about, and constructed policies and interventions about others.

The notion of public/private, independence/dependence, collective/individual "rights" and "freedoms" defined ways in which the social state administered reason and conduct; these global governing discourses included, and *excluded* through the ways in which they influenced the construction of identities and conduct as normal/abnormal, civilized/uncivilized, modern/not modern, or developed/undeveloped. The rationalities of the social state administered freedom, the liberal and social welfare state divisions between public and private, and the divisions of state, civil society, and family (see Hindess, 1996; Popkewitz, 1998; Popkewitz, 2000a; Rose, 1999).

The global and the local

The spread of global discourses about a universal, normal childhood, families, and nations were new governing discourses that homogenized childhood, families, and nations, while also differentiating those children, families, and nations that were considered abnormal, uncivilized, or different. While the terms modernization and economic theories of developing nations, according to Escobar (1995), emerged only after World War II, the eighteenth- and nineteenth-century cultural and racial theories that constituted what it meant to be civilized, modern, or developed constructed progress, science, and industrial development as the essential conditions for constructing rational, civilized modern individuals, families, and nations. As evolutionary theories made civilization and progress normal and appear inevitable, the new technologies were used to assess and categorize difference and to intervene and normalize those populations and individuals constructed as primitive, uncivilized, or deviant. As Europe and the United States colonized nations, technologies of the social state were created and circulated as global reasoning about modernity and normality. As O'Malley (1998) suggests, the translation process that was integrated with indigenous governance systems, however, was complex, varied by location, and was not unidirectional.

Governing of the parent, family, and welfare through education and care of the other: Hungary, Senegal, and the United States

In the early twentieth century, the new governing strategies became prevalent within the United States in the form of the US Children's Bureau (a government institution), in foundations that supported research on child development and the family (e.g., the Laura Spelman Rockefeller Foundation), and through the establishment of a variety of research centers and publication sources in universities to study the child and family (Bloch, 1987). These centers focused on fields from psychology to sociology, from education to political science. The new field of social welfare, and the growth of social workers, was combined with the growth of philanthropically organized centers for the spread of American ideas (see for example, Addams, 1910/1960). At Clark University, G. Stanley Hall convocated groups of teachers and psychologists to study the child in order to incorporate new scientific knowledge into teaching. New schools for young children—nursery schools, kindergartens, and day nurseries for the children of working mothers—were developed, using the expertise of psychologists and sociologists on the family and child to guide the work of teachers, teacher educators, social workers, doctors, and parents (see Bloch & Popkewitz, 2000; Popkewitz & Bloch, 2001). The concepts of the universal child, evolving naturally through different stages of development, as nations were to develop toward being modern, spread through the emergence of kindergartens (Wollons, 2000) and nursery schools in the late nineteenth and first third of the twentieth century. As this enthusiasm for social and educational science as the backbone for policymaking based on the *truth* of science spread, the discourse of science and universal truths of developmental norms and educational pedagogies were also spread globally (e.g., Popkewitz, 2000b; Wollons, 2000).

In Eastern Europe, in the early twentieth century, world war, revolution, and colonization went hand in hand with a carving-up and a reconstruction of maps, as well as notions of identity and citizenship. Countries were forced to reorganize their borders, and their sense of national- and self-identity (Chakrabarty, 2000; Hanak & Held, 1992). At the turn of the century, poverty, religious differences, massacres, war, and new political ideas caused mass immigration. After World War I, the Austro-Hungarian Empire was broken into parts, with, for example, much of what was known as the Magyar Hungarian national population being separated into multiple countries beyond what became Hungary.

At the turn of the century, Hungary was still part of the Austro-Hungarian Empire. It was an important independent nation within the large empire, relative to neighboring countries—e.g., Romania, Macedonia, Galicia, Serbia, Bosnia-Herzegovina, Montenegro. Budapest was the center of bourgeois activity, normal cultured family life; art, music, literature, and pure and high culture were critical factors in discriminating good families from others and establishing different degrees of difference (Lukacs, 1988).

Until World War I, bourgeois families were able to participate in schooling that was based upon not only Hungarian high cultural norms, but also on a construction of the Western and European model of elite and cultured families. This hybridized model embodied norms for "civilization" against which non-Hungarian nationals were judged.

After World War I, the eventual break-up of the Austro-Hungarian Empire was tied to an increase in racialized and nationalistic discourses related to normal education, childhood, and families. Hungarian family and child life was governed in relation to Western European norms of the autonomous, rational individual, and child-rearing in relation to Western European norms of educated and cultured families. In addition, the growing threat from the East after the 1917 Russian revolution resulted in new ways of thinking about family and nation in relation to the formation of USSR.

Looking both to the West, toward their own nation built out of a variety of disparate ethnic groups, and toward the East, Hungarian national norms for the modern family and child integrated bourgeois family values and liberal ideas of the West with Magyar ethnic family customs and Magyar languages to construct a pure cultural conception of the Hungarian subject. A national socialism evolved that used a constructed, racialized conception of sameness, and that also defined difference simultaneously. It used a notion of cultural purity that integrated European values of freedom and rationality with a collective care for families and children, particularly if they were normal, or could be normalized through different types of cultural interventions such as schooling. As births declined in bourgeois families, new governing ideas about the need to define populations statistically, and increase population (at least in Magyar/national families) became important. Simultaneously, social eugenics theories, similar to those used in the United States, were used to rationalize racial, class, and sexualized/gendered discursive social interventions into the lives of poor women and their families.

In order to regulate nationality, languages taught in schools were Western European and Hungarian; children from different primary language backgrounds were required, as were their parents, to learn at least Hungarian to assimilate into Hungarian society. While there was resistance from many non-Magyar (Hungarian) nationals within the borders, pre–World War I technologies categorized Hungarian nationals as cultured and civilized, while others, although living within Hungarian borders, were different, uncultured, or uncivilized (see Anderson, 1991; Hanak & Held, 1992; Hupchick & Cox, 2001; Lukacs, 1988).

In French West Africa, as one result of the 1884 Treaty of Berlin, the French took over Senegambia, keeping Senegal as part of Francophone Africa, along with its other colonies—the Ivory Coast, Guinea, Mali, Chad, Niger, Mauritania, Morocco, and Algeria. The cities of Gorée, St. Louis, Dakar, and Thiès were made official parts of France, with citizenship given to the populations within these urban areas; the rural and majority population in Senegal was still to be assimilated to French culture and civilization, but its members were not given legal citizenship. This difference, along with the provision for French schooling for the elite in Senegal, and for the children of the rich, or noble families within selected cities, provided the foundation for the assimilation policy that characterized French policy throughout Francophone Africa (Young, 2001). While these governing strategies were met with an indigenous translation of ideas and practices, the first generation of highly educated French West African politicians were schooled through this process. French and Senegalese citizen and poet Leopold Senghor, president of postindependence Senegal, Humphrey Boigné, president of postindependence Ivory Coast, and Sékou Touré, president of postindependence Guinea, were schooled in the same boarding school in Senegal during this period.

Colonial governing of family life integrated the French assimilationist policy that was aimed at normalizing African citizens and making them European with a policy that treated rural African families as abnormal and unworthy of modernization or schooling. While the French language became the language of instruction in schools, and the official national language for political and economic participation in the economy or government affairs, the vast majority of the children throughout the highly rural, agricultural, and pastoralist societies in Senegal and in French West Africa were not able to go to school, or to learn French.

The new governmentalities about what constituted culture and modernity, who was civilized, and who was civilizable, therefore, included the conception that some Senegalese could achieve an education that would make them seem French and European, while others could not. The others were still, however, part of the economic French system—as farmers, traders, and workers in the new government bureaucracies, helping to establish a market for French goods within Senegal, and the desire to become like the French.

In addition, there were other discourses making the governing of citizens/noncitizens within Senegal and the French colonies more complex. Islamization had occurred much earlier. Thus, Koranic schooling and religious practices added additional layers to the construction of family life, good education, children's activities, language education, and moral conduct. In some ways, these became hybridized with French/European logic, mentalities, ways of acting, and belief systems. In addition, indigenous ethnic languages and customs that preceded the French and the Moslem influence also formed the background for a complex and hybrid system of cultural reasoning about family and village social and cultural living patterns, child-rearing customs, and a collective indigenous system of care of self and others.

The resulting hybridization of different governing discourses resulted in many different patterns of "family" and "child-rearing" within the Senegal of the colonial period. To the present, the majority of the country is characterized as Moslem. Families were matriarchal or patriarchal, depending upon ethnic group. Families were (and are) polygynous, particularly in rural areas. Taboos about birth control, and the rationale for large families to assure family subsistence, related to the combination of local customs, Moslem laws, and French laws that limited the number of wives to four, and encouraged reproduction from puberty to menopause.

Child-rearing, which underwent much change in relation to place and custom, and to time in the early twentieth century, was still done by an extended family. Only as families moved to urban areas did the nuclear family, and a concept of individuality reminiscent of Europe, begin to govern what the family should look like.

Women and men's roles in cities and rural areas were far different from the European norm of family, as autonomous, individualistic, and nuclear. West African women participated heavily in agricultural work in rural areas, and were well-known for their participation in economic trading activities. Education of younger and older children took place within the extended family, with oral literacy and knowledge of the Koran important parts of the indigenous and Islamic systems of education. Girls from an early age participated in household labor, childcare for younger children, and agricultural work, while boys participated in agriculture and herding.

Within cities, where European influence was greatest, family organization for the Senegalese still remained similar to that of the rural Senegalese. However, the model of civilization and modernity for women gradually was related to a femininity and a Western bourgeois mentality that identified being at home, being schooled, and *not* doing agricultural or commercial trading as European and modern. It became fashionable for elite or bourgeois Senegalese men to provide support for their own large families, and to reduce collective care and financial support for the extended, rural family. However, the assimilation of French norms also integrated indigenous ethnic and Islamic customs into individual conduct and family educational and social identities. Different beliefs and customs represented the multiplicity of identities on display at particular moments in and out of bureaucratic, economic, and school settings.

While the process of colonization and assimilation under the French was critically important to changes in the Senegalese economy, politics, laws, and policies during the years of colonial rule, it is important to recognize the complexity of the hybridization of identity and conduct that was occurring—a hybridization that was highly contingent on place, ethnicity, religious origin, caste, and class. These patterns of social administration were related to the complexity of strategies for governing in the neocolonial period following the independence of Senegal in 1962 and the ascendancy of rule by President Leopold Senghor.

Post–World War II remappings of political space

Discursive shifts in governing emerged during the 1930s in Africa, the United States, and Europe. The education of a core of African leaders in the West, participation in Europe's world wars, and new demands for civil rights and independence led to a remapping of Africa into independent nations, and to the rise of new governing discourses related to social rights, needs for equality of opportunity, local empowerment, and participation. The modernization of underdeveloped nations and individuals also began as an aftermath of the Marshall Plan, as well as the development of new international organizations that were to reconstruct, refabricate, and redevelop Europe and guide the development and modernization process in other independent nations. While the discursive shifts were important, the practice of a socially administered *freedom* and the governing of normality remained essential aspects of cultural reasoning systems. In the United States, new governmentalities of the victor, in relation to the rise of Cold War animosities with the Soviet Union, led to a scramble to intervene into and govern nations. The voices of independence, empowerment, and civil rights within former colonies of Europe were matched by new demands for local empowerment, community activism, voice, and political, economic, and social rights for inclusion in the United States.

The new technologies of governing included the rise of new institutions that were to use science and technology to bring progress, modernization, industry, and markets to underdeveloped nations (see Escobar, 1995). The United Nations, the World Bank, the U.S. Agency for International Development (USAID), and, for one example from Europe, the Swedish International Development Agency (SIDA) competed with Soviet socialist agendas to increase economic links with different newly independent nations. Within Senegal, conceptions of a free market, and links with capitalist countries that would help to develop nations economically, were contrasted with a desired independence laced with a romantic nostalgia for an imagined precolonial collective socialism, and rationalities of governing linked with economic/social relations with the Soviet Union. But the political and economic rationalities were also cultural systems of power and knowledge that governed the souls, identities, and conduct of the children, families, and citizens of the new nations.

Regions were unified by language, thus limiting and controlling who gained access to appropriate forms of literacy and schooling, the use of certain ideas within textbooks, and the spread of customs through university education abroad (Escobar, 1995; Ghandi, 1998; Smith, 1999). Development became a globalized economic, political, and cultural/social mentality that became reasonable at both the individual level (to be like the French, in Senegal, for example, was an extreme good for some), and national/continental levels.

While the colonization and the carving up of Africa took place during the nineteenth and twentieth centuries, by the late 1950s, the new African nation-states, using virtually the same divisions of the European colonialists based on the Treaty of Berlin mapping of Africa, demanded assistance to develop nations. Governments desired economic links with Europe, Western schooling, and imitated, at least in many urban areas, the European in dress, language, and customs. In Senegal, strong ties with the French were retained through the presence of school systems, fellowships to go to French universities, and French bureaucrats in Senegal who helped run the government. To be cultured and developed still was related to "acting" French, and to privileging assimilation to French culture and customs over rural life and tradition, despite Senghor's poetry and narratives, which romanticized the importance of tradition and pride in being African (e.g., Anderson, 1991; Bhabha, 1994; Escobar, 1995; Young, 2001).

Senghor, who had been carefully schooled at a Senegalese-based French boarding school and later at a university in France, formed the independent new government of Senegal, based on a politics of *negritude,* a return to a nostalgic collective and "pure" African past, that was a hybridized vision of Senegalese who were also French, with the ability and desire to be modern and European. While different countries chose different pathways, Senegal retained strong ties with the French. It was only in the mid- to late 1980s when those international organizations referred to above, and including the International Monetary Fund and units of the United Nations (UNESCO), became so influential. Bank loans and other aid were used to develop the country by modernizing industry, building new schools, and highlighting the need for training in agriculture, industry, and bureaucracy. But structural adjustment policies were also required to repay international loans. These stressed neoliberal rationales of governing privileging the private over public and the autonomous individual over the collective. The effects of these policies were complex, and often related to increases in food prices, school fees, and decreased access to social services.

In Eastern Europe, the remapping of territories in the post–World War II period effectively carved up countries and remade the map of Europe (once again). This time, however, the Soviet Union, with socialist, peasant, and communist parties in Hungary in place, took over the government. The resulting universalist modern and evolutionary socialist policies related to families and children in Hungary, and elsewhere under Soviet domination were linked with mentalities about universal provisions, full employment, and equity of provisions. Soviet discourses of equity and a revolutionary social policy hid the regulation and forced homogenization of language, custom, voice, and nationality, constructing dissent as abnormality. The public was merged with the private for national good; but the discursive belief that there was a private space in families and civil society remained.

The Remapping of Public Welfare Discourses through Geography and Imagined Nations/Citizens: The Demise of the Social State

Blurring the boundaries of governing dichotomies

Governing discourses construct what appear as real or natural, boundaries between the state and nongovernmental organizations in civil society, including the family. As suggested earlier, this is related to an artificial and historical Western construction of the public and the private that rests at the very foundation of liberal philosophies of the state, which form the rationality of all modern welfare states (see Fraser, 1997; Hindess, 1996; Rose, 1999; Schram, 2000). The rise of uncertainty related to postmodernity, postindustrialized societies, globalization of economies and communication systems, and what appears to be a postsocialist, postcolonial period has led toward a distrust of the central state as a solution to global, community, and individual economic and social problems. However, the recent calls for Third or Fourth Ways of governing (e.g., see Giddens, 1998) place greater reliance on a civil society and the autonomy of individual and group actors and action that reinscribes the false notion that freedom, community, individual autonomous action, and privacy are not only possible, but desirable (Rose, 1999). Instead, the governing discourses that socially administer freedom, privacy, and the conduct of conduct continues through a "steering from a distance."

However, there are new global discourses that are circulating as the reasoning of Family-Child Welfare within international welfare policy texts. These are evident in the texts and practices associated with Third Way welfare reforms of New Labour in Great Britain (Giddens, 1998), in the Welfare to Work reforms in the United States, and in reforms embedded in Europe's social policy imperatives as well as those of international agencies such as the World Bank, or the International Monetary Fund. The new discourses emphasize *self-sufficiency, local responsibility, privatization, independence/autonomy, local and individual involvement, entrepreneurialism, motivation, quality, efficiency with equality, choice,* and *hard scientific evidence* that will prove that the new rationalities of governing produce progress and truth.

These new discourses, which tend to float from country to country and region to region, require analysis and a critical interrogation in terms of the ways they govern reason, identity, and conduct. As in earlier examples, the governing discourses appear inclusive and good and necessary for everyone, while they also hide governmentalities that exclude many from choices that appear to offer ways to think about oneself and others.

Welfare and care in a late modern or postmodern period: The nonsocial state

In place of the social state that tried to legislate the care of citizens, Rose (1996, 1999) suggests that the nonsocial state is a new way to govern through the flow of ideas, global knowledge societies, multinational institutions and corporations, and international laws and rights groups that circulate global ideas of family and child welfare. Postindustrial, postsocialist, and postcolonial "risk" societies, communities, and individuals are to take care of each other—with little expectation of broad social subsidies from any government or international organization.

The new governing patterns construct us as self-governing communities and individuals, where local and community levels are to be *caring communities* that take care of each other. The new discursive framing accepts the desirability for decentralization and privatization of care and responsibility, a rise of community and individual autonomy, entrepreneurial activity, and the flexible, problem-solving and responsible self. The discourses also underline the lack of faith in state policy as adequate to effect the complex array of economic, political, and technological changes and uncertainties that are part of the process of global and international change.

The discursive practices relate to the importance of decentralization, localization, the importance of community and the private sector, individual responsibility and self-sufficiency, reflection and participation, local action and democracy, flexibility and choice. They are also expressed in notions of uncertainty and instability, rather than closure or the ability to ensure care or education for all members of a nation (citizen/noncitizen). New forms of power/knowledge, disciplinary and biopower (Rose, 1999) govern desire, motivation, action, and conduct, and, particular to the focus of this chapter, the way we reason about caring for ourselves and for others.

Within this framing of welfare and care, again, there are inclusions and exclusions. The imaginations of citizenship, welfare, and care may be globally available, and held up as ideals during this current period of uncertainty. But the ideals, in fact, are neither shared universally as ideal norms of citizenship (democracy, autonomy, choice, universal laws and rights) nor are they equally available. Thus, while norms may be present for a Third Way (Giddens, 1998), or as a welfare-to-work model that will create autonomous and responsible citizens in the United States, or self-sufficient citizens elsewhere, the reasoning is about the universal family and child, community, and nation, while the rationalities continue to be targeted at interventions for only some. As in earlier governing discourses at the beginning of the twentieth century, it is the abnormal parent, community, child, and nation that are to be made normal through targeted interventions into their lives and subjectivities. This reasoning promotes the idea of choice, efficiency, and equal opportunity through a combination of free market and democracy around the world. But as we see in so many places, choice is a discourse that embodies the rationalities of a true individuality, autonomous action, and an equality of political and economic participation that is not possible for many individuals, nations, and regions that are poor, or differ from the universal imaginaries of the well-developed and normal modern child, family, and nation. Following are examples that illustrate these points.

The universal rights of childhood—Leaving No Child Behind

The recent UNESCO document, *A World Fit for Children* (2002), addresses "National Programmes of Early Childhood Development [that] focuses on emotional, social, physical, cognitive, and healthy development," with other foci on the "Girl Child," and "No Child Left Behind." These discourses echo other texts and documents from around the world in the way they frame policy (policing) as natural, and point to children and families who are abnormal, as the "No Child" who is to not be left behind (while other normal children are ahead). The Children's Defense Fund, an advocacy organization for poor families and children in the United States, has now copyrighted its logo "Leave No Child Behind" (Children's Defense Fund, 2001, inside front cover page), as this phrase has been taken or translated into reform discourses in education by the Bush administration, and has appeared in World Bank, UNESCO, and UNICEF documents, and government documents of other nations, such as those obtained from Senegal. *(BASIC) Education for All* (UNICEF, 2000, 2001), a reform program adopted by UNICEF and by a variety of other organizations and nations as a reform goal, using that precise phrase, is another example of circulation of discourses that, while appearing to be inclusive and democratic—*Education for All*—points toward those considered to be different and in need of remediation, development, or modernization through a variety of strategies, including the modern primary or elementary school, adult literacy programs, and nonformal community-based health and family planning programs.

It is this notion of constructing conduct and subjective identity for *all* that is critical as one looks at the discourses as they travel in somewhat different ways in the late twentieth century and beginning of the twenty-first century. The technologies that are used to fabricate commonly understood ideas about the child and family travel as universalist notions across nations, through the exportation of ideas in research, the sharing of expertise, and the amalgamation of new ideas for the globalized and knowledge-based economies of an increasingly postindustrial world. The importance of human capital formation in conjunction with the governing discourses about how to become modern and developed have resulted in the spread of mass schooling around the world. There is a truth and authority about the need for *hard evidence* (see the "Shonkoff" report, 2001, for example, or the National Research Council report edited by Shavelson & Towne, 2002), which is supposed to equal objective, experimental, hard scientific data, and the voice of expertise and evidence that inform our policies—apparently globally, despite so many needs for understanding contingent and hybridized differences.

The ideas of universality are laced with an understanding of community and new nationalisms and a remapping of regional identity (e.g., the European Union, or "old" and "new" Europe), while at the same time, the discourses of what is good for the child and family (e.g., developmental and educational child development programs, parent education, marriage, birth control—particularly for adolescents—and employment) translate similar messages. These discourses circulate through cultural systems of knowledge (over the internet, in books, at conferences, etc.), and through the microactivities of everyday interactions and communications by experts—in doctor's offices, in schools, government expert reports, and researchers, and teacher educators' ideas. The globalization of the modern, scientific citizen was and is still being transmitted in complex, circulating manners from West to East/East to West, North to South/South to North. While there is nothing inherently new about this circulation, the new discourses of the child and family and the good citizen have commonalities and contingent particularities that relate to history, culture, and space and place.

Making progress through "development" and basic "education for all"

Many of the discourses of the twenty-first century sound similar to those at the turn of the previous century. They embody notions of linear natural/biological development (as individuals go through early childhood, middle, and adolescent childhood stages prior to that responsible adulthood, followed by becoming elderly), and, in relation to nations, still involve moving from less developed to more developed nation-status—terms evident in recent World Bank documents as methods of categorizing progress.

The new governing discourses that travel, however, are similar to those from the early twentieth century, yet they embody critical differences. The neoliberal economic and social welfare discourses about state/nonstate that need to be broken apart, encourage decentralization of government and enhancement of civil society as though these were *natural,* untarnished-by-state governing, spheres of influence. The discourses focus on localization, bringing in the community, empowerment of marginalized voices, and bringing partners and new voices to the table. They privilege discourses of privatization (parent involvement, private-public partnerships) and the realm of private lives, while continuing to administer that privacy. The discourse of dependency/independency has moved from the desirability of families to be dependent on state support for raising families, to a new desire for responsibility and self-sufficiency from the state; at the same time, dependent relations between the social state and corporate welfare has become necessary in uncertain economic times. The new global governing discourses of citizen responsibility and service are to promote responsible families and encourage philanthropic, religious-based, nongovernmental, and private volunteer services that, as a collective, are (again) asked to partner in the salvation of individuals, family, and nations. This is an enhanced citizen responsibility and independence in the name of self-sufficiency and a critical and self-reflexive form of self-governing in the name of democracy that, discursively, forms the new cultural *raison d'etat.* Again, the new governing discourses are both cultural/material in their strategies and effects. They are about the construction of women/men, adults/children, families/civil society/state, and the lives, conduct, and subjectively/materially organized identity and experience of individuals, groups, and nations.

In the United States, the new needy are constructed as those clients or consumers or citizens (women/children/poor families) who need help to reach the goals of responsibility and self-sufficiency, with offers of subsidies for childcare, food stamps, and some aid for transportation. The new needy in the United States have case workers to help find employment, and five years of limited availability of welfare support, if jobs cannot be found or kept. Those who are needy, and targeted by *education for all* educational reforms, for example, are also those who are abnormal in that they are perceived as not being capable of self-sufficiency, or those who have a difficult time achieving in homes, schools, and communities that are resource-poor and segregated from "rich" America. However, these families are also families that seek good care and education for their children, but have historically been constructed as different, abnormal, and in need of special interventions to normalize their lives.

Others seen as abnormal are those adolescents who drop out by eighth grade or those in high school who are assisting in childcare and family/economic work (Fuller, Kagan, & Loeb, 2002). Those who resist the government's definitions of what they must do and "leave the welfare system"–who refuse the surveillance, punishment, panopticanism, and regulations of the discourses that enter their souls (Foucault, 1977; Rose, 1989)—are constructed as the most different, the most irresponsible, the most abnormal, for resisting state and private help.

Within these framings are also the discursively organized practices of laws, rules, regulations, and the organization of social provisions and benefits within the United States. In earlier periods, these framings have organized "choices." They continue to constrain choices, conduct, and subjective identities related to good parenting and quality education. They also constrain possibilities that relate to the availability of options, as well as the difficulty in finding or keeping jobs, or good childcare when families are *responsibly* employed. This is particularly so for those with infants 12 weeks or older who are required to work 30–40 hours per week by the new welfare legislation, but for whom there are few quality childcare provisions for young children (see, for example, Holloway, Fuller, Rambaud, & Eggers-Pierola, 1997/2001; Lowe & Weisner, in press; Weisner and Fieldwork Team, 2000; also see Dahlberg, Moss, & Pence, 1997 on the discursive construction of "quality"). According to recent research by Edin and Lein (1997), and earlier research by Stack (1974), families find support in extended family systems, the informal childcare system used by families throughout the world, particularly when choices of quality childcare, at affordable prices, are not available.

In Senegal, a recent World Bank document on reforms ("Quality Education for All Program," World Bank, 2000) spoke of new reforms related to girls' education, bilingual instruction, and early childhood initiatives focusing on Early Childhood Development programs in local settings (nutrition, informal education programs), and in community-based preschool programs. Whereas before the government of Senegal had promoted public provision of preschool programs—and had only reached the 5 percent coverage level in the early 1990s—now the programs are to be increased but with greater "authorization of private preschool establishments" that will "bring education closer to parents and the community."

Reform documents suggest that "Government is shifting from a centrally managed teaching force to one that is locally managed. Lower salaries (for teachers) will permit more rapid expansion"; lower training for teachers, lower salaries and benefits will also increase "education for all," but the "all" who will receive these less well-trained and less well-paid teachers are the rural and the poor, the others, while expenditures on education for rich children is increasing through private sources. In World Bank, Unicef, and UNESCO documents, early childhood initiatives and primary education opportunities are to be enhanced and evaluated in terms of their cost effectiveness and the ways in which they improve the efficiency of education (in terms of enrollment and retention, cost containment, local entrepreneurial investment, local responsibility and input). Reforms are directed to communities to build their own schools with virtually nonexistent private resources, and to parents to participate more actively in their children's education, despite heavy agricultural and commercial workloads to ensure subsistence family survival.

Families are to enhance girls' participation—which is linked to lower fertility rates, and a reduction of "lone" motherhood—a new gendered and racialized discourse that is circulating in the reports of international organizations about others to be cared for, surveyed, and regulated. Parents are to be educated and become involved, by paying increased fees for their daughters' and sons' public schooling, or for localized community private efforts to find any schooling at all. While structural adjustment policies (SAPs) since the 1980s from organizations such as the World Bank are often blamed for the neoliberal orientations that have pushed many into increased poverty, have reduced public schooling, and increased school fees, etc., the cultural reasoning systems (that families can afford to pay school fees; that privatization of public services is good) and new knowledge/power relationships (Foucault, 1980) have allowed certain senses of what is reasonable/unreasonable to become natural and good, while alternative ideas, conduct, identity, and possibilities are excluded.

Despite the growth in poverty, international reports for many countries highlight the importance of early brain development, healthy cognitive, social, and emotional child development, and the growth of *quality* early child development programs in communities and in group care settings. UNICEF (2001) estimates that in 2001, 8.1 percent of boys and 7.6 percent of girls from 36 to 59 months were in preschool (public and private) in Senegal. In their texts, they promote pedagogical attention to young children's sensory-motor and pre-operational stages of development (labels directly from Piaget, *Petite Enfance—periode sensori-motrice;* 2–7 *years—phase pre-operatoire*), while also advocating localized, community-developed preschools (private sector), *and* reduced training/salaries for teachers. While this may be a necessary policy for a resource-poor country such as Senegal (see for example, Swadener and Wachira's description in this volume of the Harambee movement in Kenya that increased preschools through a similar set of ideas), the decentralization to poor rural areas is a move backward for a country that, in 1990, had tried to increase public subsidization of preschools and primary schools, required teacher training, and civil service jobs with decent salaries for trained teachers.

In Senegal, present community and local welfare or care systems, inclusive of both traditional and modern governmental systems, have been hybridized over centuries with a variety of forms of welfare for families. This indigenous governance has encountered the new governing discourses by responding to some with initiative and enterprise, and by responding to others with a traditional self-help mentality built on frustration and a history of exclusionary discourses that have been present since the late nineteenth century (and earlier). Early indigenous governing related to "welfare" highlighted collective self-help strategies that have been used to safeguard ethnic groups and to organize the cultural systems through local laws and policies, customs and rules from Islam, are hybridized with colonial and postcolonial laws, regulations, and discursive practices. These hybrid discourses are translated contingently. They frame current definitions of "marriage," rules of childhood adoption, attendance at Western or Koranic schooling, participation by children, women, and men in local labor opportunities, and the development of and maintenance of local agricultural production in relation to local, national, and international globalized markets.

The hybridization of different governmentalities in relation to the care of family, the definition of a universal or a local conception of childhood, notions of child labor as good or bad—in local eyes or in terms of universal rights of childhood—depend upon a variety of complex ideas about the worth of schooling, different forms of literacy, the need for children to participate in family economies, and the importance of religion in determining conduct and beliefs. With 90 percent of the country Moslem, and the growth of Western schooling combined with Koranic schooling, requirements for French and local languages and the needs for traditional family customs to combine with world customs of health and labor, many practices are debated. With global uncertainty, constant prescriptions about what one should do, and incessant poverty in relation to the very visible first world, the education for children, family, and childhood/childcare and education, prescribed by international agencies, is far from certain to be taken as a natural good, particularly in rural areas. At the same time, the need to enter into a global system where schooling appears to matter makes this a difficult and uncertain time, and provides difficult choices for families. As the modern Western public school and globalized languages continue to govern rationalities of logic, reason, and desire, an increase of Senegalese children in primary and secondary, and possibly early childhood, education appears predictable.

In Hungary, and other Eastern/Central European countries and in the newly independent states of the former Soviet Union, the new governing discourses that encourage a targeting of resources, and privatization aligned with a shrinking of the social state and its universalist subsidies for families and children, are systems of reasoning that are new over the past decade; the growing acceptance of poverty and unemployment as possibilities, or necessities, are ideas or expectations that were nonexistent (or at least unacknowledged) ten years ago. The new set of governmentalities, therefore, must be examined in terms of the way it constructs what is possible, acceptable, and normal, who and what is included, what or which groups are excluded. A society in which parents must work in low-wage jobs to be self-sufficient and responsible, that kicks parents "off welfare" if they earn just too much (the targeting system exported from neoliberal market countries such as the United States and Great Britain by the World Bank and the International Monetary Fund), or if they spend time with a sick child and not at a job, is one in which the new governmentalities about how to care for young (and old) citizens must be recognized, and seen as problematic.

In Hungary and in Bulgaria, and in other Eastern and Central European countries (Galasi & Nagy, 2001; Gancheva & Kolev, 2001), poverty has increased particularly among children, and particularly in ethnic minority groups (Turks, Roma gypsies). Nonetheless, expenditures for children in the richest groups in each country have increased, while, oddly enough, expenditures for children in the poorest groups have decreased. Universal subsidies for families, ranging from health care, childhood immunizations, parental leave, family subsidies, and virtually free childcare, have been erased from public discourses. This involves the creation of new understanding, but also a forgetting of what had been expected or natural, or normal. Rather than a linear shift in thinking, then, this embodies a rupture in the discourses that form and construct truths and cultural systems of what is taken for granted. Today, the family "in need" has not been successful as an entrepreneur in the new societies of East-Central Europe. The privatized companies, often brought into partnership with Western countries that have disbanded social services such as daycare, have fostered the new rich, and made it acceptable (one might even say normal) to be poor. The new policies turn on where to draw the "poverty line" or where to target those in need of social subsidies. They highlight the problematic growth of "lone parents," terminology that was largely unknown as a social problem prior to this past decade. While nearly 80 percent of children age 3–5 are attending preschool programs, according to recent reports (Mickelwright, 2000), spaces in childcare centers are unused, and many programs have been closed due to lower fertility rates. As in other countries, high quality public programs have lost support from the state and are turning toward increased public-private partnerships, and increased parent fees and involvement (Bloch & Blessing, 2000).

Governing "development" and "normal" children and families in new ways

In conclusion, the many documents that present hard "evidence" of the importance of quality early childhood development programs, of the importance of early prenatal and infant-toddler experiences to brain development (Shonkoff, 2000), of the importance of teacher training and reasonable salaries in the making of the "good teacher" and the dearth of high quality developmentally appropriate programs, even in the richest of nations, are also the documents that accept the growth of child poverty as inevitable—for now (see Bradbury, Jenkins, & Micklewright, 2001). They normalize privatization, community partnerships, and enhanced public-private involvement at preschool levels and at all other levels of education. As documents of the world organizations blame single motherhood, and encourage marriage and employment as the way to normalize family, they also speak of poverty, insecurity of health care, and the growth of inequality in income around the world as acceptable—whether in the United States, in Hungary, or in Senegal. These same global discourses also accept dependencies between the state and multinational corporate welfare.

Within these frameworks of globalization of discourses and of individual cultural histories, one needs to question the ideas of "universal rights for childhood" and for women, and reforms that call for leaving No Child Behind, or Quality Preschool Education for All. The complexity of the ways in which ideas circulate and construct allowable imaginaries of how nations (and the world) care for a globalized, fluid, migrating citizenry requires vigilance, examination, and continual critique. How, in an era of uncertainties and possibilities, can we find an ethical and material way to reconstruct new imaginaries of social rights and responsibilities, and new ways to think, act, and to reason?

The cultural knowledge systems, or "reasoning of the state" include certain ways of acting—based upon global governing discourses—while they exclude other possible identities and conduct from "reason." The cultural and material effects, particularly as we look at racialized, colonial, and gendered "reason" about the welfare of families, children, care and education that are circulating globally, force us to look for the local contingent and historical ways in which groups have responded to new governmentalities. The spread of the modern, Western school is one example that has taken hold around the world. However, the targeting of deficiency and abnormality that is included within the inclusionary discourse of early childhood or basic education for all, the choice to go to private schools, or the building of community schools must also be interrogated. Who and what constitute the "all" targeted for inclusion in these new policies? What do they get, and what identities and "reason" must be given up in relation to new governing patterns in the early twenty-first century?

Circulating Discourses—Global and local contingent differences that are historically bound

As I have tried to examine specific localities, the necessity of seeing governing as a "translation," between indigenous discursively organized governing and a circulation of global governing discourses, has been important. Conceptions of development, childhood, childcare, education, schooling, and welfare are concepts that are not bounded by Western or European reason. Neither are they pure ideas within locally contingent contexts. Governing involves an examination of the way discourses "travel" or circulate between global and local governing systems, the ways in which subjectivities of self and nation are formed. While power/knowledge relations embedded within discourses include and exclude many possibilities from reason and conduct, they frame the possibilities of local, historically and culturally contingent, systems of meaning. The resulting cultural reasoning systems embody complex, multiple boundaries related to self and other; they also represent new opportunities and possibilities for thought, action, and continuing critical inquiry.

Notes

1. I want to thank members of the Thursday group for reading and responding to different drafts of this manuscript, including Devorah Kennedy, Sabiha Bilgi, Dori Lightfoot, Dar Weyenberg, and Stephen Thorpe. In addition, Tom Popkewitz, Kerstin Holmlund, and Ingeborg Moqvist have each given me very helpful comments. I want to thank Tom Weisner and the UCLA "Health and Culture Project," and Gunilla Dahlberg at the Stockholm Institute of Education for providing me with their ideas, comments, and the space and time during my sabbatical to do this work.
2. Discourses can be seen as "practices that systematically form the objects of which they speak. . . . Discourses are not about objects: they do not identify objects, they constitute them and in the practice of so doing conceal

their own invention." (Foucault, 1974, p. 49; discussed by Watling, December 6, 2002, as appearing first in Ball, 1991, *Foucault and education: Disciplines and knowledge.* London: Routledge; in the reference list, I give the Foucault citation. I'm grateful for the discussion by R. J. Watling, December 6, 2002 on the Foucault list serve; see Foucault-sig@lists.education.wisc.edu).

3. Research on the U.S.A., Hungary, and Senegal presented here is based on new textual analyses of recent documents, as well as research projects done in Senegal since 1975, research on the history of U.S. education (e.g., Bloch, 1987, Bloch & Popkewitz, 2000), and two projects in East-Central Europe in the 1990s (see, for example, Bloch & Blessing, 2000).
4. This was true in other cities throughout the nation; however, the specific idea of "missions" and "settlements" was brought to my attention by Dar Weyenberg.

References

Addams, J. (1910/1960). *Twenty years at Hull House.* New York: The New American Library.

Anderson, B. (1991). *Imagined communities: Reflections on the spread of nationalism.* London: Verso Press.

Baker, B. (2001). *In perpetual motion.* New York: Peter Lang.

Bhabha, H. (1990). *Nation and narration.* New York: Routledge.

———(1994). *The location of culture.* New York: Routledge.

Bloch, M. N. (1987). Becoming scientific and professional: An historical perspective on the aims and effects of early education. In T. S. Popkewitz (Ed.), *The formation of school subjects: The struggle for creating an American institution* (pp. 25–62). New York: Falmer Press.

Bloch, M. N., & Blessing, B. (2000). Restructuring the state in Eastern Europe: Women, child care, and early education. In T. S. Popkewitz (Ed.), *Educational knowledge: Changing relationships between the state, civil society, and the educational community* (pp. 59–82). Albany: State University of New York Press.

Bloch, M. N., & Popkewitz, T. S. (2000). Constructing the parent, teacher, and child: Discourses of development. In L. D. Soto (Ed.), *The politics of early childhood education* (pp. 7–32). New York: Peter Lang.

Bradbury, B., Jenkins, S. P., & Micklewright, J. (Eds.). (2001). *The dynamics of child poverty in industrialized countries.* UNICEF and Cambridge, UK: Cambridge University Press.

Cannella, G. (1997). *Deconstructing early childhood education.* New York: Peter Lang.

Chakrabarty, D. (2000). *Provincializing Europe: Postcolonial thought and historical difference.* Princeton, NJ: Princeton University Press.

Children's Defense Fund (2001). *The State of America's Children.* Washington, D.C.: Children's Defense Fund.

Dahlberg, G., Moss, P., and Pence, A. (1999). *Beyond quality in early childhood education.* London: Routledge Press.

Darwin, C. (1859). *On the origin of the species by means of natural selection.* London: John Murray.

Donzelot, J. (1979/1997). *The policing of families.* Baltimore: Johns Hopkins University Press.

Edin, L. and Lein, L. (1997). *Making ends meet: how single mothers survive welfare and low-wage work.* New York: Russell Sage Foundation.

Escobar, A. (1995). *Encountering development.* Princeton, NJ: Princeton University Press.

Esping-Anderson, G. (1990). *The three worlds of welfare capitalism.* Princeton, NJ: Princeton University Press.

Esping-Anderson, G. (Ed.). (1996). *Welfare states in transition: National adaptations in global economies.* Thousand Oaks, CA: Sage Publication.

Foucault, M. (1965). *Madness and civilization: A history of insanity in the age of reason.* (R. Howard, Tras.). New York: Pantheon Press.

———(1977). *Discipline and punish: The birth of the prison.* (A. Sheridan, Trans.). New York: Vintage Press.

———(1979/1991). Governmentality. In G. Burchell, C. Gordon, & P. Miller. (Eds), *The Foucault effect: Studies in governmentality* (pp. 87–104). University of Chicago Press.

———(1980). *Power/Knowledge: Selected interviews and other writings by Michel Foucault.* (C. Gordon, Ed. and Trans.). New York: Pantheon.

———(1988). *The history of sexuality: The care of the self.* (Vol. 3). (R. Hurley, Trans.). New York: Pantheon Press.

Fraser, N. (1997). *Justice interruptus: Critical reflections on the "postsocialist" condition.* London: Routledge University Press.

Fuller, B., Kagan, S., Loeb, S. (April 2002). *New lives for poor families: Mothers and young children move through welfare reform.* Wave 2 Findings. Technical Report. University of California, Berkeley: The Growing Up in Poverty Project.

Galasi, P., & Nagy, G. (2001). Are children being left behind in the transition in Hungary? In B. Bradbury, S. P. Jenkins, & J. Micklewright (Eds.), *The dynamics of child poverty in industrialized countries* (pp. 236–253). UNICEF and Cambridge, UK: Cambridge University Press.

Gancheva, R. and Kolev, A. (2001). Children in Bulgaria: Growing impoverishment and unequal opportunities. *Innocenti Working Papers.* No. 84, Innocenti Research Centre, Florence, Italy: UNICEF.

Ghandi, L. (1998). *Postcolonial theory: A critical introduction.* New York: Columbia University Press.

Giddens, A. (1998). *The third way: The renewal of social democracy.* Cambridge: Polity Press.

Gordon, C. (1991). Introduction to governmentality. In G. Burchell, C. Gordon, & P. Miller (Eds.), *The Foucault effect: Studies in governmentality.* Chicago: University of Chicago Press.

Gupta, A., & Ferguson, J. (1997). *Culture, power, and place: Explorations in critical anthropology.* Durham, NC: Duke University Press.

Hanak, P. & Held, J. (1992). Hungary on a fixed course: An outline of Hungarian history. In J. Held (Ed.), *The Columbia history of Eastern Europe in the twentieth century.* New York: Columbia University Press.

Hindess, B. (1996). Liberalism, socialism, and democracy: Variations on a governmental theme. In A. Barry, T. Osborne, & N. Rose (Eds.), *Foucault and political reason: Liberalism, neo-liberalism, and rationalities of government* (pp. 65–80). Chicago: University of Chicago Press.

Holloway, N., Fuller, B., Rambaud, M. & Eggers-Pierola, C. (1997, 2001). *Through my own eyes: Single mothers and the cultures of poverty.* Cambridge, MA: Harvard University Press.

Hupchick, D. P. and Cox, H. E. (2001). *The Palgrave concise historical atlas of Eastern Europe.* New York: Palgrave Press.

Lascarides, V. C., and Hinitz, B. F. (2000). *History of early childhood education.* Source Books on Education, vol. 55, Garland Reference Library of Social Science, Vol. 982. New York: Falmer Press.

Locke, J. (1763/1988). *Two treatises on government.* (Edited and with notes by Peter Laslett). New York: Cambridge University Press.

Lowe, E. D., & Weisner, T. S. (In Press). "You have to push it—Who's gonna raise your kids?": Situating child care and child care subsidy use in the daily routines of lower income families. *Children and Youth Services Review.*

Lukacs, J. (1988). *Budapest 1900: A historical portrait of a city and its culture.* New York: Grove Press.

Martin, J. R. (1986). Redefining the educated person: Rethinking the significance of gender. *Educational Researcher, 15* (6), 6–10.

Micklewright, J. (February 2002). Social exclusion and children: A European view for a US debate. *Innocenti Working Papers.* No. 90. Innocenti Research Centre, Forence, Italy: Unicef.

O'Malley, P. (1998). Indigenous governance. In Dean, M. & Hindess, B., *Governing Australia: Studies in contemporary rationalities of government.* England: Cambridge University Press.

Orloff, A. (1996) . *Gender and the welfare state.* Institute for Research on Poverty Discussion paper no. 1082–96. University of Wisconsin–Madison.

Popkewitz, T. S. (1998). The culture of redemption and the administration of freedom in educational research. *Review of Educational Research, 68* (1), 1–34.

———(2000a). Rethinking decentralization and the state/civil society distinctions: The state as a problematic of governing. In T. S. Popkewitz (Ed.), *Educational knowledge: Changing relationships between the state, civil society, and the educational community* (pp.173–200). Albany: State University of New York Press.

———(Ed.). (2000b). *Educational knowledge: Changing relationships between the state, civil society, and the educational community.* Albany: State University of New York Press.

Popkewitz, T. S., & Bloch, M. N. (2001). Administering freedom: A history of the present—Rescuing the parent to rescue the child for society. In K. Hultqvist &

G. Dahlberg (Eds.), *Governing the child in the new millenium* (pp. 85–118). New York: RoutledgeFalmer.

Rose, N. (1989). *Governing the soul: The shaping of the private self.* London: Routledge.

———(1996). The death of the social: Refiguring the territory of government. *Economy and Society, 25,* 327–366.

———(1999). *Powers of freedom: Reframing political thought.* Cambridge, UK: Cambridge University Press.

Rousseau, J. J. (1762/1979). *Emile.* (A. Bloom, Trans.). New York: Basic Books

———(1762/1993). *The social contract and discourses.* London: J. M. Dent.

Said, E. (1978). *Orientalism.* New York: Vintage Press.

Sainsbury, D. (1999). Gender, policy regimes, and politics. In D. Sainsbury (Ed.), *Gender and welfare state regimes* (pp. 246–275). New York: Oxford University Press.

Schram, S. (2000). *After welfare: The culture of postindustrial social policy.* New York: New York University Press.

Shavelson, R. J. and Towne, L. (Eds.) (2002). *Scientific research in education.* Washington, D.C.: National Research Council.

Shonkoff, J. P., and Phillips, D. A. (2000). *From neurons to neighborhoods: The science of early childhood development.* Washington, D.C.: National Academy Press.

Smith, L. T. (1999). *Decolonizing methodologies. Research and indigenous peoples.* London: Zed Books, Ltd.

Stacey, J. (1996). *In the name of the family: Rethinking family values in the postmodern age.* Boston: Beacon Press.

Stack, C. B. (1974). *All our kin: Strategies for survival in a black community.* New York: Harper and Row.

UNESCO (2002). *A world fit for children.* Paris and Geneva: UNESCO.

UNICEF and Governement du Senegal. (Decembre, 2000). Rapport de l'enquete sur les objectifs de la fin de decennie sur l'enfance. (MICS-II–2000). New York and Dakar, Senegal: UNICEF.

UNICEF (Draft, June 2001). Analyse de situation de l'enfant et de la femme au Senegal–2000. New York and Dakar, Senegal: UNICEF.

Weisner, T., & The Fieldwork Team (2000). Understanding better the lives of poor families: Ethnographic and survey studies in the New Hope experiment (pp. 10–13). *Poverty Research News.* Northwestern University/University of Chicago: Joint Center for Poverty Research.

Wollons, R. (Ed.). (2000). *Kindergartens and cultures: The global diffusion of an idea.* New Haven, CT: Yale University Press.

World Bank (March 20, 2000). Project appraisal document on a proposed credit in the amount of SDR 36.7 million (US$50 million equivalent) to the Republic of Senegal for a Quality Education for All Program in support of the first phase of the Ten-Year Education and Training Program (PDEF). Washington, D.C.: The World Bank (Human Development II, Country Department 14, Africa Region)

Young, R. J. C. (2001). *Postcolonialism: An historical introduction.* Oxford, UK: Blackwell.

CHAPTER TEN

GOVERNING CHILDREN AND FAMILIES IN KENYA

LOSING GROUND IN NEOLIBERAL TIMES

Beth Blue Swadener and Patrick Wachira

Introduction

In contrast to many of the late capitalist/industrialized European and North American states discussed in this book, Kenya is not often constructed as a "welfare state." We would argue that most Kenyans would not agree that they now live in, or ever lived in, a welfare state. Yet, a number of governmental programs were initially designed and intended to promote the well-being of the population or provide various forms of protection and services to citizens. These include an array of health-related programs (e.g., child and maternal health promotion programs, malaria treatment, family planning, and nutrition/agricultural education) that are no longer as well funded or as widely available, as in the early years after independence. It should be noted, however, that universal policies such as health care or extended family/maternity leaves, commonly associated with European welfare states, were never state policy in Kenya.[1] Government programs also include a centralized system of public education, administered by the Ministry of Education, which originally provided textbooks and other school supplies and covered the costs of examinations and other fees. These government subsidies have been dramatically affected by cost-sharing "austerity measures" of structural adjustment programs

(SAPs). Such changing patterns of governing children and families, mirroring postsocialist and neoliberal global policies, are the focus of this chapter.

While Kenya did not follow a socialist path to postindependence nation-building, as did several African nations (e.g., Tanzania and Mozambique), its early postcolonial policies did reflect principles of universal access to education, healthcare, and other services often associated with the "welfare state." For example, for many (if not all) Kenyans, basic education, primary healthcare, and associated goods and services were free for several years after their nation gained independence from the British in 1963. Like many sub-Saharan and Latin American nations, Kenya accrued large-scale debts to several multinational development banks, most notably the World Bank and the International Monetary Fund. Debt restructuring conditions and pressures from external donors have led to a number of cost-cutting "austerity measures" that have undermined what remains of a "welfare state" in Kenya and directly affected the majority of families raising children (Gakuru & Koech, 1995; Swadener, Kabiru & Njenga, 2000; Weisner, Bradley & Kilbride, 1997). Additionally, pressures of corporate globalization and free market–based trade liberalization, combined with urbanization and associated family dislocation, rising unemployment, corruption leading to economic mismanagement, and a worsening national infrastructure, have adversely affected Kenyan families. These patterns reflect larger patterns of neoliberal policy, as summarized by Tikly (2001):

> The fragility of the African state in the context of international relations and the postcolonial status quo has ensured that many African states are more susceptible to global forces than are wealthier nations. This susceptibility provided the conditions of much of the imposition from the early 1980s of a new neoliberal orthodoxy in the political economy that has disrupted indigenous postcolonial hegemonic projects and accumulation strategies . . . this orthodoxy has severe implications for all areas of social welfare, including education, and has served to exacerbate social stratification. (p. 165)

In using a postcolonial lens to analyze impacts of structural adjustment policies and other neoliberal policies on children and families in Kenya, we draw from Hall (1996), who asserts that a central concern of using postcolonial critique is to "re-narrativize" (Hall, 1996, cited in Tilky, 2001) the globalization story in a way that "places historically marginalized parts of the world at the center, rather than at the periphery of, in this case, education and globalization debates" (Tilky, 2001, p. 2). Tilky (2001) further asserts that:

> postcolonial critique is concerned with the continuing impacts of education of system of European colonialism and with issues of race, culture, and language, as well as other forms of social stratification including class and gender in postcolonial contexts. Thus, postcolonial critiques draw attention to the transnational aspects of globalization and of social inequalities and seek to highlight forms of resistance to western hegemony. (p. 2).

A number of contexts for this chapter cannot be discussed in depth, but deserve mention. These include social and cultural contexts, background on Kenya's precolonial, colonial, and postcolonial history, and an understanding of frameworks for educational and social policy in Africa. We should also note that writing the only chapter in this volume focusing exclusively on one African state (also see Bloch's comparative discursive analysis that includes a focus on Senegal, this volume) added to our sense of responsibility to provide sufficient contextual information, without essentializing forms of the "welfare state" and its increasing dissolution in the wake of restructuring policies and privatization in Africa. We are aware that there already exists a large body of literature that constructs the African political and economic climate as in crisis. As Parpart & Staudt (1989) put it over a decade ago, "Development first preoccupied Africanist literature, but crisis is now the dominant theme" (p. 1). It could be argued that the dominant theme in the new millennium is *economic globalization* and its impacts on Africa. We would problematize both dependency theory and statist perspectives as insufficient for understanding the local contradictions and complexities of family life in Kenya or any other sub-Saharan nation. Thus, in our analysis of the impacts of neoliberal policies, we recognize the agency, resistance, and creativity that children and families in "difficult circumstances" in Kenya and other sub-Saharan nations bring to daily life. We also acknowledge the complexity and contradictions of locally enacted popular culture practices in contemporary African contexts, as we frame a postcolonial critique of neoliberal policies in Kenya.

We agree with Popkewitz and Bloch (2001) and others in this volume, who have warned against overuse of a structural social policy analysis that blames neoliberalism or uses an essentialized treatment of national history, and who advocate the use of critical structural and poststructural theoretical framings in historical policy analysis. In this chapter, we find value in standpoint feminism and do not rule out the integration of critical theory with poststructuralism, as advocated by Fraser (1997):

> [We] might posit a relation to history that is at once antifoundationalist *and* politically engaged, while promoting a field of multiple historiographies that

> is both contextualized *and* provisionally totalizing. Finally, we might develop a view of collective identities as at once discursively constructed *and* complex, enabling of collective action and amenable to mystification, in need of deconstruction and reconstruction. In sum, we might try to develop new paradigms of feminist theorizing that integrate the insights of Critical Theory with the insights of poststructuralism. Such paradigms would yield important intellectual and political gains, while finally laying to rest the false antitheses of our current debates. (p. 219)

In describing a shift from redistribution to recognition in postsocialist theory and policy, Fraser (1997) defines the liberal welfare state as providing "surface reallocations of existing goods to existing groups" (p. 27), in order to meet basic or universal human needs. This process "tends to support group differentiation and can also generate backlash misrecognition" (p. 27). She further defines socialism as a "deep restructuring of relations of production," a process that can "blur group differentiation" and "help remedy some forms of misrecognition" (p. 27). Fraser argues that, "with the decentering of class, diverse social movements are mobilized around crosscutting axes of difference" (p. 13). Although the politics of difference is not in the widespread vernacular of contemporary Kenya, per se, "tribalism" or "*majimboism*" is still apparent in terms of the ethnic makeup of the opposition parties, organized in the early 1990s under mandated multiparty reforms, and state-sponsored violence, officially referred to as "ethnic clashes," which has left many Kenyans homeless or refugees in their own land (Human Rights Watch, 1997b). Yet, at the time of this writing, presidential elections are scheduled in the near future and coalition-building efforts of opposition parties have cut across tribal/ethnic and party lines. A decentering of class, however, remains a distant vision in Kenya.

In contrast to tenets of "the welfare state," neoliberal policies emphasize free trade, deregulation, privatization and decentralization of government programs, and replacing the idea of the "public good" or universal rights of citizens with individual responsibility and governmental accountability. The related discourse of blame, or "pathologizing of poverty" (Polakow, 1993; Swadener & Lubeck, 1995) in the United States and elsewhere (Polakow, Halskov, & Jorgensen, 2001; Sibley, 1995), has added to a recurrent deficit discourse that constructs those in poverty as having only themselves to blame. The radical restructuring/devolution of welfare policy and cutbacks in entitlement programs in the United States is well documented (e.g., Cannella, this volume; Ehrenreich, 2001; Gordon, 1994; Mink, 1998; Polakow, 1993; and Schram, 2000). The U.S. discourse has changed to reflect such "post-welfare" policies, including some states renaming social workers "self sufficiency coaches" and some state social service agencies adopting the slogan, "zero tolerance for unemployment." All such changes evidence a shift from entitlements to employment (though rarely providing a living wage or benefits), and away from education and work-related training to immediate job or volunteer placement. A later section of the chapter will problematize the discourse of dependency.

Additionally, and much in evidence in Southern Hemisphere contexts such as Kenya, the cutting of public expenditures for social services and national infrastructure (e.g., roads, water, energy, and environmental protections), while de-unionizing and reducing workers' rights and wages, are all hallmarks of neoliberalism. In Kenya and across Africa, user fees and privatization of former state-run free services and resources are frequently part of SAPs. Particularly troubling has been the privatization of water. Downsizing the civil service has also contributed to unemployment, which had reached over 60 percent at the time of this writing.

In terms of theoretical framings of our discussion of child and family policy and the "welfare" state in Kenya, we draw from postcolonial theories (Bhabha, 1994; Dimitriadis & McCarthy, 2001; Gandhi, 1998; Mignolo, 2000; N'gugi wa Thiong'o, 1993; p'Bitek, 1986; Spivak, 1999; and Willinsky, 1998) and theories of "postsocialist" conditions, including work by Fraser (1989, 1997), Gordon (1990, 1992), Polakow (1993), and Polakow, Halskov, & Jorgensen (2001). An understanding of Kenyan governing patterns and education policies is particularly informed by hybridity theory (e.g., Bhabha, 1994; Spivak, 1999; McCarthy, 1998), which deconstructs cross-migratory patterns, relationships between colonizers and colonized, and hybrid forms of information, policy, and shifting, mutually influenced, practices. We also briefly raise possibilities for "decolonizing research methodologies" (Mutua & Swadener, 2003, and Smith, 1999) in the context of collaborative research. Throughout the chapter, we reflect on the notion of *governing the child,* framed by Rose (1999), who asserted that "childhood is the most intensively governed sector of personal existence," and that "the child—as an idea and a target—has become inextricably connected to the aspirations of authorities" (p. 123). Similarly, we have appreciated the recent work of Hultqvist and Dahlberg (2001) on governing the child "in the new millennium," and agree that there is "need for a continual critical scrutiny of the past, not for the sake of the past but for the sake of the present" (p. 6). As these authors further state, "[T]oday's discourses on the child reassemble past discourses in new patterns and inscribe different assumptions about the child" (p. 6).

This framing has led to several questions we continue to reflect upon as they pertain to African contexts, particularly Kenya. How might the village raising its children differ from, and in, postmodern governmentalities, including the regulation of the child and family? How have increasingly globalized, Western influences and related hybridity issues in postcolonial Kenya influenced social policy and governing patterns as they pertain to children and families?

Historical Contexts for Kenyan Education Policy

In the following sections, we discuss the various ways, indirect, covert, or overt, in which the British colonial legacy continues to impact upon and influence the social, political, economic, and cultural lives of the people of Kenya. We focus on education and early childhood policy, or governing patterns, during precolonial, colonial, and postcolonial times, all framed in relation to present time. We also discuss gender roles as they relate to changing patterns of governing children and families. Finally, we discuss impacts of structural adjustment policies on contemporary families in Kenya and on the growing number of street children in Kenya. We conclude with a critique of dependency theory and argue that interdependence reflects persistent/prevailing African values.

As Ines Dussel reminds readers in her chapter (this volume), arguing with Foucault (1980), it is challenging to provide national and historical context without "pronouncing 'origins' and 'foundations.'" Yet, there are still periods of time in which markedly different types of formal and nonformal education tended to be enacted among different cultural and income groups in Kenya. Similarly, it is difficult to avoid the common divisions of history that tend to be framed in terms of precolonial, colonial, and postcolonial periods. While we use these distinctions to organize the next few sections of the chapter, we acknowledge that many aspects of colonial and (particularly among more traditional or "persistent pastoralist" groups) precolonial cultures, identity dynamics, and policy-related discourses persist, even in "postcolonial" Kenya.

Indeed, we would argue that any construction of the "postcolonial" still embodies much of what is termed "colonial," similar to ways in which postmodernity embodies modernity. We adopt Quist's (2001) use of the term "postcolonial," rather than postcolonial, which can be read as "carrying the idea of a linearity and chronology, signifying one period followed by another" (p. 299). As McClintock (1992, p, 85) asserts, " . . . the term post-colonial is haunted by the very figure of linear development . . . and marks history as a series of stages along an epochal road from pre-colonial, to colonial to the post-colonial." We adopt Quist's (2001, p.299) use of the term postcolonial to suggest that continuity, a back and forth relationship, a constant between the past and present-day cultural and sociopolitical relations with implications for the future. This affords the opportunity of engaging with the continuing complex interrelationships among factors and forces that simultaneously impact the postcolonial situation.

Education in Precolonial Kenya

Precolonial Kenya was comprised of over 40 ethnic groups or tribes, which were relatively isolated geographically and culturally. Although there was no formal indigenous schooling system, there was considerable educational activity both formally and informally. Traditional education was primarily social/functional and was achieved informally by the parents and elders. Traditional education played an important role in fostering and preserving the cultural values and ideals of each community. Early childhood education was integrated into daily life in the form of stories, fables, riddles, and games, often shared with young children by grandparents. Communal cooperation and respect of the environment were emphasized, in contrast to Western/colonial knowledge. Ngugi wa Thiong'o (Cantalupo, 1995) argues that the relative prosperity enjoyed in precolonial Kenya could be explained by the generally harmonious relationships between its people and nature, as contrasted to the colonial impacts of dislocating peasants from their privileged position in the natural cycle, including displacing them from grazing corridors and fertile farmland or *shambas.*

"Formal" education was achieved primarily through gender-differentiated instruction and initiation rituals through which individuals in an age-set graduated from one level of social status to another after acquiring new knowledge. Formal education was also given to young people through apprenticeships to craftsmen or *fundis* (experts), in which individuals could acquire skills as carpenters, blacksmiths, or medicine men (traditional healers) through practical application. The gender separation of the "formal" education was not for exclusionary reasons but rather for practical purposes. In contrast to colonial education, traditional education valued knowledge possessed by various groups, including women. Indeed, in some societies in precolonial Kenya, women played a more visible role in education, as they spent most of their day with children, in their traditional child-rearing role. Traditional education was, therefore, fundamental and included everyone in the learning process. With the tribal societies thus isolated, traditional education served to perpetuate a stable, albeit often stratified by age and gender, system of social relations—until the coming of Europeans, first as missionaries, affected the entire fabric of traditional cultures.

Education in Colonial Kenya

Colonial education in Kenya, as in other African contexts, was essentially Eurocentric, exploitative, discriminatory, and hegemonic. Its central goal was the subjugation of the indigenous learning system. The earliest written accounts on the history of education in colonial Africa describe Christian or Islamic foreign missions, which dominated the education sector for many years (Brock-Utne, 2000). Formal, colonial education in Kenya started with the establishment in 1846 of a school at Rabai, in the Coast Province, by the Church Missionary Society with the main purpose being the promotion of Christianity. As Fafunwa (1982, p. 21) asserts, "education to win African souls for Christ was made a central objective of mission education in colonial Africa." Missionary education obviously relates in very literal ways to notions of "governing the soul" (Rose, 1999). Education was viewed as an important instrument for providing skilled and unskilled labor in their farms/plantations and clerical workers for the growing colonial government and, therefore, the colonial government started to subsidize a small portion of education, appropriating many of the schools already established by the missionaries, who continued to meet most of the education budget. Thus, the missions were viewed as agents of colonialism in the practical sense and the government allowed them a degree of flexibility in the provision of education.

Colonial education was racially and culturally stratified, with different curricula for different races. Far greater resources were devoted to the education of Europeans, followed by Asian and Arab education, while education for the Africans came last. Similar to other colonial education strategies, European settlers provided only minimal education for Africans, designed to produce workers capable of taking instructions. As Fafunwa (cited in Brock-Utne, 2000) argues: " . . . the volume and quantity of education the colonial administrators were willing to give to the Africans were the barest minimum necessary for such auxiliary positions as clerks, interpreters, preachers, elementary teachers and so on" (p. 19). The British colonial argument was that the different "races" present in Kenya at the time (e.g., Africans, Arabs, Asians, and Europeans) needed the kind of education that was deemed suitable to and "appropriate" for their position in colonial life (Kiluva-Ndunda, 2001). Rodney (1976, as cited in Kiluva-Ndunda, 2001) termed colonial schooling as "education for under-development." This was not an educational system "designed to give young people confidence and pride as members of African societies, but one which sought to instill a sense of deference toward all that was European and capitalist" (Brock-Utne, 2000, p. 20).

Kenya's colonial early childhood history is a segregated one, mirroring primary education, with separate preschools for European, Asian, and (where preschools existed) African children. Initially, preschools for Kenyan children were intended to be nonacademic, nonteaching, childcare settings—a view that persisted until the early 1970s (Swadener, Kabiru & Njenga, 2000). Custodial childcare was also provided on Kenya's many tea, coffee, and sisal plantations, as it is today.

Education after Independence: **Harambee** *and Beyond*

At independence in 1963, an education system was needed to replace the underdeveloped and racially segregated system that Kenya inherited from the colonial era. The new education system was intended to help meet the social, economic, and political goals of the newly independent country, foster a sense of nationhood and national unity, promote social equality and respect, and restore the cultural heritage of Kenyan people, which had diminished as a result of the imposition of the "alien" European culture (Eshiwani, 1990). In order to meet the new challenges of "Kenyanization," new curricula and a coordinated national program were required; thus, the central government assumed responsibility for a secular educational system, taking over from the missions (Eshiwani, 1990; 1993). A highly centralized system of education was established in which government control was exercised through financing and regulating education. The government, however, continued to welcome the participation of the missions and other volunteer organizations in the provision of education.

Shortly after independence, however, it became increasingly evident that the government could not provide the necessary schools to meet the population's demand for greater access to education. Thus, the *Harambee* schools were established, building on a grassroots spirit of education and childcare from the "*Mau Mau*" independence struggle. These schools were community self-help projects in the spirit of *Harambee,* meaning to "pull together" for development. Grassroots communities, self-help groups, and people (*wananchi*) pooled their financial and material resources and provided the labor to put up a building, donate a cow, or provide other materials for a local school or preschool.

Education policy in Kenya, as in most African nations, took various forms in its early, postindependence days. Cutting across political ideologies at the time of independence and shortly after, most emerging African nations had phenomenal growth in education and gave funding priority to education while always within resource-limited governments and governing strategies. Since independence, education in Kenya accounts for the largest portion of the government's expenditure, with approximately 30 percent of the national budget spent on education, compared to 14.6 percent after independence in 1963 (Nieuwenhuis, 1996), although less than 1 percent of the national budget goes to early childhood education and development. Significant "progress" was made in providing universal education to Kenyans, with the national enrollment rate in primary education reaching 95 percent in 1991, compared to 50 percent at independence. The high enrollment in primary education can be largely attributed to the combined efforts of the government, parents, local communities (*Harambee* movement), and nongovernmental organizations (NGOs). Postcolonial education in East Africa has been critiqued as reflecting the myth of modernization through schooling and Western education (Ferguson, 1999; Vavrus, in press). As Vavrus reflects, "the belief in a linear progression from underdevelopment to development by building more schools and increasing literacy rates has undergirded both socialist and liberal development policies since independence in 1961" (p. 1).

Primary schools did not historically charge fees, although cost-sharing measures related to Structural Adjustment Programs (SAPs) have included fees for enrollment, testing, uniforms, books, building funds or *Harambee* fundraisers, and other incidental costs. The government meets the operational costs of running public schools through the Ministry of Education, Science and Technology. It is also the government's responsibility to train teachers and pay their salaries, through the Kenya Teachers Service Commission (TSC). Government spending on the development of school facilities, including teachers' housing, is minimal, and the responsibility is left largely to the Parent Teachers Associations (PTAs), school committees, and the local community through *Harambee* or self-help schemes.

Turning to early childhood, in 1971 the government (with the assistance of the Bernard Van Leer Foundation, from the Netherlands) established the Preschool Education Project, based at the Kenya Institute of Education (KIE). Prior to that time, early childhood education (pre-primary education for children 3 to 6 years of age) was the responsibility of local communities, NGOs, churches, and other volunteer organizations. By 1980, the Ministry of Education took over full responsibility from the Ministry of Culture and Social Services, creating preschool sections at MOE headquarters and the inspectorate. In 1984, the National Centre for Early Childhood Education (NACECE) was established, in part for training preschool teachers, developing and disseminating appropriate curriculum, and coordination with external partners and other government agencies (Swadener, Kabiru, & Njenga, 2000). Preschool teachers are not hired through the Kenyan government, although their training is facilitated by NACECE and the DICECE (District Centres for Early Childhood Education). Most rural preschools, for example, function on a *Harambee* basis, with a local community hiring the teacher, putting up the building, and providing other needed resources (e.g., a feeing program). This reflects a frequent (international) division in governing children, between preschool and primary in which preschool programs and teachers' employment are private and locally governed.

Thus, the regulation and monitoring of preschools in Kenya (through national guidelines, district/local school inspectors, and DICECE trainers) reflects an interesting mix of indigenous and universal, mainly Western, assumptions about child development, "quality," and advocated "best practices." An emphasis on building national identity in a culturally diverse society and joining the international community reflects growing globalization—even as it is reflected in governing the youngest citizens and their preschool teachers and programs. Assumptions about universal "best practices" in early childhood education also permeate Kenyan early childhood guidelines and training, although most are balanced with traditional childrearing information (e.g., the values of using traditional weaning foods, using mother tongue stories with children, benefits of intergenerational care, etc.). The Kenyan Preschool Guidelines were based on earlier UNICEF documents (Kabiru, personal communication), again demonstrating the hybridity of policy development and influence of global donors. The guidelines tend to reflect the discourse of many other nations' early childhood planning documents, underscoring the shared discursive framing of the governing of childhood that has been part of national/international or global/local discourses.

Gender Issues: Economic and Social Marginalization

Women in precolonial Kenya played a prominent role in the family in terms of economic production (Kiluva-Ndunda, 2001), and had substantial right to control the means of production and ownership of what was produced. Women have lost much of their traditional authority and autonomy in colonial and postcolonial Kenya. Colonialism introduced urbanization and intensified class and gender differences, as existed in precapitalist societies (Robertson & Berger, 1986). In postcolonial Kenya there is persistent gender segregation in the workforce. This can be traced, in part, to postcolonial policies in which career training institutionalized a gender-segregated workforce (Kiluva-Ndunda, 2001).

In contemporary Kenya, women continue to bear the main responsibility for the welfare of Kenyan families, and one third of Kenyan households are headed by a female (Adams & Mburugu, 1994; Kilbride & Kilbride, 1990), typically a single mother. Paid labor continues to be gender-segregated in postcolonial Kenya, with women frequently being dominated and exploited. Males, for example, account for 79.1 percent of the total formal employment with women accounting for only 20.9 percent in 1989 (Kiluva-Ndunda, 2001). Women also earn significantly less than males in all occupations. In addition, girls and women are often oppressed in other ways, especially those dependent on their husbands, who yield more power economically and socially in Kenyan society. Women cannot inherit property, and wife-beating, rape, wife inheritance, and forced circumcision are all aspects of life among several of the ethnic groups in Kenya (Kilbride, 2000, p. 135; Ombour, 2001). Kilbride, Suda, and Njeru (2000) observed that having a baby before marriage may expose the girl to punishment by the parents or relatives and expulsion from school.

Against this backdrop of economic and social disadvantages for women, the Children's Act (2001) considers children born out of wedlock the responsibility of the mother alone. There is *no legal responsibility on the part of the father* to support and maintain his illegitimate children unless he wishes to accept that responsibility, and applies for such responsibility in court (Amisi, 2001). Some child advocates have argued that the Children's Act should have incorporated the substance of the 1959 Affiliation Act, which became an Act of Parliament at independence. This legislation enabled the mother of a child born out of wedlock to seek a maintenance court order against the father in a range of circumstances, but was repealed in 1969. At the time of this writing, a mother and a child rights group had helped a two-year-old child challenge a section of the Children's Act in High Court, saying that the new law discriminates against children born out of wedlock (Koome, 2002). As Martha Koome, chairperson of the International Federation of Women Lawyers in Kenya, stated: "It is very disheartening to know that child maintenance issues may have been more advanced in 1959, when the Affiliation Ordinance came into force, than today, when we have a Convention on the Rights of the Child and The African Charter on the Rights and Welfare of the Child" (Koome, 2002).

Widespread rural to urban migration, due to poverty and devaluing of Kenyan agriculture, has also contributed to the number of single-parent families, a phenomenon often described as dislocation (Kilbride & Kilbride, 1990). Many female-headed households live in the sprawling slums of Kenya's major cities, particularly Nairobi, or remain in the rural areas, with the father leaving to find work in a town or city. Austerity measures and a worsening economy have greatly affected mothers, as they often have little education (due to discriminatory policies in provision of education), high dropout rates due to exorbitant fees (in which parents would rather educate boys in the case where there is not enough money to educate both), high pregnancy rates, and a school curriculum not sensitive to girls and diverse cultural beliefs (e.g., women should be dependent on men).

"Children in Debt": Impacts of Structural Adjustment

Similar to other African countries, reductions by the Kenya government on spending for subsidized food, healthcare, and school-related expenses has meant that the cost of these basic necessities has been passed on to families, leaving them with fewer resources to devote to the education of their children. As a mother in Kisumu Municipality (near Lake Victoria) put it, "Books, uniforms, building fund, admission fees are all required, and if you don't have them, children are sent home!," while a father in rural Embu District (near Mt. Kenya) commented, "life is very demanding—it is just living hand-to-mouth. . . . We are supporting the (education) system rather than benefiting from the system and there is no going back!" (Swadener, Kabiru, & Njenga, 2000, pp. 176, 247).

Two widespread results have been the increasing rates of school dropouts and lower school enrollment, as families cannot afford fees and more children engage in income-generating activities to contribute to their family income or simply for their own survival. Ironically, this trend is corroborated by a World Bank (2001) report, which stated that, "Poverty-related deprivation contributes to low education attainment in Africa. Poor children spend more time than other children contributing directly or indirectly to the household income. As a result, they are less likely to spend out of school hours on school work . . . and more likely to be tired and ill-prepared for learning" (World Bank, 2001, p. 25). Studies in a number of countries in Africa indicate that school enrollments declined in countries that adopted structural adjustment policies. As Reimers and Tiburcio (1993, as cited in Brock-Utne, 2000, p. 23) state: "It is clear that the adjustment programs supported and promoted by the World Bank and IMF during the 1980s have not worked for many countries. . . . International financial institutions are supposed to be part of the solution, not part of the problem, and their record has to be assessed by the number of success stories they can claim, not by whether they can or cannot be blamed for the failure." Studies by UNICEF and other research teams have also documented the negative impacts of structural adjustment programs on children and other vulnerable groups in Africa (Bradshaw, et al., 1993; Kilbride & Kilbride, 1997; Kilbride, Suda, & Njeru, 2000). James Grant, the former Executive Director of UNICEF, described structural adjustment as having a "human face," often that of a child (Grant, 1993) and the first author has documented ways in which such macroeconomic policies and related local dynamics are directly linked to the quality of life experienced by families and the opportunities afforded their children (Swadener, Kabiru, & Njenga, 2000). In fact, UNICEF used the phrase "children in debt" for several years to convey the strong correlation between third world debt, structural adjustment policies, and children's increased risk. UNICEF estimates that if just one of every five dollars Africa pays for debt servicing instead went to primary education, there would be a place in primary school for every child (UNICEF, 1996, p. 3.).

The impact, at the level of family existence and economics, of global recession and the related debt crisis is unevenly, though increasingly, documented (Bradshaw, et al., 1993). Associated policies have included greater community cost sharing, higher prices to consumers, increased unemployment (Hancock, 1989), and dislocation (Kilbride & Kilbride, 1990; 1997). Walton and Ragin (1990), in discussing impacts of over-urbanization, note that, "[t]he urban poor and the working class are affected by a combination of subsidy cuts, real wage reductions, and price increases stemming from devaluations and the elimination of public services" (p. 877).

The relationship between structural adjustment programs and a rise in child mortality and malnutrition rates in some parts of Kenya has also been noted by Gakuru & Koech (1995). This, combined with the sharp increase in female-headed households—many living in urban slums in extreme poverty far away from family supports—underscores the threat to a so-called welfare state in this and other sub-Saharan African nations. However, lest we portray Kenyan families as passive victims of global economic policies, we would agree with Weisner, Bradley, & Kilbride (1997, cited in Swadener, et al., 2000, p. 265) that: "African families face serious crises today. They are under economic, demographic and political pressures of all kinds; yet, families are not mere hapless victims of global change. They are proactive, resilient agents and creators of change."

Governing Street Children and Other Out-of-School Youth

The number of out-of-school children and youth has grown dramatically in the last decade, and has contributed to the growing number of street children in Kenya. Estimates of the number of street children in Nairobi, for example, range from 10,000 to over 30,000, and children can often be seen begging or trash-picking on the streets of smaller towns as well. Similar to other countries, the majority (as many as 80 percent) of "street children" return home to sleep frequently enough to be considered children "on the street," versus homeless children, "of the street" (Kilbride, Suda, & Njeru, 2000). School-related expenses are quite high in Kenya (e.g., $250–350 annually for primary and typically much higher for secondary, particularly if it is a boarding school). When compared to the annual per capita income of $260 (down from $340 in 1991), the financial sacrifice for children's education is starkly apparent. This has also contributed to an increase in informal and nonformal education programs aimed at the "rehabilitation" of street children and usually run by NGOs and religious organizations.

Street children are "governed" in a number of ways and there are mixed viewpoints on governing—or rehabilitation—of street children from the churches and *wananchi* (Kenyan citizens). Some churches have distanced themselves from involvement with street children, claiming that it is the government's responsibility. Other churches have pointed out that programs or agencies who feed and clothe street children are simply *encouraging dependency,* a familiar neoliberal argument, not unlike those regarding welfare reform in the United States and several European nations. Still other churches, however, have sponsored an array of community-based programs to tackle the issue of street children, with some parallels to current "faith-based initiatives" promoted by Bush and other conservative leaders in the West. The government generally views street children as a menace and a threat to security and to Kenya's number one source of revenue—tourism. The government policy "criminalizes" (Kilbride, et al., 2000, p. 145) street children, often arresting and charging them with vagrancy in the case of boys, and "loitering with intent" (prostitution) for girls. The street children are often rounded up and forced into desolate and crowded government-run remand homes and "approved schools." While in police custody or in remand homes, most children have reported incidences of beatings, abuse, and unsanitary conditions (Kilbride, et al., 2000, p. 146; Human Rights Watch Africa, 1997, pp. 62, 68).

Approaches to tackling the issue of street children are quite complex. There have been many attempts to provide solutions to the problem but none of them has been effective. The government, like most upper-class citizens in Kenya, feels that most of the children on the street are there on their choosing or with the encouragement of their parents or guardians. Similar to neoliberal policies in the United States and parts of Europe, there is reluctance to provide financial assistance to these poor families, or even find sponsors for their children, as this is viewed as encouraging dependence. We would argue, however, that the government, in conjunction with globalization discourses and international donor economic, political, and cultural policies, has failed to address the social and economic factors that have left most families incapable of earning a decent living to feed and house, let alone educate, their children, thus, leaving children with no choice but to fend for themselves. As discussed earlier, the Kenya government and many of its more privileged citizens have resorted to discourses of blame and pathologizing of poverty that constructs those in poverty as having only themselves to blame, a familiar neoliberal argument, in contrast to culturally grounded solutions that would have been applicable in a more communally sensitive society.

Global Policies/Local Lives: Impacts of Sociopolitical Change on Child-rearing

In a national study documenting impacts of rapid social, economic, and cultural change on child-rearing practices and early childhood education, the first author, together with NACECE and DICECE collaborators, interviewed over 460 parents, grandparents, preschool teachers, children, and community leaders in Kenya. Data were collected in 8 districts and a cross-section of locations, including rural, urban, plantation, and traditional/pastoral settings (Swadener, Kabiru, & Njenga, 1997; 2000). Although the study did not set out to document impacts of neoliberal, "postsocialist" policies, including structural adjustment measures, many of the narratives reflected such themes. Across these varied settings, parents and others concerned with the care and early education of young children described services that had previously been available to families that either had a cost share (e.g., a range of school-related costs) or were unavailable (e.g., basic medications or feeding programs in preschools). In other words, families were increasingly *governed* by policies stipulated by donors, including austerity measures and payment for services, most of which had previously been free.

Although education remains public in Kenya, it is no longer "free" (Buchmann, 1999; Mutua & Dimitrov, 2001; Mutua & Swadener, 2001). In most cases poor families and, increasingly, middle-class families in Kenya cannot afford a public education. This issue was cited by families as one of the greatest economic hardships they faced—particularly when they were forced to decide which children they would educate, if any. Since preschool is still primarily a private, community-based program, many parents also discussed their desire for more government or NGO assistance in order to make preschool more affordable and reinstate health and nutrition services previously available. The following brief narrative summaries, drawn from contrasting parts of Kenya, provide a glimpse into this manifestation of "children and the State," particularly in terms of impacts of neoliberal policies on child-rearing and early education.

> *Unemployment is almost complete—some don't own land, look for casual labor, squat on relatives' land, and are living a hand-to-mouth life most of the time.*
>
> —Mother in Embu District

Responses to questions about social and structural changes affecting child-rearing and concerning the major problems facing families (in Kenya) were typically quite similar, often overlapping, with a discussion of changes leading directly into a list of social and economic problems. The most common themes in this regard were the overarching issue of increasing poverty and an array of related economic problems. First among these, in terms of the frequency with which financial problems were mentioned, was the cost of living; second was the rapidly rising cost of educating children in Kenya. The cost of living had several dimensions, including the loss of purchasing power for basic necessities for families or, as one Maasai mother put it, "the higher cost of everything." As the Kenyan press frequently laments, the gap between the day-to-day realities of the majority of citizens (*wananchi*) living in poverty and the distanced and donor-dependent economic policies of its government is widening (Swadener, Kabiru, & Njenga, 2000, p. 266).

Housing Constraints and Safety

> *I have stayed in this slum since 1982 and sell charcoal. My parents died when I was young and I never went to school. I have five children and am a single mother. On a typical day I am just trying to get enough money for food and caring for my youngest children. I get some help from my children and often take them with me when I am selling charcoal. Before, people here were all living in cartons, then in 1984 they all burned. Now we are in tin houses. We only got a dispensary recently, and there is only one pay toilet for many families. My children and I are often hungry.*
>
> —Single mother in Mukuru, Nairobi

Another aspect of the high cost of living was housing. Rents had become much higher, and, as families grew, their living conditions were more crowded. A number of problems were associated with this, ranging from discipline difficulties to the rise in the number of street children (as older out-of-school boys, for example, were often subtly encouraged to leave their mother's home, at least to find food or work on the street during the day). There were also fewer open spaces, playgrounds, and other recreational settings for children—particularly in urban areas. Those interviewed in rural areas also described more squatters, some of whom were living on relatives' land and others just starting a small *shamba* (farm) on another's property. Some of these squatters were reported to be a source of "compound kids," the rural equivalent to "street kids" in urban areas. Such children were idle during the day, going home to get food and occasionally doing casual labor, and were seen as a "bad influence" on other peers.

Access to housing was also a growing problem in urban areas, with rapid expansion of slums, estates growing through the addition of illegal extensions to existing housing units, and crowding frequently mentioned by the parents interviewed in both Nairobi and Kisumu. Such uncontrolled growth frequently meant that few services were available (e.g., water, sewerage or latrines, trash removal or rubbish burning pits, etc.), making environmental hygiene a major problem. This, in turn, led to outbreaks of disease, including dysentery, which was particularly problematic in one of the Nairobi slums sampled in the study. Even when people were able to arrange temporary or semipermanent housing in such slum settings, the possibility existed of entire neighborhoods being bulldozed by the City Council or burning down in mysterious fires (*Daily Nation,* September 12, 2001), rendering hundreds homeless in a single night.

Problematizing "Dependency" in Public Policy Discourses

We would agree with Fraser and Gordon (1997) that "dependency" is a highly stigmatized, ideological term, which has many registers of meaning (p. 123), including social, economic, psychological, political, and metaphorical meanings. In neoliberal discourse, critiques of welfare dependency are pervasive and often serve to pathologize single mothers who are not self-sufficient. The rhetoric of welfare reform in the United States and elsewhere provides much evidence of this negative construction of dependence, including the name of the 1996 U.S. welfare reform legislation, The Personal Responsibility and Work Opportunity Act. This legislation was intended to "end welfare as we know it," though not to eliminate poverty.

When viewed in African contexts, discourses of dependency and self-sufficiency are complex and often contradictory. We would argue, however, that in Kenya (as in much of sub-Saharan Africa), discourses and assumptions of interdependence are far more culturally grounded and "realistic." Values of cooperative economics and the widespread expectation that the family member(s) making the most money will pay the school fees of others in the family are examples of interdependence as socially and culturally accepted. Whether at the level of family or nation, the dependence on donors or "sponsors" is also generally accepted (if resisted) and pervasive. Kenya is dependent to a large degree on foreign donors, whether multilateral agencies and development banks or NGOs. As Buchmann (1999) states:

> Innovations in the educational sector can come from National governments, but they are just as likely to spring from the abundant networks of African organizations that are usually small in scale and local in orientation. Indeed the emergence of national non-governmental organizations, local community groups and grassroots organizations as providers of basic educational services may be the most promising route to extending basic education to Africa's poorest children. (p. 512)

The interdependence encouraged by free trade and corporate globalization, however, is likely far more problematic, as reflected in the persistent colonization of Africans by the West, through so-called free market globalization and outcomes of hybridity in the course of Kenya's "development." Another complexity of global discourse is what Dale (1999, cited in Tikly, 2001) refers to as the "installation of interdependence," which refers to "the spread of environmental, human rights and peace issues by the new 'global civil society' and NGOs." Such global mechanisms are also seen as influencing education policy in Kenya and other African nations, no less than the imposition of SAPs in education by World Bank and IMF policies, as discussed throughout the chapter. Kenya, like other African states, is often forced to negotiate policy agendas of multiple global agencies, which may be contradictory (Tikly, 2001, p. 166). These complex global discourses, their national negotiations, and local impacts have clearly affected the way education is governed and financed. Such persistent and contradictory global interdependencies further complicate the discourse of dependence and serve to mock calls for independence.

Final Reflections

We agree with Scheper-Hughes and Sargent (1998) that:

> The cultural politics *of childhood* speaks on the one hand to the public nature of childhood and the inability of isolated families or households to shelter infants and small children within the privacy of the home or to protect them from the outrageous slings and arrows of the world's political and economic fortunes. On the other hand, the *cultural policies* of childhood speak to the political, ideological, and social uses of childhood. (p. 1)

Such intersections of childhood with larger social, economic, and ideological discourses is at the heart of governing children and families and is in constant flux in Kenya, as elsewhere in the world. As James and Prout (1990, p. 1) state, "any complacency about children and their place in society is misplaced, for the very concept of childhood has become problematic during the last decade." Similarly, Stephens (1995, p. 8) argues that, "[a] focus on childhood—and on other domains previously differentiated from the realm of political economic—is thus important, insofar as it breaks the frame of dominant models of transformation in the world system." Stephens also argues that child and family researchers should rethink their studies "in the light of social and historical macroperspectives" (p. 8). We would advocate a constant mixing of analysis of such macroperspectives, with research focusing on the rapidly changing local perspectives of children, parents, and caregivers, in order to better understand the complex dynamics of governing children and families in Kenya or other Southern Hemisphere settings.

While analyzing the many ways in which the state and larger, multinational banks and other bodies govern children and families in Kenya, we have been struck many times by the persistent patterns of colonialism and the contradictory spaces of "postcolonial" life, that subtly resist colonial power relations. To quote Dimitriadis & McCarthy (2001), "Our period of intense globalization and the rise of multinational capital has played a large part in ushering in the multicultural age—an age in which the empire has struck back . . ." (p. 117). We strongly support calls for decolonizing research and moving beyond Western, imperialist models, to an analysis from a growing number of indigenous scholars (e.g., Gandhi, 1998; N'gugi wa Thiong'o, 1993; Smith, 1999; San Juan, 1999; Spivak, 1999). We recognize the struggles and attempts to deal with the [im]possibilities inherent in carrying out decolonizing work, and agree that the work of such scholars stands at the center of the "beginning of the presencing" (Bhabha, 1994) of a disharmonious, restive, unharnessable knowledge that is produced at the ex-centric site of neo/post/colonial resistance. Such authors serve as transgressive authorizing agents whose positions at once marginalize and singularize the totalizing metatexts of colonial/Western knowledge (Mutua & Swadener, 2003).

We are also concerned about "overdetermined discourses" that construct the third world and contribute to false binary categories of difference, often defined by Western scholars (Mohanty, as cited in Gandhi, 1998). While we have not wanted to overemphasize the economic margins, we recognize that there are gross social inequities that contribute directly to marginality and difficult circumstances for growing numbers of children and families in Kenya. Thus, in emphasizing the daily experiences of Kenyan families whose lives have been deeply affected by neoliberal global policies and growing assumptions of normative and universal discourses of development and early education, we have attempted to take the position of allies, or "allied others" (Rogers & Swadener, 1999). Though policy analysis is critical, we agree with Schram (2000), that a more radical response is also required:

> In the end, these considerations remind us that social justice is still contingent on all families being able to access basic social welfare entitlements. All families should be able to practice a 'politics of survival.' Parents should have nutritional assistance, housing, schooling, and the like. Parents should be able to have access to the basic services needed to raise their children: health care, to receive the education and job training they need not only to be effective parents but also productive citizens. Whether these universal entitlements should be guaranteed all at once under some comprehensive family policy or whether they should be built up one after another was decided a long time ago. The time for incrementalism to get radical and radicalism to get incremental is long overdue. (p. 182)

We believe, with Dimitriadis & McCarthy (2001), that, "A strategy of alliance might allow us to produce new anti-discriminatory pedagogies that will respond to this fraught and exceedingly fragile moment of globalized, postcolonial life." Such alliances should foreground the voices of children and families and should challenge the pervasive assumptions of late capitalism with indigenous/hybrid sensibilities and solutions.

These alliances at the local level can create more inclusive spaces for the formulation and enactment of policy by increasing the possibility of those at the margins having a voice, so that policy- and decision making is not left in the hands of the privileged few. Emerging indigenous governance structures should be supported, including policies that resist neoliberalism's emphasis on "self-sufficiency." As this chapter documents and as Tikly (2001, p. 166) asserts, structural adjustment programs have undermined the state and civil society, pointing to the need for policy change (p. 166). We would agree with Ngugi wa Thiong'o, that there is a need to move the center toward correcting the economic and political structural imbalance between the West and third world nations, and actively resist the social and structural imbalances between the few who control the resources and the silent majority living in poverty (Cantalupo, 1995; Ngugi wa Thiong'o, 1993). We join members of growing grassroots movements in advocating that governing patterns of children and families in Kenya be decolonized and democratized.

Epilogue

This chapter was written prior to the general and presidential elections in Kenya in December 2002. The elections ushered in a new government, after 39 years of Kenya African National Union (KANU) party rule. Since that time, a number of policy issues affecting children and families have changed. In particular, the new government pledged to provide free primary education to all public school students. The policy to provide free and compulsory education (also referred to as Universal Primary Education, UPE) by the new government is in line with NARC's (National Alliance Rainbow Coalition) campaign pledge and with the Children's Act, passed in 2002. The latter calls of the provision of free and compulsory education, conform to the international charters that Kenya has ratified, including the Declaration on Education For All (EFA) at the World education conference in Jomtien, Thailand in 1990 and the 2000 World Education Forum in Dakar, Senegal.

The free primary education is set to benefit more than 3 million children eligible for school who had been out of school due to numerous levies charged as part of mandated cost-sharing. Primary school enrollment stood at 85 percent in 2002, down from 95 percent in 1990 (Rugene & Njeru, 2003). Prior to the reintroduction of free education in early 2003, 6.2 million children were enrolled in Kenyan public primary schools (Siringi, 2003). With the new policy, more than 1.2 million more children have enrolled. There are still another 2 million more out of school (Wanyonyi, 2003). Early Childhood Education (ECE) enrollment stands at 35 percent of all eligible children but may be decreasing with less support from the Ministry of Education. In addition, parents are questioning paying preschool fees in the "new era" of free primary education.

The new government is also moving forward with policies addressing the plight of street children. The government has placed many of them in temporary homes that are providing rehabilitation programs with the purpose of helping the children make the transition to formal education (Gitonga, 2003). The older out-of-school youth are being "rounded up" to join the National Youth Service for training that can prepare them for a better future (Gitonga, 2003). While conditions in this program have been critiqued, this represents one of the most systematic government actions on behalf of street children to date.

We would also like to clarify that at the time the data cited in this paper were collected, Structural Adjustment Programs (SAPs) was the phrase used to refer to the World Bank/IMF debt restructuring and austerity measures in place in Kenya. This has since changed; aid recipient countries are now required to prepare and submit Poverty Reduction Papers

(PRSPs) to the IMF/World Bank for approval. The international financial institutions introduced The Poverty Reduction Strategy (PRS) as a participatory country-led mechanism to more sharply focus countries' poverty reduction efforts (World Bank, 2000). This was in response to concerns there was a stalemate in the fight against poverty, thus prompting an intense reexamination of lenders' own development and debt strategies (World Bank, 2000).

The result of this reexamination was an agreement at the World Bank/IMF 1999 Annual Meeting, to link debt relief to the establishment of a poverty reduction strategy for all countries receiving World Bank/IMF concessional assistance (World Bank 2000). The Poverty Reduction Strategy Papers (PRSPs) are to provide the basis for assistance from the World Bank and the IMF, as well as debt relief. While Kenya has prepared and submitted its PRSP (Ministry of Finance and Planning, 2003) to the lending institutions, it is important to note that aid to Kenya was suspended in the year 2000 (Orina, 2003) after the government failed to meet some of the key conditions for further assistance. Donor countries and international institutions did resume aid to Kenya with the new government in place, after the new administration promised to fulfill the conditions required for its resumption (Orina, 2003).

The World Bank/IMF PRS initiative is welcome in that it has the potential to have an impact on the most vulnerable part of society, the poor, especially women and children, by making poverty reduction the primary goal guiding the use of resources from debt reduction and loans. It also gives civil society a meaningful participatory role in the design and implementation of a national strategy for poverty reduction by moving away from reforms dictated by creditors and donors. PRSPs, however, are still based on neoliberal policies that focus on budget austerity, economic growth, and liberalization.

Note

1. Although there are maternity leave policies affecting public sector employees and some private sector jobs, there are many workers whose rights to such leave are greatly restricted—e.g., casual laborers, hawkers, and, increasingly, workers in private sector jobs.

References

Adams, B. & Mburugu, E. (1994, June). *Women, work and child care.* Paper presented at the Second Early Childhood Collaboration Training Seminar, Nairobi.

Amisi, O. A tale of hard work, courage and generosity. *Daily Nation,* September 12, 2001.

Bhabha, H. K. (1994). *The location of culture.* London: Routledge.

Bradshaw, Y. W., Noonan, R., Gash, L. & Sershen, C. B. (1993). Borrowing against the future: Children and Third World indebtedness. *Social Forces, 71* (3), 629–656.

Brock-Utne, B. (2000). *Whose education for all? The recolonization of the African mind.* New York: Falmer Press.

Buchmann, C. (1999). Poverty and educational inequality in sub-Saharan Africa. *Prospects, 29* (4), 503–515.

Cantalupo, C. (1995). *The world of Ngugi wa Thiong'o.* Trenton, NJ: Africa World Press.

Children's Act (2001). *Kenya Gazette Supplement No. 95 (Acts. No.8).* Nairobi: The Government Printer.

Dale, R. (1999). Specifying globalization effects on national policy: A focus on mechanisms, *Journal of Education Policy, 14,* 1–17.

Dimitriadis. G. & McCarthy, C. (2001). *Reading and teaching the postcolonial: From Baldwin to Basquiat and beyond.* New York: Teachers College Press.

Eshiwani, G. S. (1990). *Implementing educational policies in Kenya.* Washington, D.C.: The World Bank.

———(1993). *Education in Kenya since independence.* Nairobi: East African Educational Publishers.

Ehrenreich, B. (2001). *Nickel and dimed: On (not) getting by in America.* New York: Holt and Company.

Fafunwa, B. A. (1982). African education in perspective. In Fafunwa, B. A. & J. U. Aisiku (Eds.) *Education in Africa: A comparative survey* (pp. 9–28). London: Allen & Unwin.

Foucault, M. (1980). *Power/knowledge: Selected interviews and other writings, 1972–1977.* C. Gordon (Transl.). New York: Pantheon Books.

Fraser, N. (1989). *Unruly practices: Power discourse and gender in contemporary social theory.* Minneapolis: University of Minnesota Press.

———(1997). *Justice interruptus: Critical reflections on the "postsocialist" condition.* New York: Routledge.

Gakuru, O. N. & B. Koech (1995). *The experiences of young children: A contextualized case study of early childhood care and education in Kenya.* Nairobi: KIE/NACECE.

Gandhi, L. (1998). *Postcolonial theory: A critical introduction.* New York: Columbia University Press.

Gordon, L. (Ed.). (1990). *Women, the state, and welfare.* Madison: University of Wisconsin Press.

———(1994). *Pitied but not entitled: Single mothers and the history of welfare.* New York: Free Press.

Grant, J. (1993). *The State of the World's Children Report.* New York: UNICEF.

Hall, S. (1996). "When was postcolonial"? Thinking at the limit. In I. Chamber & L. Curtis (Eds.), *The postcolonial question: Common skies, divided horizons.* London: Routledge.

Hancock, G. (1989). *Lords of poverty.* London: Macmillan.

Hultqvist, K. & Dahlberg, G. (Eds.) (2001). *Governing the child in the new millennium.* London: Routledge/Falmer.

Human Rights Watch/Africa (1997). *Juvenile injustice: Police abuse and detention of street children in Kenya.* New York: Human Rights Watch.

———(1997b). *Failing the internally displaced: The UNDP displaced persons program in Kenya.* New York: Human Rights Watch.

James, A. & Prout, A. (Eds.). (1990). *Constructing and reconstructing childhood.* Basingstoke, U.K.: Falmer Press.

Kenya Institute of Education (2000). *Guidelines for preschool education.* Nairobi: KIE/ NACECE.

Kilbride, P. & Kilbride, J. (1997). Stigma, role overload, and delocalization among contemporary Kenyan women. In T. Weisner, C. Bradley, & P. Kilbride (Eds.), *African families and the crisis of social change.* Westport, CT: Bergin and Garvey.

Kilbride, P. & Kilbride, J. (1990). *Changing family life in East Africa: Women and children at risk.* University Park: Pennsylvania State University Press.

Kilbride, P., Suda, C., & Njeru, E. (2000). *Street children in Kenya: Voices of children in search of a childhood.* Westport, CT: Bergin and Garvey.

Kiluva-Ndunda, M. (2001). *Women's agency and educational policy: The experiences of the women of Kilome.* Albany: State University of New York Press.

Koome, M. (2002). Spare a thought for the "fatherless" child. *Daily Nation,* August 20.

McCarthy, C. (1998). *The uses of culture: Education and the limits of ethnic affiliation.* New York: Routledge.

McClintock, A. (1992). The angels of progress: Pitfalls of the term post-colonialism. *Social Text, 31 & 32,* 84–98.

Mignolo, W. D. (2000). *Local histories/global designs: Colonialist, subaltern knowledges, and border thinking.* Princeton, NJ: Princeton University Press.

Mink, G. (1998). *Welfare's end.* Ithaca, NY: Cornell University Press.

Mutua, N. K. & Dimitrov, D. M (2001). Prediction of school enrollment of children with intellectual disabilities in Kenya: The role of parents' expectations, beliefs and education. *International Journal of Disability, Development and Education, 48* (2), 179–191.

Mutua, N. K. & Swadener, B. B. (Eds.). (2003). *Decolonizing research in cross-cultural contexts: Critical personal narratives.* Albany: State University of New York Press.

N'gugi wa Thiong'o. (1993). *Moving the center: The struggle for cultural freedoms.* Nairobi: East African Educational Publishers.

Nieuwenhuis, F. J. (1996). *The development of education system in postcolonial Africa: A study of a selected number of African countries.* Pretoria, RSA: Human Sciences Research Council.

Ombuor, J. (2001). Rape and terror rule over the land. *Daily Nation,* September 12.

Parpart, J. L. & Staudt, K. A. (Eds.). (1989). *Women and the state in Africa.* Boulder, CO: Lynne Reiner Publishers.

P'Bitek, O. (1986). *Artist the ruler: Essays on art, culture and values.* Nairobi: Heinemann Kenya.

Polakow, V. (1993). *Lives on the edge: Single mothers and their children in the other America*. Chicago: University of Chicago Press.

Polakow, V., Halskov, T. & Jorgensen, P. S. (2001). *Diminished rights: Danish lone mother families in international context*. Bristol, UK: The Policy Press.

Popkewitz, T. & Bloch, M. (2001). "Bringing the parent and community back": A history of the present social administration of the parent to rescue the child for society. In Hultqvist, K. & Dahlberg, G. (Eds.), *Governing the child in the new millennium*. London: Routledge/Falmer.

Quist, H. (2001). Cultural issues in secondary education development in West Africa: Away from colonial survivals, towards neocolonial influences. *Comparative Education, 37* (3), 297–314.

Reimers, F. & Tiburcio, L. (1993). *Education, adjustment and reconstruction: Options for change*. Paris: UNESCO.

Rogers, L. J. & Swadener, B. B. (1999). Reflections on the future work of anthropology and education: Reframing the "field." *Anthropology and Education Quarterly, 30* (4), 436–440.

Rose, N. (1999). *Governing the soul: the shaping of the private self.* (2nd Ed.) London: Free Association Books.

San Juan, E., Jr. (1999). *Beyond postcolonial theory*. New York: St. Martin's Press.

Scheper-Hughes, N. & Sargent, C. (Eds.). (1998). *Small wars: The cultural politics of childhood*. Berkeley: University of California Press.

Schram, S. (2000). *After welfare: The culture of postindustrial social policy*. New York: New York University Press.

Sibley, D.(1995). *Geographies of exclusion: Society and difference in the West*. London: Routledge.

Smith, L. T. (1999). *Decolonizing methodologies: Research and indigenous peoples*. London: Zed Books.

Spivak, G. C. (1999). *A critique of postcolonial reason: Toward a history of the vanishing present*. Cambridge: Harvard University Press.

Stephens, S. (Ed.). (1995). *Children and the politics of culture*. Princeton, NJ: Princeton University Press.

Swadener, E. B., Kabiru, M., & Njenga, A. (1997). Does the village still raise the child? A collaborative study of changing child-rearing in Kenya. *Early Education and Development, 8* (3), 285–306.

Swadener, B. B., Kabiru, M. & Njenga, A. (2000). *Does the village still raise the child?: A collaborative study of changing childrearing and early education in Kenya*. Albany: State University of New York Press.

Swadener, B. B. & Mutua, N. K. (2001). Mapping terrains of "homelessness" in post-colonial Kenya. In Polakow, V. & Guillian, C. (Eds.), *Homelessness in international context*. (pp. 263–287). Westport, CT: Greenwood Press.

Swadener, B. B. & Lubeck, S. (1995). *Children and families "at promise": Deconstructing the discourse of risk*. Albany: State University of New York Press.

Swadener, B. B. & Mutua, N. K. (2001). Mapping the terrains of homelessness in post-colonial Kenya. In V. Polakow & C. Guillean (Eds.), *International perspectives on homelessness* (pp. 263–288). Westport, CT: Greenwood Press.

Tikly, L. (2001). Globalisation and education in the postcolonial world: Towards a conceptual framework. *Comparative Education, 37* (2), 151–171.

UNICEF (1996). *The progress of nations.* New York: The United Nations Children's Fund.

Vavrus, F. (in press). "Educated girls know English": Myths of modernization and development in Tanzania. *TESOL Quarterly.*

Walton, J. & Ragin, C. (1990). Global and national sources of political protest: Third world responses to the debt crisis. *American Sociological Review, 55,* 876–890.

Weisner, T. S., Bradley, C., & Kilbride, P. K. (Eds.). (1997). *African families and the crisis of social change.* Westport, CT: Bergin & Garvey.

Willinsky, J. (1998). *Learning to divide the world: Education at empire's end.* Minneapolis: University of Minnesota Press.

World Bank (2001). *World development indicators 2001.* www.worldbank.org

World Bank (2001). A chance to learn: knowledge and finance for education in sub-saharan Africa. http://www-wds.worldbank.org/pdf_content/00009494601032-905304794/multi_page.pdf.

Epilogue References

Worldbank (2000). Partners in Transforming Development: Approaches to Developing Country-Owned Poverty Reduction Strategies. http://www.worldbank.org/poverty/strategies/prspbroc.pdf Retrieved 4/6/03.

United Nations Center for Human Settlements (Habitat)(2001). From Structural adjustment programmes to poverty reduction strategies: Towards productive and inclusive cities. http://www.unchs.org/programmes/ifup/conf/HabitatPresentation-English.PDF. HABITAT. Retrieved/6/03

Ministry of Finance and Planning (2003). Poverty Reduction Strategy Paper (PRSP) http://www.treasury.go.ke/overview.htm. Retrieved 4.15.03

Gitonga, L. (2003). Diary . . . Ready for school. *Daily Nation,* March 23, 2003 (Nairobi).

———(2003). Marked change as donations increase. *Sunday Nation,* March 30, 2003.

Orina, E. (2003). IMF to resume lending to Kenya in July. *Daily Nation,* January, 17, 2003

Rugene, N. & Njeru, M. (2003). Free primary school starts next week. *Daily Nation,* January, 4, 2003.

Siringi, S. (2003). Schools act on intake. *Daily Nation,* January 10, 2003

Wanyonyi, T. (2003). Don't neglect indigent secondary students. *Daily Nation.* April 15, 2003

SECTION IV

LIMITING THE BOUNDARIES OF REASON: NEW POSSIBILITIES/ IMPOSSIBILITIES

CHAPTER ELEVEN

PEDAGOGY AS A LOCI OF AN ETHICS OF AN ENCOUNTER

Gunilla Dahlberg

> *There is a whole race of judges, and the history of thought is like that of the court, it lays claim to a court of pure Reason, or else Pure Faith. This is why people speak so readily in the name and in the place of others, and why they like questions so much, are so clever at asking them and reply to them.*
>
> —Gilles Deleuze

The images of an autonomous, self-motivated, flexible, and competent child, and a teacher who is an interacting and reflective practitioner, who empowers and speaks in the voice of this child by opening up for the child's independence, self-managing, and choice, have traveled around the world in educational practices and child research in the late twentieth century and still travels in early-twenty-first-century educational discourse. These are images of the child and the teacher that also seems to be bound with a new understanding of governing and citizenship and a return to ethics. As a matter of fact, it seems like the self-governing child is a citizen who is governed through ethics. In this chapter I will scrutinize and contest some discourses related to these images of the child, and its related pedagogical practices and ethics. It may sound strange that I will contest discourses related to this child—a child who symbolizes a free and active subject, an independent child with a desire to learn. However, what is seen as the normal and the "truth" about the child, and done in the best interests of the child, is not something ethically neutral, natural, and objective. Our knowledge of the child, as well as of the teacher and pedagogy, are historically bounded constructions, and as such they are social and historical artifacts, and as such they cannot be placed outside time, space, and relations. Due to that they are also provisional and open for change.

My purpose with this contestation is twofold. The first purpose is to critically examine the different discourses and technologies, which form a type of common sense, or reason, through which this child and teacher are inscribed. By using research from studies of the *history of the present,* studies that have drawn inspiration from Michel Foucault´s work, I will problemetize how we, in this historical period, make ourselves into subjects, and how this making can appear so natural.

Studies of the history of the present start out from an unease with the values of the present, and by historicizing and denaturalizing the taken-for-given notions, practices, and values of the present, such studies can open up a space out from which one can revise and reformulate other possible ways of reasoning and practicing pedagogy. Nicholas Rose (1999, p. 282), writing about studies of the history of the present, suggests that their aim is "to help maximize the capacity of individuals and collectivities to shape the knowledges, contest the authorities and configure the practices that govern them in the name of their nature, their freedom and their identity." Accordingly, research within a history of the present can be seen as a form of critical engagement of the present, as making the production of discourses open for scrutinity and denaturalization also makes them open for revisions (Popkewitz, Franklin, & Pereyra, 2000).

Hence, historicizing the autonomous and self-motivated child and the interacting and reflective teacher, a child and teacher who are supposed to open up for diversity and pluralism can open possibilities for other choices to be made. To open up such possibilities I will especially examine what kind of ethic the *norm of autonomy* embodies, a norm which today seems to be taken as the Truth of pedagogy, and that is related to a "new" form of constructivism and interactionism producing an intense and continuous self-scrutiny and self-evaluation (Rose, 1999).

On the base of this historicizing and destabilizing of the present, the second purpose of the chapter is to explore if there are other possible ways of conceptualizing the child, teacher, and pedagogy than in terms of the norm of autonomy. To explore other possibilities and other ways of making ourselves into subjects, and to rethink pedagogical practice and ethics, I will use Emanuelle Levinas´ thoughts on *alterity and welcoming* and Jacques Derrida´s thoughts on *hospitality.* Levinas and Derrida have played a major role in formulating the contemporary account of the ethical within poststructural discourse, and their notion of ethics diverges significantly from that found in Anglo American analytical philosophy (Taylor, 1987).

To rethink pedagogical work I will also draw on the experiences of *a pedagogy of listening* from the pedagogical practice of the city of Reggio Emilia in Nortern Italy. Together with Levinas´and Derrida´s thinking I have found that the Reggio Emilian experience can help us to reconceptualize the scene of pedagogy as *a loci of an ethical practice—as an ethical encounter,* instead of as a site of the production of an autonomous and independent child. As a way of thinking pedagogy differently, the Reggio Emilian experience will also be related to Gilles Deleuze´s perspective on the image of thought.

Provoking a new enlightenment?

Usually when ethics is discussed it is the "ought" question that is seen as the value-based question, and this question is often answered in relation to a Kantian universal moral. From a normative and universal perspective one is asking what a human being is, can be, and ought to be. This is a perspective that has been questioned by so-called poststructural thinkers. Michel Foucault, Emanuelle Levinas, and Jacques Derrida, among others, have argued that the universal and normative thinking, which is related to the enlightenment, has been brutal to many, which does not mean that they displace the enlightenment; rather, they try to open up for *a new enlightenment.*

As an answer to Kant's essay "What is enlightenment?," Foucault (1991, p. 42) writes that the enlightenment should not be seen as faithfulness to doctrinal elements, but rather as a form of reactivation of an ethos that he describes as a permanent critique of our historical era. For Foucault, deconstructing and rethinking taken-for-given assumptions is *an ethical attitude,* a form of affirmation that opens up for alternative constructions and for the possibility to think differently. It opens up for the Other, for the issue of *diversity, difference, and otherness,* and for new possibilities and potentialities—issues that seem to be the most pressing in a globalized world, as they relate to questions of integration and inclusion, but also to questions of exclusion, domination, and repression.

To provoke a new enligtenment, poststructural thinkers have deconstructed the privilege of the human subject of modernity, a subject that we today take as a given, but that they argue has a tendency to totalize alterity and otherness, and hence is a danger for responsiblity, for decision, for ethics, and for politics (Caputo, 1997). Descartes' thinking has been decisive for the construction of this subject–a subject that constitutes and defines itself through its own constructive activity. Taylor (1987) states that in modernity the autonomous and self-conscious subject has been the locus of certainty and truth and the first principle from which everything arises and to which all must be returned. He states: "What appears to be a relationship to otherness—be that other God, nature, objects, subjects, culture, or history—always turns out to be an aspect of mediate self-relation that is necessary for complete self-realization in transparent self-consciousness" (p. xxii).

It is this autonomous subject that poststructural thinkers have challenged through what has been called *the linguistic turn* and that has resulted in talk about *the death of the subject*. A talk that surely has given rise to the many subversive reactions toward poststructural thought. The death of the subject should not be taken literally. It has to be seen in the context of what poststructural thinking is trying to transgress. In this process of deconstruction the authentic, coherent, autonomous, and transparent subject of the project of modernity has been in focus for transgression. Michel Foucault (1982, p. 208), for example, describes his project as an interest in "the different modes by which, in our culture, human beings are made into subjects" and in his studies he has identified those practices and technologies by which we are making ourselves into subjects. He also states that knowledge and power is intertwined in regimes of Truth, regimes that normalize human beings for the modern institutions of which they are a part. Via these dominating discourses and practices, human beings are continually constructing themselves, a process Foucault does not understand as a relation of domination, but rather as *subjectification*. It is through these discourses that persons become subjects, at the same time as they become bound into corporeal and affectual relations with certain truths and authorities. Authorities or expertise of human conduct such as psychologists, lawyers, doctors, criminologists, and so forth have lent their vocabularies of explanation, procedures of judgement, and techniques of remediation, to others such as teachers, nurses, parents, managers (Rose, 1999, p. 92). This means that we are continually making ourselves, but are also made, through an active embodying of dominating discourses and practices in which we as human beings take part—in the family, the preschool, school, etc.

However, discourses are arbitrary constructions, and hence, not value-neutral or natural, which means that there are possibilities to border-cross the fixed and transparent subject. It is for this reason that Foucault talks about a *knowledge/power regime.* Dominating discursive regimes function through the concepts, classifications, conventions, and categories that we use to represent reality, and it is through these discourses that we know what is seen as normal or nonnormal, right or wrong, and they become productive through creating procedures of inclusion and exclusion. We govern ourselves through these norms, so "what begins as a norm implanted 'from above', such as the universal obligations of literacy or numeracy, or the adoption of appropriate patterns of conduct in child rearing, can be 'repossessed´ as a demand that citizens, consumers, survivors make of authorities in the name of their rights, their autonomy, their freedom" (Rose, 1999, p. 92).

By deconstructing and problemetizing the inclusions and exclusions on which professional competences and authority are based in the pedagogical field, one also gets to know the practices they produce and make possible. Showing that which seems to be "natural" and taken for given, how knowledge has become *a norm,* opens up for other possibilities and choices to be made. Valerie Walkerdine (1984), for example, has in one of her works analyzed how developmental psychology and its concepts and classifications such as universal developmental stages, can be seen as a form of language that has constructed what a child is, can be, and should be in the project of modernity. Hence, making some practices possible and others not. She proposes:

> the understanding of the "real" of child development is not a matter of uncovering a set of empirical facts or epistemological truths which stand outside, or prior to, the conditions of their production. In this sense developmental (as other) psychology is productive; its positive effects lies in its production of practices of science and pedagogy. Developmental psychology, therefore, can be seen as a language, which starts living its own life through *processes of normalization,* and hence also constructs pedagogues and children and their respective expectations and social practices. (p. 163)

Governmentality and the autonomous, flexible, and self-managing child

What kind of reasoning is, then, constituting the discourse on the autonomous and flexible child with a desire for self-realization and self-managing through choice? From Foucault´s perspective of *governmentality* these discourses can be seen as new techniques and as a rupture in the social administration and government of the child, and as such they are new ways of understanding ourselves and acting upon ourselves as subjects of freedom (Rose, 1999; Fendler, 2001; Hultqvist & Dahlberg, 2001; Popkewitz & Bloch, 2001). Nicholas Rose (1999) has shown how the discourse of the humanistic subject, since the Enlightenment, has produced a thinking and acting that implies that we are all the time striving and longing for freedom. Accordingly, freedom has become a form of *technology of government*—a form of social administration—within institutions of modernity, such as early childhood institutions and schools.

In relation to this, the construction of an autonomous, flexible, and self-managing child can be viewed as connected to a "new" form of social administration of the self that has taken shape, alongside former images of the self, at the end of the twentieth and the beginning of the twenty-first century. It is a new way of understanding the child as a subject of freedom. This child Popkewitz and Bloch (2001) have named the *constructivist cosmopolitan child,* and although there are overlapping discourses, this child differs from the image of a *cosmopolitan child* constructed at the turn of the twentieth century—the cosmopolitan child they have historicized is in relation to the government of the social citizen, a citizen constructed at the turn of the twentieth century, when the progress and the betterment of society and the individual was seen as one project. In this project the child was supposed to be prepared for a form of pluralism where nationhood and the citizen were aligned into a universalized and globalized "American," "British," etc. This cosmopolitan self was seen as no longer bound to a sense of identity built through geographical location and face-to-face interactions. It was a move from gemeinshaft to a form of gesellshaft, and where, as Popkewitz and Bloch (2001, p. 89) write "individuality was embodied in multiple and anonymous social relations formed through, among others, urbanization, the new capitalist industrialization, a generalized Protestant view of salvation, and liberal political rationalities." In this process the "soul" was the site of the administration of freedom displacing earlier religious themes with secular discourses that made inner dispositions, sensitivities, and awareness of the individual the site of individual salvation. And a "salvation story was now told in the name of the new, secular citizen in the early twentieth century whose subjectivity embodied the obligations, responsibilities, and personal discipline aligned to liberal democratic ideals."

Although the idea of pluralism represented children from diverse ethnic and class backgrounds, the register of social administration connected the child, family, and community, by involving overlapping universal discourses of state policy, philanthropy, social science, health, medicine, and morality. To provide for social cohesion children´s dispositions were classified through norms drawn from universal values, something that at the same time as it included also excluded, as categories used were not universal, but historically particular categories dividing children through particular binaries such as whiteness/blackness, male/female, civility/savagery

(Baker, 1998). Hence, this shows *a particular normalization of diversity*—a normalization that was ordered through universal norms of child development, the family as the primary group and a legislative expertise (Popkewitz & Bloch, 2001, p. 99).

If the cosmopolitan child of the late nineteenth century had to be prepared for a pluralism constructing a universalized identity, the constructivist cosmopolitan child, Popkewitz and Bloch argue, has rather to be prepared to be a global citizen/worker. The soul is still the site of governing, but now the child is supposed to be an active, flexible, and adaptable child, ready for diversity and uncertainties both in school, family, and work. The discursive practices of the constructivist cosmopolitan child, a child who is an empowered, problem-solving individual capable of responding flexibly to problems that have no clear set of boundaries or singular answers, is built on a form of pedagogical constructivism joining a particular cognitive psychology and symbolic interactionism. This joining of psychological technologies has given concrete form to practices of reforms such as cooperative learning, and group or peer learning. This valued image of the child has its counterpart in the globalized economy, with its valorization of the entrepreneurial self, innovation, and national competetiveness. So, embodied in the "new" form of constructivism, prevalent in the field of education today, is an *entrepreunerial self*—a flexible "actor," ready to respond to new eventualities and empowered through self-reflections, self-analysis, including the incorporation of the voices and actions of local communities. Popkewitz and Bloch also argue that today´s decentralized forms of governing and the stress on individual choice and the local context is related to the need for individuality and flexibility in the face of uncertainties over freedom, democracy, and interdependencies in the globalized culture and economies in the twenty-first century.

Rose (1999, p. 174) talks about this change in the political role of the state as *a double movement of autonomization and responsibilization*—as a process where the social state gives way to that of the facilitating state and where the individual, as well as organizations and other actors, are being set free to find their own destiny, and at the same time being responsible for that destiny as partners. He states that "politics is to be returned to society itself, but no longer in a social form: in the form of individual morality, organizational responsibility and ethical community." In this process '*community*' becomes instituted as a sector of government, as a government through community, and as such it is brought into existence and can be "mobilized and deployed in novel programmes and techniques which encourage and harness active practices of self-management and identity constructions, of personal ethics and collective allegiances."

Following Rose´s thinking, in this process, the former universal and uniform citizen of the nation-state becomes retrieved from a social order of determination into a new ethical perception, as an individualized and autonomized actor, who has got unique, localized, and specific ties to her or his particular family and to a particular moral community. Rose suggests that in this new process of citizen formation one can discern the emergence of a new diagram of power operating in a field that he has termed *ethico-politics.*

In this process of change the child is supposed to adopt a new relation to her/his self in its everyday life as the self is more and more to be an object of knowledge, and autonomy is to be achieved through a continual enterprise of self-improvement, through the application of rational knowledge and techniques for self-reflection and self-management. To live as an autonomous individual who is constructing an ethical life is to have learned these knowledgeable techniques for understanding and practicing upon oneself. The striving to live an autonomous life "and in order to discover who we really are and to be able to realize our potentials and shape our life styles, we become tied to the project of our own identity and bound in new ways into the pedagogies of expertise" (Rose, 1999, p. 93). Hence, *the norm of autonomy* produces an intense and continuous self-scrutiny and self-evaluation. By referring to Wagner, Popkewitz and Bloch (2001) argue that this entrepreneural and cosmopolitan constructivist self can be seen as a hybrid of different discourses in which new patterns of calculative routines are to "make" the citizen who manages his own personal ethics—it is an ethics that involves the constitution of new forms of authority in the government of conduct.

Processes of normalization

Do the prevalent ideas of the autonomous, competent, and self-managing child, a child who is enabled to govern herself in the name of freedom, actually open up for *diversity* and *respect for the Other,* or do they still constitute *universal sameness?* Questions that follow from this question are: What kind of ethic does the *norm of autonomy* (Rose, 1999), which we today seem to take as the Truth of pedagogy, embody? A norm that seems to be related to a "new" form of constructivism and interactionism and producing an intense and continuous self-scrutiny and self-evaluation. What ways of reasoning and what rationales and strategies are we using for making ourselves into these researching, motivated, flexible, and free-choosing actors, who are ready to respond to new eventualities and uncertainties in the face of the twenty-first century? And how are we making ourselves into subjects through the prevalent strategies and techniques of self-evaluation and self-inspection, such as documentations, portfolios, and examinations of dispositions and learning styles? And what will it mean for a child to be governed as an autonomous individual and an active consumer, who must actively construct its own life through the choices s/he makes about her/his conduct and who must take individual responsibility for the nature and consequences of the choices made?

In a genealogical study Lynn Fendler (2001) examines if these strategies can change *universal sameness.* Her argument is that the so-called autonomous child is still governed through *the normalizing gaze* of the classifications and categorizations of psychology—through what she has called *developmentality.* A gaze in which children´s learning and well-being is shaped and valued according to procedures that are justified in terms of psychological norms of development and in which the social administration of the child and its life is produced through research discourses such as cognitive psychology, medicine, biology, brain research, etc. This gaze is related to ways of understanding children, teachers and their work, which constructs a power/knowledge complex by representing, classifying, and normalizing the child and the teacher through the concepts related to these scientific discourses—a power/knowledge complex where lack and needs still seem to become the constitutive elements of the child, as well as of the teacher.

Similar thoughts are presented by Popkewitz and Bloch (2001). They argue that, although it might seem contradictory, the *constructivist cosmopolitan self* involves images of the universal child—an image that intersects "scientifically derived age norms with the normalization of the dispositions and sensitivities of a 'problem-solving´ child or person." According to this they propose that developmental knowledge is still the mainstay or foundation for best practice in these new reforms, and that scientifically guided principles based on generalizations that are sufficiently reliable are universal and scientific sign-posts for "who" the child is, and how to guide his/her progress or development. Their conclusion is that many of the same norms detailed in psychological research studies from the 1940s still guide teachers practices today, with few adaptations to broaden norms to include other children. However, today the notions of development intersect with *a new normality of the child*—a child who will be flexible, and who is developmentally ready for the uncertainties and opportunities of the twenty-first century. Popkewitz & Bloch (2001) write that:

> the scientific knowledge of the self, this "new" pedagogical contructivism, which "in itself is a hybrid of different discourses consisting of an amalgam of particular cognitive psychologies and symbolic interactionist approaches joined with political discourses of reform"—seems to be related to notions

of universal development which "overlap with flexibility, problem-solving, and uncertainty to provide new salvation themes. (pp. 14–15)

Reducing the Other into the same?

According to the above, generalizations and classifications are not innocent and neutral. They are productive as they inscribe intentions and purposes into pedagogical practices and, hence, creating boundaries for what becomes possible to think and how to act. Following Popkewitz and Bloch the autonomous and flexible child, who is developmentally ready for the uncertainties and opportunities of the twenty-first century, seems to have become a new normality of the child. Although all talk about difference and diversity, once again general and fixed categories are used to define the child. However, putting everything that one encounters into premade categories implies that we *make the Other into the Same;* as everything that doesn´t fit into these categories, that is unfamiliar and not taken-for-given, has to be overcomed. Hence, alterity disappears. This betrays the complexity in children´s lives and closes down possibilites to view the child in relation to multiplicity and change. In one of his books Gregory Bateson (1988) warns that we live with the illusion that the map is the territory or landscape, and that the name is the same as that which is named. The following quotation from Lewis Carroll´s (1893/1973) book "Sylvio and Bruno Concluded" shows the dangers with this kind of mapping:

> Mein Herr looked so thoroughly bewildered that I thought it best to change subject. "What a useful thing a pocket map is!" I remarked. "That's another thing we´ve learned from *your* Nation," said Mein Herr, "map-making. But we´ve carried it much further than *you*. What do you consider the *largest* map that would be really useful?" "About sich inches to the mile." "Only *six inches!*´ exclaimed Mein Herr. "We very soon got to six *yards* to the mile. Then we tried a *hundred* yards to the mile. And then came the grandest idea of all! We actually made a map of the country, on the scale of *a mile to the mile!*" "Have you used it much?" I enquired. "It has never been spread out, yet,´ said Mein Herr: "the farmers objected: they said it would cover the whole country, and shut out sunlight! So we now use the country itself, as its own map, and I assure you it does nearly as well." (pp. 556–557)

Vitalizing ethics through constituting a space for alterity and radical difference

Is it, then, possible to elevate the child, the Other, beyond objectification and totalization, beyond the universal sameness that has been one side of the Janus-faced project of the Enlightenment? How can pedagogy be conceptualized if one does not understand it in relation to the self-reproduction of a sovereign and autonomous subject? And how can we find configurations that are open for the constitution of a changeable child with multiple subjectivities without running into ready-made mappings and fixed and uncontestable moral codes of conduct? Rose (1999) has stated that if the ethico-politics only operates at the pole of morality, there will only be a shift in the loci of authority, decision, and control in order to govern better. Hence, we have to be made aware of our own allegiance to the ideas of autonomy and community by evaluating the technologies and authorities that seek to govern us, as free individuals, through ethics. What kinds of relations of truth and power are we governed by and govern ourselves through?

Emanuel Levinas and Jacques Derrida have made important contributions to the question of how one would be able to develop an ethico-politics that operates closer to the pole of ethics instead of morality. Their notions and *vitalization of ethics* can help to scrutinize and contest some of the discourses constructing the cosmopolitan constructivist child, teacher, and parent. For Levinas, ethics is the first philosophy and he proposes that ethics has to precede all thought. He also questions the certainties, logocentrism, and egocentrism characterizing the modern Western society, arguing that they cannot constitute ethical relations. In his work he examines how the striving toward autonomy and independence can even be violent and oppressive. To rethink ethics Levinas has had to challenge the sovereign and knowing subject of the Enlightenment, by arguing that one has to think "otherwise than being"—beyond essence and the *autonomous and rational self* (Kemp, 1992; Levinas, 1969). The autonomous self, Levinas argues, in its will to master and make the world comprehensible, through an abstract and universal system of knowledge and truth, assimilates and makes the Other into the same or negates otherness. This means that the autonomous self can "understand" the Other, but when this takes place, the Other is made into the same, and is then also dispossessed the possibility to be an Other. Levinas has called this the *Logos*—a form of transparent understanding that reduces the unknown to the known. To describe the process through which this takes place Levinas uses the metaphor *grasp.* In order to retain its freedom, the knowing subject, with claims to mastery, grasps the Other, and the stranger is then made familiar and intimate and *made into the same,* into something that the "free thought" can grasp and have at its disposal (Kemp, 1992). Thought, then only becomes self-reflection and philosophy and has become what Levinas calls an egology. In this process thought neutralizes the objects and everything becomes neutral and objectified quantities, instead of unique phenomenas. Hence, alterity disappears and novelty is excluded.

To contest totalities and universal sameness Levinas bases his ethic on *Absolute Alterity*—an alterity that is premised on the *responsibility for the other.* It is responsibility for the other, rather than autonomy, independency, and rights of the self, that constitutes an ethical relation for him. This is an Other that I cannot represent and classify into a category and hence not totalize. It is only if one can affirm the Other and one´s own responsibility for the Other that, according to him, freedom can be reached. For Levinas it is the face-to-face encounter that ruptures my ego, and in this encounter the Other is absolutely Other and I have to take responsibility for the Other. This relation is a *welcoming*—a welcoming of the Other, *the stranger.* I hear the *call* of the Other, and this call is an ungraspable call as the Other is an absolutely other, which I cannot comprehend and contain. Respecting the Other as the call of the Other—as *Absolute Alterity*—means elevating the Other beyond objectification, beyond the confines of the knowing subject, as in the *face-to-face relation* the capacity of the self to possess and master the Other dissolves (Kemp, 1992). Levinas (1969, p. 9) writes, "The Other is the stranger that disturbs intimacy and whom it is impossible to reduce to myself, to my thoughts and my possessions."

To think otherwise than being, Levinas reformulated the ancient notion of the infinite. For him the infinite is the "trace of the other" that disrupts and dislocates human subjectivity. It is never comprehensible by or in finitude. The Other always goes beyond my understanding of her. Levinas (1969, p. 204) says that receiving the idea of infinity can be seen as an incessant overflowing of the self. "The Other is infinity, and we can conceptualize infinity, but we can never comprehend infinity" (Taylor, 1987, p. 196). From this follows that ethics for Levinas is not about prescribing rules for what is right or wrong—how another individual should behave or ought to do. It is a way of affirming the irreducible alterity of the world we are trying to construe. This is not empathy and tolerance. Empathy and tolerance is what Levinas contests; from his perspective, empathy and tolerance mean that you can represent the Other in your thought or get insight into the Other, which would be the same as reducing the Other into an intellectual object without concrete reality (Kemp, 1992). Deconstructing the idea of totality also dissolves the idea of the whole child, a child who is defined through homogeneous and coherent identities and fixed in time and place. For following Levinas' thinking, the child must rather be seen as a decentered child, who is changeable and with multiple identities.

Pedagogy as a dissymetrical and endless relation built on welcoming and hospitality

To think of another whom I cannot grasp is an important shift and it challenges the whole scene of pedagogy. It poses other questions to us as pedagogues. Questions such as how the encounter with Otherness, with difference, can take place as responsibly as possible—as something that the so-called free thought cannot grasp through categories, classifications, and thematizations.

Following Levinas we then need to vitalize and intensify ethics so it can be possible to open ourselves for alterity, by *a welcoming of the stranger.* From this perspective teaching and learning have to start with ethics—with receiving and welcoming—and it is the receiving from the Other beyond the capacity of the I, which constructs the discourse of teaching (Derrida, 1999, p. 18), a teaching that interrupts the philosophical tradition of making ourselves as the master over the child. This requires restructuring of subjectivity through challenging the universal subject. To be able to hear the ungraspable call of the child, and to have the capacity to relate to absolute alterity, one needs to interrupt totalizing practices that give the teacher possibilities to possess and comprehend the child. We have to dare to open up for the unexpected and affirm what is to come.

The word "welcome" is a very frequent word for Levinas (1969) in one of his main works, *Totality and Infinity.* In his book *Adieu to Emmanuel Levinas,* Derrida (1999, p. 93 and p. 46) says that the use of the concept of welcoming in *Totality and Infinity* prepares the treshhold, for opening the way for the arrival of the word *hospitality.* Although that word was not frequently used or emphasized by Levinas in that book, Derrida shows that in the concluding pages hospitality becomes the very name of what opens itself to the face, or what welcomes it. Hospitality is the "yes" of the Other. Derrida then proposes that Levinas, as well as himself, redefines subjectivity as hospitality. From this follows that instead of viewing pedagogy in relation to the autonomous and sovereign subject who privileges grasping and gathering, one can view *the subject as a host and pedagogy as a relation*—an encounter—but not as a harmonious relation and encounter. It is *a dissociating relation* that can leave room for the radical otherness and singularity of the Other—a dissociating relation that is characterized by inevitable and endless uncertainty, dissensus, dissymmetri, ambiguity, interruptions, or even, as Derrida (1999) has said in relation to Levinas´ ethics, "impossibilities." This relation, which opens up to the right of the stranger, is also the chance of human togetherness according to Levinas and Derrida, a right that is the condition for love and friendship. So here we have a paradox: *being-together* presupposes infinite separation and dissociation. Or as Derrida (1999, p. 92) so intriguingly expresses it: "The social bond is a certain experience of unbinding without which no respiration, no spiritual inspiration, would be possible."

An ethics of learning, taken from this thinking, opens for the unique in the encounter with the Other, but not unique in an essentialist way, as welcoming and hospitality *assume radical separation*—absolute otherness. Instead of a society built on autonomy, Levinas refers to a society that can be understood in terms of *dependency* and built on *a network of obligation*—but not in its prescribing and imperative form. This is not a dependency with monetary value, an exchangeable value; it is more like Lyotard's (1984) description of *a dissensual society,* where the social is an effect of a duty toward others, which we cannot wholly understand. In contestation with the modernist idea of future and progress this is not a plannable and programmable future, the future of strategic planning, neither the relative future, but "the absolute future, the welcome extended to an other whom I cannot, in principle, anticipate. This is the "*tout autre,"* which Levinas also talks about, and who disturbs the complacent circles of the same. It is a gift that gives itself without return, whenever the occasion calls for it" (Caputo, 1997, p. 200).

Relating to the child as the Other, then, something incalculable comes on the scene, and for Levinas this kind of relation to the Other is justice. Teaching thus becomes "answerable to the *question of justice* rather than to the criteria of truth" (Levinas, 1969, p. 89). Prior to any question, Derrida (1999) states, there is a more original *affirmation,* a "yes" to the Other, to the neighbor and the stranger, a *yes, yes, yes* that comes before the question, before science, critique, and research. This yes is for him responsibility and engagement in a future to come. We then need spaces where children, together with adults, can speak and be heard. Spaces, where adults also can become surprised and where pedagogues can see the possibilities in uncertainty and dissensus. This possibility means "the impossiblity of controlling, deciding, or determining a limit, the impossibility of situating by means of criteria, norms, or rules . . ." (Derrida, 1999, p. 35). However, then we have to *respect and pay attention to singularity,* to the radical heterogeniety of an individual and an event, as totalities invalidate the uniqueness in the Other´s existence and hence *sacrifice difference.*

Pedagogy as an ethical space

Are there any existing pedagogical practices that can give inspiration for the kind of shift that Levinas´ and Derrida´s thinking would imply in relation to teaching and learning? As said before, to think an Other whom I cannot grasp is an important shift that challenges the whole scene of pedagogy. From my own experience I think that the municipal preschools in Reggio Emilia in Northern Italy can give a lot of inspiration for such a shift and for making new practices possible. Although at the same time I think, to make such a shift requires a huge endeavor, which not only implies bringing the school-system in, but the society as a whole.

When described in international contexts, the pedagogical practice in Reggio Emilia is mostly seen as an experience built on the idea of "the hundred languages of children" and as a pedagogical practice relating to children with great respect and having an image of the child as competent and with lots of potentialities, and the teacher as a reflective and participating practitioner. They are also known for working with children´s problem-solving in longer term project-work, where children´s individual and cooperative learning-processes are documented and reflected on. Into the schools they have also brought the *atelierista,* a pedagogue coming from another discipline than education and who can function as a challenger for bringing in new languages into the everyday practice. They have also got what they call *pedagogistas,* who function as pedagogical consultants for a group of schools (for a further description of the Reggio Emilian experience see especially the books published by themselves; see also Edwards, Gandini, & Forman, 1993; Dahlberg, Moss, and Pence, 1999; Project Zero/Reggio Children, 2002).

To view children as competent instead of immature and incompetent, and as such disqualified, excluded, and marginalized, is in itself a great shift. Especially when you can see how the intelligent child, who is all the time trying to make meaning of the world, is vivid in their everyday pedagogical practice. The same holds for the way the teachers document and continually visualize and reflect on children´s learning-processes and the way the parents and the community participate in the preschool institutions in Reggio Emilia. From what has been said above, the Reggio Emilian child appears to be a perfect model for the autonomous and flexible child—a child suited for the uncertainties over freedom and democracy and for the entreprenuerial working life facing us in the twenty-first century. This then is a constructivist cosmopolitan child that the researchers referred to above have examined as a child constructed through the norm of autonomy and developmentality. However, if one goes deeper into the Reggio Emilian project, such a view can be contested. In Reggio Emilia, at least from my reading of their practice, the pedagogues together with the children, the parents, and the community, all the time are trying hard to deconstruct and border-cross what here has been called *the norm of autonomy* and the universal child, by transgressing prescribed categories of developmentality. This means that their practice cannot be seen as a programmable approach. Even the concept of approach seems to be misleading. It is more of a modest, tentative, and experimental relation that they have. A thought that resists application, but has to be used (Grosz, 1995).

By reference to Levinas and Derrida one could say that they have opened up *an ethical space,* which they have called *a pedagogy of listening.* In their pedagogical work the pedagogues are all the time posing questions on how the encounter with Otherness, with difference, can take place as responsibly as possible—as something that the so-called free thought cannot grasp through categories, classifications, and thematizations. Or in the words of Loris Malaguzzi (1993), the philosopher who was the first head of the communal preschools in Reggio Emilia: "I have saved my world as I have always tried to change it." In relation to this he argued that "pedagogy is not generated by itself, it is only generated if one stands in a contesting and loving relationship with other expressions of the present."

The idea of a pedagogy of listening is based on relations, and it is closely related to their constructions of the child, the pedagogue, and the parent as co-constructers of knowledge and learning. Carlina Rinaldi (1997), one of the consultants of Reggio Children, has in a speech described the pedagogy of listening in the following way:

> When we talk about listening we talk about an active as opposed to a passive listening, and about receiving and welcoming. Listening is a welcoming of the other and an openness to the difference of the other. The difference of the other is foremost a right to be different. . . . Difference can never be positive or negative in itself. It becomes positive or negative in relation to what kind of receiving and welcoming of difference the single preschool can give. It depends on whether the preschool or society wants to give to its children or citizens likeness or anonymity or if it wants to emphasize difference. A preschool or society emphasizing anonymity is much easier to govern.

The pedagogues in Reggio Emilia have worked hard to be able to construct a responsible and respectful affirmation of the child, by cultivating a critical attitude and through constructing and reconstructing the child, as well as themselves as pedagogues (Lenz Taguchi, 2000). Doing this one could say that they have tried to challenge images of the child and its identity as homogeneous and coherent and as fixed in time and space. This critical attitude has helped them, to a considerable extent, to keep their pedagogical practice alive as well as to resist totalizing procedures and to address the child, the Other, without reducing alterity.

Starting out from documentations of pedagogical processes, they constantly reconstruct and reinvent, through rereading, relistening, revisiting, revisioning, relearning and reinterpreting. This way of unfolding, taking apart, as well as folding, seems to have been a productive way of transgressing what has been folded over by tradition in the everyday practice of the schools. By fixing meaning provisionally and tentatively through researching, negotiations, and commentaries, they release the possibility of new practices, which have got the possibility of challenging the dualistic and binary thought of mind and body, nature and culture, private and public, etc. In this process the child, the pedagogue, and the parent, as well as the whole institution, are constructed anew in each moment. This also gives Reggio its movement, a movement that implies that it cannot be imitated and mastered in any simple way.

My reading of Reggio Emilia is that they are working with what I would conceptualize here, with inspiration from Samuel Weber (Readings, 1996, p. 32), as a form of *deconstructive pragmatism*. In their everyday practice they continuously struggle to problemetize the inclusions and exclusions on which their own professional competence and authority is based in order to open up for new possibilities. Doing this they have been able to study how dominant regimes establish processes of normalization and surveillance and how these processes go hand in hand with science. That has helped them to refuse to codify the children into prefabricated developmental categories and hence they have to a great extent been able to transgress the idea of a lacking and needy child. Loris Malaguzzi (1993), in one of his speeches, stated that we have been enslaved by science. Truth is science. Otherwise there is no Truth. By this statement he did not want to neglect science, but he wanted to make us aware of the enormous impact of science, while experiences from pedagogical practices are not listened to. So, from my understanding, the pedagogues in Reggio in their everyday work, constantly rework their own spaces of action through experimentation, questioning, and pragmatism.

One could say that the work in Reggio Emilia could be understood as a certain kind of *in-ventionalism in the every day practice*—an in-ventionalism that tries to respect the particularities of each situation and event. This in-ventionalism challenges representations and puts new possibilities into work, but only provisionally and tentatively. But at the same time as they question, they also affirm what is to come, as they have inserted this reconceptualizing and critique into *an ethical space* of open-ended, experimental, nonprogrammable reflection and self-questioning. And through experimenting they try to introduce new possibilities and new lines of thought, new potentialities, into their pedagogical practice. So the pedagogy of listening turns to what is to come, by inviting yes, yes, yes and the pedagogues are trying hard to listen to the child from its own position and experience and not to make the child into the same. This is not so much a processes of labeling, identification, recognition and judgement—it is more about *an ethics of an encounter;* an encounter in which the child and the teacher changes themselves along with the process of experimentation.

To open up and take apart can from this perspective be seen as an ethical attitude and a way to create an opening to the coming of the Other. Hence, in a pedagogy of listening the child becomes an absolute stranger for whom I have got responsibility as a teacher. Such a relation to the child is a form of affirmation of the everyday lived world and it requires attentiveness and responsiveness to the singularity of the particular situation or event. In this there is always the possibility not to be so governed, but at the same time, the in-vention of the Other may be dangerous. Something that Carlina Rinaldi (1994), one of the consultants of Reggio Children, often has said: "This can be dangerous and we have got no guarantees!" This very Foucauldian expression does not mean that they open up for total chaos and relativism, as that would be the same kind of "death" as total order.

Related to this critical attitude of the pedagogues, they have also given up the idea of the teacher and the researcher as the Great Liberators (Lather, 1991; Dahlberg, et al., 1999), as the source of what can be known and should be done, the Masters of Truth and Justice. Instead the pedagogues talk about choosing from many possible uncertainties and perspectives–a problem, a hypothesis, a doubt, a zone of nonknowledge, so as to hold the unexpected alive and affirm the coming—the surprise and the wonder (Rinaldi, 1994).

Knowledge as "a tangle of spagetti"—a rhizome

Project work in Reggio also differs from what it is usually seen as—as a work where the pedagogue gives the children problems to solve, but where the pedagogue already knows the answers and has preplanned not only the process but its conclusion. Reggio Emilia project-works more often take the form of a dialogue, in which children become partners in a process of experimentation and research by inventing problems and by listening to and negotiating what other children are saying and doing. In this process the co-constructing pedagogue has to dare to open her/himself to the unexpected and to experiment together with the children—in the here-and-now event. S/he challenges the children by enlarging the number of hypotheses and theories, as well as their more technical work, and hence enlarging the choices that can be made, instead of bringing it down to universal trivializations. Through this, children make themselves a specific relation to the different themes of the project, for example time, light, amusement parks for birds, communication and messages, etc. And they get a responsible relation to other children by listening, but also by negotiating in between each other. And this way of working, where the pedagogues open up the possibility for children to invent questions, as an art of constructing problems, implies that the children get the possibility to invent a problem before they are supposed to find a solution. This contests the question-and-answer pattern so common in pedagogical practices, a pattern that can easily have the effect that the child has got nothing to say as s/he is forced into a predetermined position. Or as Deleuze and Parnet (1987, p. 9) say:

> If you are not allowed to invent your questions, with elements from all over the place, from never mind where, if people pour them to you, you haven´t much to say. While encountering others, and when each child is bringing in her/his lot a becoming is sketched out between the different perspectives. Then a block starts moving, a block, which no longer belongs to anyone, but is "between" everyone. . . . Like a little boat which children let ship out and keep loose, and is stolen by others.

However, from our own work we have seen how very difficult it is for us as pedagogues and adults to stick to children's theories, hypotheses, phantasies, and dreams, as we already have got an idea of the right answer, and we habitually use the concepts and theories that we ourselves have embodied—concepts that own us, which means that we cannot become surprised (see further Dahlberg, Moss, and Pence, 1999).

To understand Reggio Emilia's idea of a project-work, I think we also have to rethink our mainstream idea of knowledge. We have to make disciplinarity a permanent question, through keeping open the question of what a discipline and what a school subject is, and ask ourselves why we have organized knowledge in a certain way. A project-work in Reggio Emilia grows in many directions without an overall ordering principle. It can be seen as a small narrative, a narrative that it is difficult to combine in a cumulative and additative way. This challenges our mainstream thinking of how knowledge is acquired—as a form of linear progression, where the metaphor is the tree. In one of his speeches, Loris Malaguzzi (1993, May) talked about their idea of knowledge as a "tangle of spagetti" and this, I think, is a view of knowledge that can more be seen as *a rhizome,* rather than the hierarchical and centralized idea of the tree of knowledge. The rhizome is a concept that Deleuze and Guattari (1999, see also Deleuze and Parnet, 1987) have used as a way of transgressing notions such as universality, question-and-answer patterns, simple judgements, recognition, and correct ideas. In a rhizome there is no hierarchy of root, trunk, and branch—it is not like a staircase, where you have to take the first step before you move onto and reach the other, which is the tree metaphor of knowledge, the aborescent idea, which is so prominent in education. For Deleuze and Guattari, thought and concepts can be seen as a consequence of the provocation of an encounter, and they view the *rhizome* as something that shoots in all directions with no beginning and no end, but always *in between,* and with openings toward other directions and places. It is a *multiplicity* functioning by means of connections and heterogeniety—but a multiplicity that is not given but constructed. Thought, then, is a matter of experimentation and problematization—as *a line of flight* and as an exploration of *a becoming*—as something different. Deleuze and Guattari (1999) propose that: "A rhizome has no beginning or end; it is always in the middle, in between things, interbeing, *intermezzo.* The tree is filiation, but the rhizome is alliance, uniquely alliance. The tree imposes the verb 'to be,' but the fabric of the rhizome is the conjunction 'and . . . and . . . and'. This conjunction carries enough force to shake and uproot the verb 'to be'" (p. 25).

In education it is often said that there needs to be a progression. But when there is a multiplicity of interconnected shoots going off in all directions this metaphor is difficult to use. Then you have more of a pluralism and multiplicity, which confronts the binaries of Western thought that have left so much of reality outside of education, such as sound, light, movement, smell. Something that Reggio Emilia has recaptured when they work with *"the hundred languages of children"* and *a pedagogy of listening.*

This image of thought also brings into question the assumption of the coherent and transparent subject. It was Nietzsche´s perverse taste, Deleuze (1995) says, that opened for him to say things in his own way, in affects, intensities, experiences and experiments:

> It is a strange business speaking for yourself, in your own name, because it doesn´t at all come with seeing yourself as an ego or a person or a subject. Individuals find a real name for themselves, rather, only through the harshest exercise in depersonalization, by opening themselves up to the multiplicities everywhere within them, to the intensities running through them. A name as the direct awareness of such intensive multiplicity is the opposite of the depersonalization effected by the history of philosophy. It´s depersonalization through love rather than subjection. (pp. 6–7)

When Deleuze and Guattari (1999) talk about depersonalization they bring in the idea of *between* things, an idea that they use as a way of escaping from the confines of the unified subject: "*Between* things does not designate a localizable relation going from one thing to the other and back again, but a perpendicular direction, a transversal movement that sweeps one *and* the other away, a stream without a beginning or end that undermines its banks and picks up speed in the middle" (p. 25).

The self here becomes defined as *a process of becoming* instead of as identity. It is a nomadic subject who enters into the domain of the virtual and the possible (Smith, 1997).When you proceed from the middle it is more a journey of coming and going rather than starting and finishing. However, Deleuze and Guattari warn us to view the middle as the average.The between is, rather, where speed is picked up, and speed is a mode of intensity and production—a collective production and creation rather than representations. From this deleuzean perspective follows, that each child's contribution can be seen as connected in a multiplicity of ways to other children's and teachers' contributions, and to the milieu that has been choreographed. The group then is not an organic whole uniting hierachized individuals, but a constant generator of depersonalization—*a collective production and creation;* something that Loris Malaguzzi perhaps had in mind when he said: "we are not only one, we are a hundred, a thousand, a million—we are many!"

Opening up for a radical dialogue

There are also a lot of similarities between Reggio Emilia´s thoughts of a pedagogy of listening and Bill Readings´ (1996) view of pedagogy as a relation and as a network of obligation. Readings argues for the importance of listening to the Other, the marginalized, which has not yet made itself heard, in which: "the condition of pedagogical practice is an infinite attention to the other. It is to think besides each other and ourselves to explore an open network of obligation that keeps the question of meaning open as a locus of debate," for "doing justice to thought means trying to hear that which cannot be said, but which tries to make itself heard" (p. 165).

To make such a pedagogy possible, he argues that we need to *decenter the pedagogic situation.* We need to view the landscape of pedagogy as a *radical dialogue* that does not resolve into a monologue, nor is controlled solely by the sender as a formal instrument in the grasp of the subject, which has been the modernist priviliging of the sender over the addressee. It is a move from transmission—as well as from the idea of determining the conditions for a dialogue. It presupposes that one does not view the addressee, the child, as an empty head, as the child´s head is already full of language. Decentering teaching begins with an attention to the *pragmatic scene of teaching,* through respecting the complexity of the obligation involved. Readings (1996) has vividly expressed this in the following way:

> I can address the Other only to the extent that there is a separation, a dissociation, so that I cannot replace the other. In each encounter the Other escapes the self. The child escapes the teacher—which means that the teacher can never comprehend the child. . . . To be hailed as an addressee is to be commanded to listen and the ethical nature of this relation can not be justified. We have to listen without knowing why, before we can know what we are listening to. To be spoken to is to be placed under an obligation, to be situated within a narrative pragmatics. We are obliged to them without being able to say exactly why. For if we could say why, if the social bond could be made an object of cognition, then we would not really be dealing with an obligation at all, but with a ration of exchange. If we knew what our obligations were, then we could settle them, compensate them, and be freed from them in return for payment. (p. 227)

This relates to Levinas' thinking, when he proposes that thought has mainly been related to the monologue—not with the dialogue—a monologue that sees itself as universal and able to comprehend everything in a totality. To describe the monologic thinking he uses Socrates as an example, something that surely is surprising, as his thinking often is used as a way of describing a dialogic thinking. Levinas argues that for Socrates the whole Truth already existed as "a reminder" in our minds. Then the teacher only had to be the deliverer—a midwife—of a human being´s knowledge. Thought, then, becomes a mirror of the self, and philosophy has become an egology, a narcissism, where the thinker only views its self (Kemp, 1992). This monologue, where the teacher claims to know and speak or explicate for the Other, is, according to Rancière (1991), a pedagogical myth and fiction, which divides the world into two: the knowing and the ignorant, the mature and the unformed, the capable and the incapable. To speak for and explicate he says is the myth of pedagogy, which rather than eliminating incapacity, in fact creates it. This is done, according to him, by establishing the temporal structure of delay. He relates this to the nineteenth-century myth of Progress, where you will see the light a little further along, a little later, after you have got a few more explanations.

Pedagogical documentation as a tool of opening up for difference, multiplicity, and ambivalence

As said at the beginning of this chapter, poststructural researchers have shown how new devices of governing, related to the autonomous, problem-solving cosmopolitan, constructivist child, such as portfolios, and self-evaluations have entered the educational agenda, and circulate globally. According to this perspective these devices can be seen as new strategies and technologies for the government of the autonomous self—an autonomous self who is grounded in the emergence of a new way of understanding ourselves and acting upon ourselves as subjects of freedom. As such they are related to shifts in the conception of identity and in the ways people relate to each other. These researchers also state that this change tends to increase those functions and variables through which the child is visualized and judged. More of the child´s personality, emotions, creativity, capacity for empathy, and cooperation are opened up for judgments, something that the English sociologist of education Basil Bernstein (1975) observed in his discussion of what he called visible and invisible pedagogy (see also Fendler, 2001). Documentation hence becomes an act of power and control—a form of visualization through the norms. A question that can be asked in relation to this research and in relation to the Reggio Emilian experience is: Does the power to see and make visible, through pedagogical documentation, always need to be related to the power to control and to the norms of developmentality? Or are there other ways of working with pedagogical documentation and portfolios that can transgress the gaze of developmentally appropriate practices? These questions are important in a time when techniques like pedagogical documentation and portfolios travel around the globe. In such a situation one has to ask what kind of reasoning is already inscribed in concepts such as portfolios and documentation? And what will these concepts and techniques *do* to us and to our pedagogical institutions?

Related to the above idea of a pedagogy of listening, which is open to difference, pedagogical documentation, as it is used in Reggio Emilia, can rather, if one follows Deleuze's thinking, be seen as a force, as an affirmation, as action and as energy and effectivity, a form of becoming. To visualize children´s theories, hypotheses, and work not only implies a way to follow children´s learning processes, it also implies an opening of a space where we can try to overcome the techniques of normalization through constructing new combinations and new assemblages. Following Deleuze, as well as Foucault, one can see power as productive and as vitality and intensity. Visualizing, then, can be seen as a form of construction—as an emotional engagement and participation rooted in the body of experience. Deleuze (1988, pp, 58–59) says that "visabilities are neither the acts of a seeing subject nor the data of a visual meaning. . . . Visabilities are not defined by sight but are complexes of actions and passions, actions and reactions, multi-sensorial complexes, which emerge into the light of day."

Seen from this perspective, pedagogical documentation becomes a form of visualization, which brings *forces* and *energies* into a project-work, forces and energies that can open us up to new possibilities, to the possibility of transformations—to difference.

In Reggio, as I understand it, they have worked hard to choreograph an ethical space by a pedagogy of listening—a space where the pedagogues have put concepts and themselves to work and through this been able to construct a more joyful pedagogy than we are used to in education. In Deleuzean terms this can be seen as a minor engagement—as a minor politics. The term "minor politics" was coined by Deleuze and Guattari, and Nikolas Rose (1999) describes it in the following words:

> These minor engagements do not have the arrogance of programmatic politics—perhaps even refuse their designation as politics at all. They are cautious, modest, pragmatic, experimental, stuttering, tentative. They are concerned with the here and now, not with some fantisized future, with small concerns, petty details, the everyday and not the transcendental. (pp. 279–280)

This kind of philosophy and research opens up a possible space and a possible way to intensify and vitalize ethics. For it makes us observant of the contingency of our constructions, and hence makes it possible to distabilize the meaning of that which we take for given and see as natural and truthful about the child. It is a constant critique and reflective attitude—a stuttering—that can be seen as *an experimental ethico-politics of life.* Derrida´s idea of a democracy to come is very close to this. He has said: "The promise of an event and the event of a promise, is the promise of democracy" (Caputo, 1997, p. 60). In relation to this I would like to see the children, the teachers, the parents, and also the politicians in Reggio Emilia as some kind of *astonished producers* (Deleuze and Parnet, 1987), and the Reggio Emilian preschools acting as a laboratory for alternative futures.

References

Baker, B. (1998). Childhood-as-rescue in the emergence and spread of the U.S. public school. In T. S. Popkewitz, & M. Brennan, M. (Eds.), *Foucault's challenge: Discourse, knowledge, and power in education* (pp. 117–143). New York: Teachers College Press.

Bateson, G. (1988). *Ande och natur. En nödvändig enhet.* [Mind and nature: A necessary unit] Stockholm: Symposion.

Bernstein, B. (1975). *Class, codes and control* (Vol. 3). London: Routledge & Kegan Paul.

Caputo, J. D. (1997). *Deconstruction in a nutshell: A conversation with Jacques Derrida.* New York: Fordham University Press.

Carroll, L. (1983/1973). Sylvio and Bruno concluded. In L. Carroll, *The Complete Works.* London: The Nonesuch Press.

Dahlberg, G., Moss, P., & Pence, A. (1999). *Beyond quality in early childhood education and care: A postmodern perspective on the problem with quality.* London: Falmer Press.

Deleuze, G. (1988). *Foucault.* Minneapolis: University of Minnesota Press.

———(1995). *Negotiations. From anti-oedipus to a thousand plateaus.* New York: Colombia University Press.

Deleuze, G., & Parnet, C. (1987). *Dialogues.* London: The Athlone Press.

Deleuze, G., & Guattari, F. (1999). *A thousand plateaus: Capitalism and schizophrenia.* London: The Athlone Press.

Derrida, J. (1999). *Adieu to Emmanuel Levinas.* Stanford, CA: Stanford University Press.

Edwards, C., Gandini, L., & Forman. G. (Eds.). (1998). *The hundred languages of children. The Reggio Emilia approach: Advanced reflections.* Greenwich, CT: Ablex.

Fendler, L. (2001). Educating flexible souls: The construction of subjectivity through developmentality and interaction. In K. Hultqvist, & G. Dahlberg (Eds.), *Governing the child in the new millenium* (pp.119–143). New York: Routledge.

Foucault, M. (1982). The subject and power: Afterword by Michel Foucault. In H. L. Dreyfus & P. L. Rabinow (Eds.), *Beyond structuralism and hermeneutics* (pp. 208–226). Hertfortshire: The Harvester Press Limited.

———(1991). What is Enlightenment? In P. L. Rabinow (Ed.), *The Foucault reader* (pp. 32–51). London: Penguin.

Grosz, E. (1995). *Space, time and perversion. Essays on the politics of bodies.* New York: Routledge.

Hultqvist, K., & Dahlberg, G. (Eds.). (2001). *Governing the child in the new millenium.* New York: Routledge.

Kemp, P. (1992). *Levinas.* Gothenburg: Daidalos

Lather, P. (1991). *Getting smart: Feminist research and pedagogy with/in the postmodern.* New York: Routledge.

Lenz Taguchi, H. (2000) "Doing Reggio?" No, doing difference in co-operative learning: Some aspects of the Reggio Emilia approach in the Stockholm project. *Forum, 42* (3), 100–102.

Levinas, E. (1969). *Totality and Infinity.* Pittsburgh: Duquesne University Press.

Lyotard, J.-F. (1984). *The postmodern condition: A report on knowledge.* Minneapolis, MN: University of Minneapolis Press.

Malaguzzi, L. (1993, May). Speech by Loris Malaguzzi, Reggio Emilia, Italy.

Popkewitz, T. S., Franklin, B. M., & Pereyra, M. (Eds.). (2000). *Cultural history and critical studies of education: Critical essays on knowledge and schooling.* New York: Routledge.

Popkewitz, T. S., & Bloch, M. (2001). Administering freedom. A history of the present. Rescuing the parent to rescue the child for society. In K. Hultqvist, & G.

Dahlberg, (Eds), *Governing the child in the new millenium* (pp. 85–119). New York: Routledge.

Project Zero/ Reggio Children. (2002). *Making learning visible: Children as individual and group learners.* Reggio Emilia, Italy: Reggio Children.

Rancière, J. (1991). *Five Lessons in intellectual emancipation.* Stanford, CA: Stanford University Press.

Readings, B. (1996). *The university in ruins.* Cambridge, MA: Harvard University Press.

Rinaldi, C. (1994, June). Seminar on documentation. Reggio Emilia, Italy.

———(1997). Seminar at the Stockholm Institute of Education. Stockholm: Sweden. Rose, N. (1999). *Powers of freedom. Reframing political thought.* Cambridge: Cambridge University Press.

Smith, D.W. (1997). *"A life of pure immanence": Deleuze's "Critique et Clinique" project. Introduction to Deleuze: Essays critical and clinical.* Minneapolis: University of Minnesota Press.

Taylor, M. C. (1987). *Alterity.* Chicago: The University of Chicago Press.

Walkerdine, V. (1984). Developmental psychology and the child-centered pedagogy: The insertion of Piaget into early childhood education. In J. Henriques, W. Holloway, C. Irwin, C. Venn, & V. Walkerdine (Eds.), *Changing the subject: Psychology, social regulation and subjectivity.* London: Methuen.

CHAPTER TWELVE

HEAR YE! HEAR YE!

LANGUAGE, DEAF EDUCATION, AND THE GOVERNANCE OF THE CHILD IN HISTORICAL PERSPECTIVE

Bernadette Baker

> The language I have learn'd these forty years,
> My native English, now I must forego:
> And now my tongue's use is to me no more
> Than an unstringed viol or a harp,
> like a cunning instrument cased up,
> Or, being open, put into his hands
> That knows no touch to tune the harmony:
> Within my mouth you have engaol'd my tongue,
> Doubly portcullis'd with my teeth and lips;
> And dull unfeeling barren ignorance
> Is made my gaoler to attend on me.
> I am too old to fawn upon a nurse,
> Too far in years to be a pupil now:
> What is thy sentence then but speechless death,
> Which robs my tongue from breathing native breath?
>
> —Thomas Mowbray commenting upon his exile from England
> in William Shakespeare's *Richard II,* Act 1, Scene III.

> Deafness is less about audiology than it is about epistemology.
>
> —Owen Wrigley, *The Politics of Deafness,* 1996: 1

Contemporary references to the welfare state often assume such a construct to be discrete and identifiable. A welfare state is thought most obvious in certain bureaucratic structures, ethical commitments, and policies that undertake to protect and administer a population or segment of the population considered vulnerable, such as children. Linked to this naturalized vision of welfare is a series of images of what a caring nation is, what constitutes a language that would convey and execute such care, and what the ideal child should be in order to be administered and protected as such; for example, through publicly funded systems of schooling.

This chapter interrupts such common senses of welfare as imagined within the construct of a nation. It does so by historicizing how discourses of governance have produced rather than revealed their objects, including the objects of the nation, welfare, a language, and the child. The historicization takes as its window onto these object-productions, the motor that the constitution of deafness, and then Deafness[1], has provided to their figuration. That is, the historical analysis is one that problematizes the naturalness sometimes attributed to particular topics, such as what welfare, the nation, a language, or the child refer to by considering how such constructs were acted on and acted back on the idea of deafness, the deaf child, and deaf education.

Mitchell Dean & Barry Hindess (1998) note four main ways that such "governmental assemablages" or "discourses of governance" can be approached in historical documents: (1) through studying the problematization of normativities; (2) through analyzing the formation of identities, capacities, and statuses of members of populations; (3) through tracing the (self)shaping and reshaping of the identities of those whose conduct is to be governed; (4) through considering the ethos of governmental practices (i.e., Are there contradictory "senses of ethics" inhabiting particular constructs being produced *through* the act of governance?).

The analysis mobilizes this approach by comparing modern texts on deaf education in a period of occidental expansion in which deaf children were historically "discovered," reconstituted, regulated as objects, and subjected to unique pedagogical experiences (roughly 1600–2000). The associated documentary focus is on English-language publications emanating from England, the United States, and Australia that comment on deaf ears, deafness, deaf education, and Deaf awareness. Their recognition as English-language and as nationally located can be understood as part of the problem brought into focus through the analysis.

Debates entered into in English, American, and Australian publications over this period are read, then, through the above four intertwined strategies that enable consideration of how ideas about welfare, the nation, language, and the child are produced through and as part of discourses of governance. The wider concern in the chapter is, therefore, not on what various nations have "done to" children who were initially ignored in early "vernacular" literature, then later labeled as deaf, and now as Deaf. Rather, what has the shifting construct and governance of deafness and "the deaf child" permitted in regard to the imagining of communities now called nations, systems now called languages, and senses of ethical responsibility now thought of as welfare?

Historicizing Welfare, Language, Nations, and the Child

Histories of nation-formation, nationalized systems of schooling, or of "the West" in general have not considered the constitutive play of inscriptions of deafness in identifying such things as "entities" in the first place. Yet in so much late-twentieth-century scholarship, implicit notions of deafness, amid binary notions of speech and hearing, often provided parameters for the theoretical insights. In *Of Grammatology,* for example, Jacques Derrida (1997) argued that a Western metaphysics of presence hinged on the way in which one heard one's self speak, a form of phonocentrism that became logocentrism. A link was forged between what was heard and what was known, and this order of things became implicit to the assertion of presence. In *Can the Subaltern Speak?* Gayatri Spivak (1995) questioned the extent to which an autonomous voice could operate in the aftermath of imperialist invasion. If the colonized "spoke back" to the colonizer in the belief of authentic resistance, were they not doing so a little naively insofar as the colonized must use a language that the colonizer can hear? In *Imagined Communitites: On the Origins and Spread of Nationalism,* Benedict Anderson (1991) asserted that it was in part around the object that speaking became that imagined communities called nations were eventually forged. The esotericization of Latin, the Reformation, and the haphazard development of administrative vernaculars made new communities imaginable.

In all three cases—deconstructivist literary criticism, postcolonial theory, and historical sociology—speech and hearing have been constituted as obvious analytical objects intimately bound to the theorization of "social groups." There is a sense in which such preoccupation with speech and hearing has both enabled striking insights regarding the theorization of "social groups" or "subjectivities" while blocking others. Such a preoccupation with speech and hearing can be both extended and interrupted by considering some historical winnows and sways that have made it possible. A historical analysis of how the constitution of deafness was imbricated in the production of national imaginaries, languages, and senses of child welfare administered through the production of govern-mental (not Government-al) assemblages is therefore one such strategy that the following sections take up.

Out of the Way? Fantasies of Deafness[2] in Early Modern "England"

> In the relentlessly aural world of early modern England—a world in which people regularly enjoyed listening to four-hour sermons—civilization was only possible through *spoken* language. (Nelson & Berens, 1997, p. 52)

In their analysis of literature that is now called English poetry, plays, pamphlets, and proclamations across the seventeenth century, Nelson & Berens (1997) state that they cannot find documents that contain "fully realized representations of deaf people." They point out that "while there are almost no representations of deaf people, there are abundant representations of deaf ears" (Nelson & Berens, 1997, p. 53).

The earlier seventeenth-century representations of deaf ears take two general and related forms. In the first form, deaf ears are a voluntary, temporary, and advantageous thing. Deaf ears appear in situations in which a god, ruler, or social superior temporarily turns off his sense of hearing in the face of an anxious speaker. In the second form of representation, deaf ears are used to constitute a perilious act of audition: the anxiety shifts from the person unsuccessfully addressing temporarily deaf ears to a "hearing person" striving to escape the world's pestilent voices (Nelson & Berens, 1997, p. 53).

The images of deaf ears are highly selective appropriations that cannot at that time construct central characters as *permanently* deaf *and* moral. Deaf ears are simultaneously metaphorized and reduced in a process that is played out in an array of literature that inscribes what it means to be moral against a backdrop of the ability to hear and speak the Word of God (Nelson & Berens, 1997).

Nelson & Berens note further the central role that a binary of speech and hearing played in narratives on morality in the early seventeenth century. For example, in Thomas Wilson's *The Arte of Rhetorique* importance is attributed to a quality called eloquence. Wilson describes eloquence as the constituting force of humanity and suggests that it is the sweetness of the speaker's utterance that will draw in the rude and ignorant, presupposing the importance of an audience who can hear (Nelson & Berens, 1997, p. 52).

The two dominant forms of representing deaf ears are not sustained across the seventeenth century: the "virtual absence of the physically deaf ends in the eighteenth century. Literary and cultural representations generally begin to occur with the realization—largely during the mid–seventeenth century—that the deaf are actually educable in English, the majority language of their country, and as a result, become 'representable' within that majority language" (Nelson & Berens, 1997, p. 53). In his study of "how Europe became deaf in the eighteenth century," Lennard Davis (1997, p. 111) concurs: "Before the late seventeenth and eighteenth centuries, the deaf were not constructed as a group. There is almost no historical or literary record of the deaf as such. We may rarely read of a deaf person, but there is no significant discourse constructed around deafness."

By the 1700s, this was not the case: "deafness was for the eighteenth century an area of cultural fascination and a compelling focus for philosophical reflection" (Davis, 1995). Nelson and Berens describe the shift in representations of those whom they refer to as the physically deaf as embodying two trajectories. The first entails a numerical difference: the previous relative absence of the physically deaf in literature can be noted because from the mid-1600s onwards there are more frequent representations and discussions. The second aspect of the shift entails a substantive difference: images change from temporarily deaf ears in "fully human," even divine, characters to later commentary that comports those assigned the label deaf and dumb as "having" a permanent, but trainable, "condition" indicative of a "lesser" humanity.

This shift was accompanied by a refiguring of imagined communities. Both Nelson and Berens and Benedict Anderson note the extent to which the invention of the printing press redefined imagined communities. For Anderson, the vagaries involved in the assertion of English as an administrative vernacular and then popular reading language are emphasized not in regard to the constitution of deafness but in regard to the possibilities for imagining the nation as a discrete object of love and service. The erosion of the sacred imagined community takes place in part through the effort to address the fatality thought to inhere in human linguistic diversity (presumed spoken). For Anderson, administrative vernaculars predated both print and religious upheaval of the sixteenth century, so they "must therefore be regarded (at least initially) as an independent factor" in the conditions of possibility for nationalism and national languages (Anderson, 1991, p. 41).

The inscription of deafness-as-a-condition in relation to the availability of print media informs the later seventeenth-century discussion of how to "deal with" a trait signaling "deficiency." Just as "savages" in the antipodes were considered savages for not speaking English, so too were children labeled deaf and dumb in England problematized in relation to what was taken as language, but doubly so, for they were not already part of some other "hearing culture." Silence is only comfortable when it is a refuge and it is only silence in relation to being able to hear: "When silence and deafness are treated as actual physical conditions in mainstream literature, these conditions are viewed as pathological and as needing to be cured if at all possible: for the English, this cure is what deaf education means" (Nelson & Berens, 1997: 66).

The works of John Wilkins (1641), e.g., *Mercury;* Sir Kenelm Digby (1644), e.g., *Of Bodies and of Mans Souls;* and John Bulwer (1648, 1707), e.g., *Philocophus: Or, the Deafe and Dumbe Mans Friend* are the first proposals to teach children labeled deaf to write and read English in order to make up for a perceived lack and to "fit in." The publications are themselves signs of a preoccupation with depicting England as monolingual—an image of homogeneity that Anderson (1991) notes was never the case. For Bulwer, the signs that he notices children labeled deaf using among themselves are gestures that, while fascinating, are ultimately to be in service to the tongue. Like Shakespeare's Mowbray, native Englishness cannot be distinguished from how the tongue is envisioned as a harp. In Bulwer, the tongue is the final goal of a "progression" from signs to speech and "lip grammar."

Nelson & Berens (1997, p. 69) refer to such lines of reasoning as a form of *linguistic colonialism.* This might be understood as a form of colonialism that operates from within to secure the impression of monolinguality and cultural solidarity in the midst of multiple forms of expressions. Not only are signs irreducible to speech and not at all derived from speech, but the proposed pedagogy in Bulwer's Utopic school insists on the imposition of a foreign language onto others as though English is a "priceless cargo" for redemption of the savage. This is a catch–22.

> Against this background, the Deaf are not fully accepted as they are, with a language of their own: only with the possibility of writing the English language onto them, of linguistically colonizing the Deaf, is there recognition in mainstream forms of literature . . . we can only "know" them through these writings (beginning in the mid-seventeenth-century), yet writing about the Deaf and educating them through the English language "shapes" them for consumption by a hearing audience. (Nelson & Berens, 1997, p. 69)

It should be noted that the cure that early texts on deaf education supposed themselves to provide was by no means considered a scientific one at the time and that the constitution of deafness as deficit and condition did not suggest medical responses. Rather, the emphasis on reading and writing English in these earlier tracts indexes a crossover point from overtly theological and moralistic to ethnic and anthropological frames of reference in the constitution of both deafness and nationhood around administrative "vernaculars." The ability "granted" to the "deaf and dumb," that is, the "realization" that a child labeled deaf can learn religion and ensure salvation through having some access to the Word was not the emergence of a new sympathy for a discrete Deaf subjectivity or distinctive populational group. It was the *constitution* and *consumption* of deaf*ness*, which Nelson and Berens note, was integral to assertions of supremacy for spoken and written forms of expression such as English and to the definition of those forms *as* languages or vernaculars.

Across the 1600s to the 1800s, the service that the constitution of deafness provided to emergent national imaginaries, the recognition of something as a language, and the ethical sense of redeeming "the deaf child" as a form of welfare thus lay in relation to how governmental assemblages were being reorganized and given new boundaries and rationales. The shift from seemingly few mentions of temporarily deaf ears in divine and clever characters to a sustained discourse on permanent deafness-as-a-condition of lesser humanity is a complicated inflection of a salvation discourse. The move from deaf ears into deafness is discernible in the pedagogic efforts of a charitable nature; a redemptive effort to transform the condition of deafness into something that can depart from the baseline of humanity which it helps to establish. In the wake of wider availability of print media, the constitution of deafness as a permanent and deficient condition in need of and open to a linguistic and religious cure upholds a scale that in turn admits "the deaf child" to a trope and hope of some (limited) improvement via education into English. In ways that are homologous to Michel Foucault's (1961) analysis of the history of madness, permanently deaf ears at first lie outside major comment, outside knowing, appearing in their absence as an elusive Derridean or Levinasian Other that cannot be analyzed, objectified, or subjected to reason and investigation. In the "inclusive" move toward participation in theological redemption, deaf ears become deaf*ness*—a permanent, trainable, investigable, and objectified condition subject to commentary, analysis, and pedagogy. It is not an overestimation to assert that in this period "the deaf child" had been "discovered" (created) as an object of salvific educational discourse amid new ethnological norms for character development, nation-formation, and vernacular elevation.

On the Way? Phantasms of Deafness *for* England

> *Herkunft* is the equivalent of stock or *descent;* it is the ancient affiliation to a group, sustained by the bonds of blood, tradition, or social status. But the traits it attempts to identify are not the exclusive generic characteristics of an individual, a sentiment, or an idea, which permit us to qualify them as "Greek" or "English"; rather it seeks the subtle, singular, and subindividual marks that might intersect in them to form a network that is difficult to unravel. (Foucault, 1998, p. 373)

The constitutive role that representations of deafness-as-a-condition, and especially as a permanent one, played in the organization of national imaginaries and administrative vernaculars changes by the mid- to late nineteenth century. The change becomes evident in the publication of textbooks for teachers, teacher trainers, and children, in the establishment of *publicly funded* schools for children labeled deaf, and in research into deafness as an unquestioned audiological topic (see Seashore, 1919).

The manualism (Signing) versus oralism (lipreading and mouthing) debates had already opened by the century's end and they could not be disarticulated from the re-imagining of nationhood around evolutionary and eugenic tropes for a stronger, purer citizenry. The quest for strength and purity had to acknowledge and indeed required recognition of the *proliferation* of communities, the upsurge in the formation of many social groups now attempting to distinguish themselves as nations. The pluralized background that other nations provide does not quite do the work of making clear what "being English" means, though, especially in relation to the child labeled deaf and dumb. In 1872, for example, Thomas Arnold had to work hard at determining what *English* deaf education actually meant, to the extent that he wrote a text for teachers and teacher trainers on the topic.

Arnold's *The Education of the Deaf and Dumb: An Exposition and Review of the French and German Systems* was published in order to ascertain which system, manualism (French) or oralism (German), ought to be employed for the education of schoolchildren who had been assigned the label deaf and dumb. The link between the constitution of deafness-as-treatable-condition and the nation-as-object-of-love for which conditions ought to be treated could not have been clearer, but in the midst of clarifying what was specifically English and "best for England," an "ambiguous and cross-corporeal cohabitation"[3] of "English" deaf education by "French" and "German" systems was required to filter out and in what would newly constitute "local" forms of pedagogical practice.

The French system as identified and described by Arnold is introduced in order to convey what is not truly English and so inhabits it as a foil.

> This system is chiefly distinguished by the use of signs though which, as a kind of intermediate tongue, the meaning of the language in general use is communicated. Signs are therefore substitutes. We all use signs. The kiss blown to a child, or the finger placed on the lips to secure silence. As a nation, however, we are sparing of them. Our continental neighbours, especially the French, indulge in them more freely. Their whole body seems to speak. (Arnold, 1872, pp. 4–5)

For Arnold, the use of signs lies outside what is designated as the English nation and further constitute something else as a language. Signs are intermediary in a normative series of expressions. They are really pictures and resemble objects. In order to keep itself as a reference point, the assertion of English *as a language* requires other foils, then, besides the expressiveness of French neighbors. Presumed distinctions between art and literature, between the abstract and the concrete, between the metaphysical and the material, between the body and the mind provide the scaffolding for determining what "language" and "higher order understanding" are: "But by what signs shall we express abstractions, purely mental states, operations, and intuitions? As none of these can be reduced to a material form it is impossible to figure them by signs. Of course, the more complex the term, the more difficult to express. The processes of the understanding cannot be described on the fingers" (Arnold, 1872, p. 6).

The French system was thus described as philanthropical but not adequate for communication or homogenization beyond specific institutions: "Each follows his fancy, so that the pupils of different institutions cannot understand each other by signs" and "Signs are deficient in number as in expression." The use of signs "is therefore wanting in all the *vallidae juncturae* and *nuances* to which composition is not unfrequently indebted for its force and perspecuity" (Arnold, 1872, p. 7).

The French system was also making a great mistake in Arnold's view by not basing signs on the French language. Signing had inverted word orders relative to the French, e.g., *a very good man* becomes *man good very a* in sign. This is considered not natural: "Signs are only the merest rudiments of a language, and at best ancillary to the acquisition of a complete language; and should not their order have been at once sacrificed to that of the French language?" (Arnold, 1872, p. 8). It is also considered unethical, uncaring, and not welfare-oriented.

> Is this how we should treat an aboriginal Australian child who had come to learn our language and make his home with us? His own tongue might be capable of the highest development as an instrument of thought, but that would not influence us in teaching him English. The construction of his would have to yield to ours, and he would learn to speak, write, and think in English, as if he had never known a word of his mother tongue. Were we even to leave him to himself, he would speedily do the principal part amongst his playmates or the servants. Teachers of modern languages know the value of this principle, and therefore as speedily as possible accustom their students to the construction of the new language by making it take the place of the old. (Arnold, 1872, p. 9)

The argument for the German system and against the French thus plays out in the linguistic colonialism that Nelson & Berens (1997) note. If the "deaf and dumb" continued in the sign system, they would "continue to think in the order of the signs, so that when they came to write or spell, they would feel like foreigners imperfectly acquainted with our language." The sense of belonging to Englishness is now being forged through how a student would write or spell as viewed by someone "hearing" (the "our") and not just in relation to interpreting a biblical passage.

Arnold argues that all of the difficulties attributed to the French method are, however, "either obviated or never encountered in the oral method" (Arnold, 1872, p. 12). So successful did he consider the system that some students he observed spoke "so clearly with tone, cadence, and accent, both French and German, that had you not been prepared you could not have suspected their deafness" (Arnold, 1872, p. 14). It was not successful without strategy, however. While the signing hands of children labeled deaf and dumb in England were far too French in the emphatic use of the body for Arnold's taste, the different kinds of strategies that he notes on the part of students who refused to use the German system indicates the limits of imposed change. The shift, from representations of deaf ears to deaf*ness*-as-a-need-for-oralism could not completely subjugate alternative ways of taking up assigned subject positions, and thus could not achieve the very homogeneity desired by reformers. Arnold concludes, however, by listing the eight reasons why "the German system is superior to the French for the education of the deaf and dumb," the first and presumably most important being that "It is superior because it relates them to language like ourselves" (Arnold, 1872, pp. 15–16).

Other educators must have agreed with Arnold because further publications followed similar lines of reasoning. "Mental culture" was presumed to be formed through a restricted avenue—exposure to a "national" language in spoken and then written configuration. Thus, a German textbook written for children was translated from the German in 1884, providing a model for deaf education in England. The book was called *Object Lessons on Object Pictures Forming an Elementary Reading and Language Book for Deaf Children for use with the Coloured Illustrations.* It was written by M. Hill, director of the *Institute for Deaf Mutes* at Weissenfells and described its purpose in the following way: "The aim of this book is manifold, but principally it is so constituted as to meet the necessarily low standard of elementary education acquired by Deaf-mutes."

Added to the book's narrative on topics such as "the garden," "the bird," and so on are pictures to match, but "the pupil must not be allowed to dwell upon the picture alone, but his attention must be directed to similar objects and circumstances in his own surroundings; in other words he is to understand the *living* world in which he finds himself; and to a proper understanding of which, the picture is only to be used as a help" (Hill, 1884, p. 111). As a test of this proper understanding, false readings have been included: "In order to avoid the thoughtless reading, which is always the danger with Deaf-mutes, and which cannot be too carefully guarded against, I have put purposely false readings, to which I would particularly call attention, as they will prove to what extent the dead letter has been animated, and will at the same time show the progress made not only in reading and speaking but also in understanding" (Hill, 1884, p. ix).

The most remarkable thing following the opening reminders to check for life, however, is that the textbook provides constant reminders of and acts as preparation for *death.* Positioned as liminal, but not-dead-yet students, the children labeled deaf-mutes constituted as the target are nonetheless in a unique relationship with death that changes the expectation of the teacher, as espoused in the book's purpose. Death is treated under so many topics in the book that the early prompts in the preface to remind the child of the animated living are mitigated against by the remainder of the content, which often instructs the child about the inanimate and dead: "The worms eat the flesh; All human beings must die"; "when I am dead my pulse beats no more"; "He [the wolf] will tear the sheep to pieces and eat it"; "Now the boy screams for he has shot his sister dead"; "Many people die when they are ill"; "Perhaps the soldier will be wounded or killed."

In a textbook written for children, deafness is presumed to preclude an understanding of life, of *living* language, of animation and action. Being called deaf is portrayed as a flat, barren, and listless canvas, verging on an arid zone that challenges existence itself. Deafness is drawn into alliance with death at the most minute pedagogical level, i.e., in both the content and in how a teacher would interact one-on-one with a student. As described above, the teacher is required to interpret a child's activity in learning to read by deceiving the child with false readings to see if the child's mind is alive in a "true" or genuine way. In the very classroom methods employed, a link is forged between spoken and written expressions, the formation of "mental" concepts, and national "evolution" of Life. This circuit would be subjected to forms of monitoring and testing that would protect against the fatalism not just presumed inherent in "the Deaf-mute," but in a wider linguistic diversity itself.

The limits of a child's humanity and educability are communicated through other foils besides the threat of death and the "lost time" of childhood. In the "German"-authored book multiple discourses circulate to resecure barriers between organisms; a child must not act like a monkey, for a monkey is sly; a child labeled deaf must recognize their freedom over a canary bird trapped in a cage; and importantly a child labeled deaf must recognize themselves in relation to other "defectives." For example:

> I, MYSELF: . . . I have 2 ears but I cannot hear. I am deaf. Since when are you deaf? The good God has given me the other four senses; seeing, tasting, smelling, feeling. Only the sense of hearing is wanting in me. Therefore what cannot you do? Now I am in the deaf and dumb school. . . . In school I learn to speak. In the past I was dumb. I was deaf and dumb; now I am only deaf not dumb, for now I can speak. (Hill, 1884, p. 218)

What had these limits of educability and humanity to do specifically with the demarcation of Englishness? In the quote above, the imposition of English language acts as the springboard to "promotion" into humanity, educability, and subcitizenship—deaf, but no longer dumb; human and linguistic, not animal and unspoken. English and England are thus mapped via the location of deafness in a series or chain of Being that establishes the grounds for sympathy and welfare for "deaf children." That sympathy is again taken up as a moral position amid the reformation of the nation's image around an administrative vernacular, but unlike Bulwer's commentary in the mid-1600s there is much greater direct focus on the "problem" of plurality, of other neighbors in the general vicinity.

What is "truly" English in the process of deaf education remains seemingly impossible to discern, however. The "German" textbook and "German" methods of oralism are used and the same pronouncements circulate in the Arnold text. It is *only* at the level of spoken and written expression, then, that England can define itself as England in such textbooks, submerging any, in Foucault's terms, "subindividual marks" in pedagogical processes, content, and systems of reasoning drawn from "elsewhere" that might complicate what could be defined as the local.

The argument over methods of deaf education thus suggest a new form of governance, not just of children labeled deaf and their teachers, but of the images that are being organized around something often called the "welfare state." They are radical arguments within the series or chain of Being at the time because they assume that the children labeled deaf are definitely trainable if not educable (i.e., able to learn written English and forms of oralism) and that now it's just a matter of how to best do it. In Arnold's book, too, they are radical arguments because they assume that children labeled deaf should be treated in some ways as other human beings are, i.e., with the same expectations and similar standards for acquired content and because he demands monies for schools for children labeled deaf just as other children have them. In this version of sympathy, then, is a new role for deafness in the constitution of nationhood: it is not that the "national system of education" has awoken to a humanitarian realization to now teach "the deaf child" who was made an object of religious pedagogy in the mid-1600s, but that "the deaf child" constitutes, in a reworked way, an object of secular administration against which the limits of "regular" education, Englishness, literacy, citizenship, and the very meaning of the term "language" can now be drawn in *measurable* form.

The above indicates how by the end of the nineteenth century, deafness is called upon in at least five ways to legitimate a sense of security in the fragile and shifting notions of nation, language, and welfare: (1) by legitimating the eugenic contouring of "children's capacities" as a form of welfare, (2) in the projection of a national discreteness against a pluralized background, (3) in the study of child mental-life as gaurantor against death and degeneracy, (4) in the elevation of infant plasticity and identity-formation as objects to be managed early, and (5) in defining the limits of humanity around the senses and around the sense of national-linguistic belonging.

Amid this discourse network, problems of forging/sustaining a national imaginary in the late nineteenth century became problems of how to secure a paradoxical sense of *shared uniqueness.* Despite every effort to cleanse and comport "foreigners" at home, the reference points that structured deaf education textbooks in the late nineteenth century suggested nonetheless that a curious and ambiguous cross-corporeal cohabitation occurred to muddy the purity and distinctiveness so desired—for what was *really* English beyond the assertion of English as a "language" and what was *really* a "language" beyond the assertion that the signing hands of a child was not? A perceived uniqueness had to be projected and spread "elsewhere" without losing its distinctiveness; but that elsewhere, rather than simply being "in the colonies" or directed at "hearing" immigrants, was constituted as the barren, unmarked, unrecognizable zones of a deafness born within.

In the Way? Seeing the Signs in Another Reconstitution of Deafness

The blank canvas of collapsed national-linguistic distinctions written onto/as deafness at the turn of the twentieth century was conjured as an uninhabited territory that someone "with the power of hearing" needed to mark. By the turn of the twenty-first century, the logical extension of and reversal within the existing systems of reasoning was achieved; it had become possible to assert that the Deaf (capital "D") constitute a transnational nation of approximately 15 million people (Wrigley, 1996).

> The global Deaf population is currently about fifteen million—on par with a modest-sized nation. Yet it is a "country" without a "place" of its own. It is a citizenry without a geographical origin . . . . Without a claim to a specific place, and without the juridical and policing agencies by which we know nations in the late twentieth century, deafness is not a recognized nation. In keeping with the medical model of the body inherited from the nineteenth century, deafness is commonly viewed as merely a "condition." But the claim of a distinct ethnic identity that has accompanied the resurgence of Deaf Awareness in the past two decades forces a reassessment of this and other identities excluded from the equation of the "normal." (Wrigley, pp. 13–14)

The conditions of possibility for asserting a placeless Deaf nation owe much to the prior formation of identity politics as "the modern form of political complaint" (Eghigian, 2000). In the "deviancies" that are produced through efforts toward normalization emerge the presumptions of commonality indicative of identity politics. Greg Eghigian argues that welfare state disability categories, formalized first under Bismarck in the Federal Republic of Germany and subsequently taken up in bureaucracies in other locales, most immediately the United States and Japan, produced entitlement mentalities. Eghighian's analysis is not dedicated to disparaging identity politics but to historicizing its emergence. He argues that the entitlement mentalities that systems of social security helped to create enabled the formation of identity politics. The categories and mechanisms of welfare, social insurance, and grievance subsequently became the main and sometimes the only way in which to address political complaints.

In the emergence of a solidified identity politics around deafness, the welfare of the child, the boundaries of the nation, and the limits of a language are reconfigured. Small "d" deafness becomes capital "D" Deafness, a different kind of inside-outside motor for the constitution of imaginaries.

Eghigian's historical analysis of German social insurance and disability policy thus has implications for places that introduced similar mechanisms. In the United States, movements called Deaf Awareness and Deaf Culturalism (both overlapping and divergent in terms of internal politics) emerged in the second half of the twentieth century. Deaf Awareness and Deaf Culturalism have brought into question the kinds of sympathy meant to be imparted to those designated as "less fortunate" or "the disabled." They have also engaged in the kinds of reversals that identity politics structures. Any simple equation of Deafness with disability, for example, has been contested. To align Deafness and disability means in some of the literature to reinforce the audist establishment, hearing-centeredness, and ableism (Brueggeman, 1999; Lane, 1997). Who "said," for instance, that "hearing" is actually an *ability?* Is better? How would *you* know? What or who is your reference point for this assumption? The disarticulation of Deafness from disability is one of the most controversial assertions issued through Deaf Awareness, in part because it could have policy, funding, legal protection, and affirmative action consequences (Lane, 1997; Wrigley, 1996).

In addition, debates have opened beyond Deaf Awareness and Deaf Culturalism in Disability Studies more generally regarding whether persons who are assigned disability labels or others who proudly take up and reinvent historically pejorative labels, e.g., Crip, Gimp, Deaf, constitute a discrete community, culture, or minority group (Rogers & Swadener, 2001). Further arguments have taken place over the definitions of these terms, e.g., Does the term minority group refer only to concepts of ethnicity or race? (Longmore & Umanksy, 2001)

New terminology has also arisen in regard to dissatisfaction with previous forms of expression. The shift from *defective* to *handicap* to *disability* to *differently abled* to *inclusive* to *Disablity* as descriptors and/or policy categories has been met by refusal to be defined in relation to any of the above, as if one "trait" essentializes all else, as if the notion of "ability" should be the lens for defining personhood, and as if "disability" exists "in" someone rather than in social relations (Danforth & Rhodes, 1997).

Contra the effort to rethink such normalizing binaries are reappropriations of nomenclature that reinvoke them in new form. As noted above, this has occurred within Deaf Awareness and Deaf Culturalism but is not a phenomenon exclusive to only one branch of identity politics. Deaf with a capital "D" can refer to several phenomena, including Deaf Culturalism, Deaf Awareness, and *the* Deaf as persons who identify as participating in a discrete linguistic community and who are proud to be Deaf (Lane, 1997; Longmore & Umanksy, 2001; Wrigley, 1996). In some of the literature in Disability Studies, Deaf Awareness and Deaf Culturalism, small "d" deaf refers to what the medical profession calls an auditory condition (Baynton, 2001), while capital "D" Deaf marks the move away from naturalizing accounts of auditory conditions into the recognition of a minority linguistic group and/or Deafness as more an epistemological construct (Lane, 1997; Wrigley, 1996). Within some arguments, this move is further marked by undermining what would be seen in medical frameworks as a technological breakthrough. The cochlear implant, for example, has been interpreted as the latest manifestation of efforts to subvert Deaf Culture, while humor has been developed to indicate the presumptions of what Campbell (2000) calls "ableist normativity." For instance,

> A married couple's conversation about the discovery of "deaf genes":
> Husband: Have you read this article?
> Wife: Yes, it is groundbreaking.
> Husband: Now we can have babies without worrying.
> Wife: That's right. With a little bit more progress of science and technology, we can avoid hearing babies. (Campbell, 2000, p. 307)

The above logical extensions of thinking and governing in terms of populational reasoning and the reversals that they invite have addressed not just the naming of someone but, circularly, the naming of something as a language; a deliberate challenge to and rethinking of the rules for articulation at the same time that they are being used. Challenges have been issued to linguists and language teachers in particular. The presumption that only spoken and written expressions constitute a language has been complemented by campaigns to have Sign languages recognized *as* languages and not as "rudiments" of a complete language as Arnold described them. Campaigns targeting the Modern Language Association have been met by scholarship that argues that *both* manualism and oralism are really just ways of colonizing the Deaf, calling into question the privileged place of Sign in Deaf Awareness and destabilizing the neat binaries on which critique, reversals, and humor often turn (Edwards, 2001).

The fact that such debates are often disseminated in English-language literature and in the United States might suggest something about the legacy of oralist deaf education, of colonialism, and of the "American" context of identity politics. The attribution to oralism's dominance, to English imperialism, and to the "fractured" nature of American society essentializes and simplifies, however, more complex problematizations and subtle, subindividual intersections that intervene in such smooth causal accounts.

In recent scholarship across disciplines, for example, new patterns of governance in relation to "What is a nation?," "What is a language?," and "What is a child?" have been discussed, especially under the term globalization. While new strategies of governmentality have been noted, constructions of Deafness and "the Deaf child" are again being marshalled to reconstitute and reinvigorate debate over wider imagined communities.

In the English-language dominated context of Australia, for instance, the premises of Deaf Awareness and Deaf Culturalism have been taken up in the kinds of research questions now being formulated in education: What do teachers of NESB children (non-English-speaking background) and LOTE (languages other than English) teachers assume as the first language of a Deaf immigrant child attending a mainstreamed school in Australia? This is a question investigated by Branson & Miller (1998).

Branson & Miller's question follows from a historical position: that schools are meant to induct children into a language that signifies a "nation"—a nation that, as for England, has never been truly monolingual. The question also follows from a Deaf Culturalist position, however, that Sign language is to be attributed the status of a language and, therefore, can be considered a child's first language.

The title of Branson & Miller's (1998) chapter speaks volumes about speaking and nationhood: "Achieving Human Rights: Educating Deaf Immigrant Students from non-English Speaking Background Families in Australia." There is a proliferation of adjectives here that captures the one-hundred-year difference between textbooks written simply for "the deaf and dumb" and new formulations of Deaf education at the turn of the twenty-first century. Whatever is considered local or specific (e.g., children from NESB families in Australia) indicates the splintering and regathering action of identity politics in defining "the local."

Branson & Miller argue that questions about what to teach children labeled as NESB and LOTE could only really emerge in regard to changes in Australian immigration law. They inadvertently draw a historical map of how a new governmental assemblage, i.e., "Deaf immigrant NESB-family students in Australia" could have become an object of inquiry.

They juxtapose two eras in immigration policy; pre- and post-1950s. The post-1950 era has seen new kinds of arrivals.

> The 1950s and 1960s brought a greater diversity of ethnic origin among the Australian population as a result of progressive removal of immigration restrictions based on country of origin, race, or color between 1949 and 1973,

> together with the extension of assisted migration schemes to non-British groups and refugees. By 1973, with the election of the Whitlam Labour Government, Australia finally rejected its White Australia policy [established in 1901], declared positive support for a multicultural Australia, and was beginning to recognize itself as part of Asia rather than Europe. (Branson & Miller, 1998, p. 91)

The authors note how immigration policies have in various ways prohibited the entry of the Deaf into Australia. Prohibitions operating through health clauses in the immigration act were and are the main form. The policies have allowed health officers, using medical grounds, to declare the Deaf a potential burden on the public purse. These restrictions may be waived "for humanitarian reasons or for the resettling of political refugees, such as those from Vietnam, East Timor, Lebanon, Burma, the Sudan, and the former Soviet Union. Such waivers have provided avenues for some deaf people to enter Australia" (Branson & Miller, 1998, p. 91). They describe these restrictions as emerging in the wake of a shift from a convict penal settlement into the imagined community of settler nationhood, where discrimination against those later to be labeled "disabled," "deaf," and then "Deaf" became established well before the White Australia policy.

The history narrated provides a segue into considering how new imagined communities could be yielded and how public schools have become a renewed site for debate over methods in Deaf education. With the emergence of Deaf Awareness and Deaf Culturalism, for instance, there has been a return to the manualism that was never fully subjugated to the point of disappearance. In 1984, UNESCO announced that Sign languages, the foremost indicators of manualism, were to be considered the first languages of the Deaf around the world, but this did not provide a formula for "dealing with" the governance of proliferation, of the child, already a welfare category, multiplied by further identity politics.

> Over the last few years, the use of Auslan as a language of instruction in schools and units for deaf students [*sic*] has reemerged. Associated with this trend is a focus on the bilingual education of deaf children through Auslan and English, with Auslan as the primary language of instruction. While the use of sign language [*sic*] in a bilingual classroom is usually seen as the most progressive move in Deaf education nationally and internationally, discussions of these processes tend to ignore a vital cultural and linguistic characteristic of Australian society, a characteristic shared by most societies. Australia is a multicultural, multilingual, immigrant society. The unconscious assumption that there are only two languages involved in dealing with lin-

guistic issues in the education of deaf children(the national spoken language, English, and the national sign language, Auslan—fails to consider the multicultural reality. (Branson & Miller, 1998, p. 88)

The pedagogical solution proposed seeks homogeneity within "the multicultural reality." The "sensitive strategies" of bilingual education for Deaf immigrant NESB children in Australia are proffered as achieving human rights. The pedagogical recommendations are not about reproductive sterilization but about controlling proliferation that the multicultural reality embodies. They are not about prenatal eradication for quality population promotion but about a perfection discourse that is still linked to quality control issues, not through biological discourse, but through a discourse of *democracy* and *human rights.* That is, the effects of power in Foucault's terms have not been disarticulated from the intensification of abilities.

So if a minority language group such as the Deaf is going to realize its human rights in Australia, it must have access both to natural sign language, to English, as well as to the language of the home, if a LOTE. Therefore, if the Deaf from English-speaking backgrounds are to achieve anything like human rights, they must have a bilingual education with sign language and English as a second language. If the NESB Deaf are to achieve these rights, they must have education through the medium of sign language plus access to both English and the home LOTE as second languages. In the case of children from a home background where the language of the home is a sign language other than Auslan (SLOA), then these children will need facilities to ensure that they can be bilingual in the SLOA and Auslan as well as acquiring the LOTE of their ethnic group and English. The agenda is daunting, but then so is the achievement of human rights (Branson & Miller, 1998, p. 97).

The authors argue that existing resources could actually be utilized to teach all children labeled Deaf and NESB two languages in Sign and two languages in written form.[4] The limit, amid the proliferation of categories, is announced in what matters most to maintaining the identity, not of the child, but of the nation: "The first priority must, therefore, be for the provision of Auslan as a medium of instruction as early as possible. Ideally, Auslan should be introduced into the home for children younger than two years and their parents" (p. 99). The rationale is a pro–Deaf Culturalist one, merged with an older version of discourse on the richness of multiculturalism—and the "unspoken" theme of fatalism-in-diversity—the threat of linguistic differences left unreconciled.

This new form of romantic pedagogy, of not seeming to interfere with the "natural language" of the home, yields a new kind of Utopia: "In this ideal world, all deaf children would, by the end of secondary school, be bilingual in two sign languages and bilingual in two written languages" (p. 99). The anticipation of chaos is stablized within a grid-like table provided. The table lays out what languages should be learned when, forging the link between nationhood and language in measurable and trackable form, and in turn making the proliferation of linguistic forms and new governmental assemblages further amenable to administration.

Restructuring Welfare and Governing Patterns of the Child: Serving Identity Politics, Postcolonial Theory, and Education

The above historicization bears out a more general observation made by Henri-Jacques Stiker (1999) in his *A History of Disability.* Stiker argues that between approximately 1600 and late 1900s in the West there was a shift in discourses of governance from charity to welfare. Medieval and early Enlightenment Christian charity in particular was distinguished by being able to situate people as vulnerable and therefore as objects through which others could do good deeds and ensure their own salvation. Welfare discourses, he argues, indicate a new kind of society and are most evident post–World War I. They refer to rehabilitation and inclusion programs that have at base a passion for sameness, normalization, and homogenization that ultimately excludes from within.

Under both discourses of governance (charity and welfare), the (problem" of "difference" is posed, albeit with unique outcomes, as one of the *integrable* rather than the *integral.* Stiker argues that while it is easy to see how systems of naked exclusion operate, it is much more difficult to interrogate how systems of inclusion work to produce subjects such as (the deaf child" as *integrable* but still as a "problem." This obscures how *integral* such subjects have been to the formation of other constructs, for instance in this analysis' case, to inscriptions of the nation, language, the child, and an understanding of the actions that the terms welfare or care refer to or can incorporate.

Thus, the above historical analysis has illustrated this major move in discourses of governance from charity to welfare: from not seeing deafness as an object of "policy" at all, to seeing deaf ears as "situate-able" but exterior and not treatable from "within," to seeing deafness and the deaf child as "situate-able" but amenable to treatment and education from "within" publicly funded institutions, to seeing Deafness as a positive linguistic minority who still need correcting (i.e., nationalizing) as a task of governance. The "problem" of the *integrable* and the *integral* is brought into focus through such specific historical examples, but what has not been considered are some of the effects of this move in the present, of how such shifts can be and are played on to problematize the conditions that made them possible.

It might be argued, for instance, that the shifting historical inscriptions of deafness, "the deaf child," and the Deaf in the period studied here have always been in service. The project of asserting nationality and delimiting something as a language have suggested themselves frequently in the governance of "the deaf child," although in different ways. What is interesting in regard to the latest inflections is the extent to which the available constructions of the Deaf and Deafness have "spoken back" to traditional forms of identity politics, to postcolonial theories of nationhood, and to the idea of literacy education, all of which challenges earlier-twentieth-century notions of welfare and service amid formulae for new ones.

First, it seems no coincidence that Wrigley entitles the opening chapter in his book *The Politics of Deafness* as "Deafness is a Big Country," and that the book inherently links identity politics to language and the concept of humanness. In noting the recent efforts of Deaf Culturalists to assert a linguistic identity, Wrigely has hit the nail on its dual and shifting heads, both embodying and problematizing the modern form of political complaint, identifying inclusion/exclusion not *within* nations, but *as* them, and simultaneously identifying the pathway to acceptance or respect as *becoming* a nation in terms of a "linguistic community." The very concepts historically used to exorcise the Deaf from within in definitions of normalcy, such as nation/citizen, are being redeployed to claim entry of a "foreign" group into a new kind of political philosophical map.

Second, as part of the new version of identity politics, Wrigley engages a striking reversal to make his point that plays on postcolonial theoretical insights. The reversal indicates ways in which Deaf Awareness and Deaf Culturalism can trouble some of the more formulaic postcolonial theories that at times barely distinguish themselves from the binaries of Marxism (i.e., rich/poor; colonizer/colonized; powerful/powerless). In the static kinds of postcolonial theory that play the discrete binary game, the liberal rationalities of governance attributed to the colonizer seek a knowledge of indigenous forms of government, such as kinship networks, to arrive at modes of regulation that work through, and that are themselves adjusted to, the internal modes of regulation of a community or cultural group. But Wrigley's strategy positions the very idea of "the nation" as the community whose internal modes of regulation need to be understood, thereby self-avowedly constructing and positioning "the Deaf community" as an organizing force, a transnational nation that can discern the kinship-linguistic systems of a different cultural group, i.e., the Hearing of various traditional nations. Pushing to the limit what would be recognized as a nation in sociology, then, Wrigley's opening lines open themselves onto a history of how constructions of deafness have occupied the metaphorical place of vestibular stimulation within various assertions of nationhood.

In addition, the insights of Deaf Awareness and Deaf Culturalism have troubled the neat categories usually assumed within postcolonial theory, i.e., how nations can be grouped. In postcolonial analyses the emphasis is often assumed to be on the colonizing nation's effects on victimized and invaded groups and what that gave back to the colonizer—e.g., England is the colonizer and beneficiary of the invasion of places now called India, Jamaica, the United States, Australia, New Zealand, etc. One superficial effect assumed in the present is that, generally, an English-speaking tourist should be able to get by in any of these countries. In pointing out how American Sign language could not be easily understood in India, Jamaica, England, Australia, or New Zealand but has much more in common with French Sign language systems because of the historical connection, Deaf Awareness literature challenges the presumptions of postcolonial theory's a priori groupings. It challenges them not just in terms of what is a nation and which nations can be historically assumed as "related," but in terms of the important processes of colonization that took place "within" rather than just without, that is, in the "diaspora" already at home.

And third, in the available systems of reasoning that now posit the Signing hands of a child labeled Deaf as able to be marked nationally—e.g., through the nationalization of Sign languages such as Auslan—whatever vaunts itself as literacy education has to, paradoxically, be "rewritten" around new understandings of what constitutes welfare. While linguists still argue over what a language is and whether English is one at all, the argument has been historically restricted to exclude any consideration of "non-phonetic" expression. In exposing the reduction of language historically to only certain modes of expression, both the Branson & Miller chapter and the perceptions of Deaf Culturalists more generally have riveted what can count as an "expression," as a system, as poetry, as literate, and as "a human power." This challenge highlights how traditional notions of "the welfare state" have been intrinsically dependent on governmental assemblages produced around restricted definitions of language, capacity, and human powers. The recent depiction of Deafness as epistemological turns on the new routes for forging human relationships that have become available out of and against old welfare policy, announcing new-ish possibilities for the governance of the child, national imaginaries, and their associated ontological investments. Argue Bauman & Dirksen (1997, p. 317) about this twist:

> The theoretical significance of "deafness" takes on new historical and metaphysical importance that pathologized "deafness" cannot. If nonphonetic writing interrupts the primacy of the voice, deafness signifies the consumate moment of disruption. Deafness exiles the voice from the body, from meaning, from being; it sabotages its interiority from within, corrupting the system which has produced the "hearing" idea of the world. Deafness, then, occupies a consummate moment in the deconstruction of Western ontology.

Finally, this analysis of the dynamic role that shifting constitutions of deafness, "the deaf child," and the Deaf have played in national imaginaries, in identifying language, and in senses of welfare indicates the limits of the very processes for building such arguments, of trying to locate history in the least-looked-for places. "National" histories in the governance of the child are difficult to trace in the places they seem most profound, in feeling and sentiment, in deafness and hearing, in anticipation of and nervousness over proliferation, and in the subindividual marks that make the idea of the national and of welfare from different perspectives difficult to disentangle. The refiguring of deafness and the deaf child indicate the extent to which, as Foucault has put it, concepts "continue to have value for us" without necessarily being *continuous.* "Finding" such histories are not heroic moments, not just because they are narrated in one version of English, which would seem to reinvoke the very things being critiqued, but because such histories can actually never be found. That is, if "the Deaf" cannot be constructed and consumed by "the Hearing" without the use of "written languages" and without essentialization occurring across a binary, then isn't this still part of the problem? If historians require "evidence" in the form of *documents* that can at some point be reduced to paper and archived in arguments about governing the child, then doesn't this miss the point of the irreducible, nonphonetic, positive excess that Sign language has now come to signify?

The very act of history as a written or oral practice becomes questionable under the burden of such reasonings. When a discipline cannot embrace the multiplicity of forms of expression that have been historically available, even if it wishes to include, master, codify, and not inevitably exclude from within, one realizes what has served it, what has helped to constitute it as a discipline while being compelled to play within its limits. It is this point of critical complicity and re-entry that

Deaf Awareness and Deaf Culturalism have helped raised awareness of, indicating how new enactments of welfare and depictions of the child are forced to engage with yet interrupt the primacy of language as spoken, written, and national.

Notes

1. I follow the convention cited in Longmore & Umanksy (2001), Davis (1997), Lane (1997), and Wrigley (1996) in the use of terminology in this paper, although that convention generates interesting historiographical/ political quandries that are not necessarily amenable to such conventions. In general, small "d" deaf refers to the dominant medical perspective of deafness as an auditory condition and thus as an adjective often used within schooling e.g., deaf children, deaf education. Capital "D" Deaf refers to the more recent reappropriation of the term through social movements referred to as Deaf Awareness and Deaf Culturalism, i.e., Deaf as belonging to a linguistic minority. For the purposes of this chapter I will be using deaf (small "d") because this is what documenting the historical senses is weighted toward, i.e., moves and moments in the history of governance through which the term deaf, (and later, Deaf) emerged as part of populational reasoning and medicalization in English language texts. In the historical works cited here by Davis, Nelsen and Berens and others, this is also the convention, i.e., small "d" deaf and deafness when referring to an historical emergence. I use throughout as well the notion of "children labeled deaf" as a way of further signaling the relatively recent appeal to the terms children or child and deaf as populational categories of welfare policy. Even when or where labels are agreed to, they are labels, and therefore culturally and historically specific, directing attention to some factors, variables, or things, and not others. Thus, as in other historical works on deafness, the predominant lower-case usage is not intended as a reinscription of the medicalized view but as an historicization of its availability to educators. Also, for the purposes of this chapter I will be using the terms linguistic and language interchangeably, although I realize to do so opens onto debate within the field of linguistics as to whether English is a language. At the level of instruction in grade schools, however, such debates seem already settled.
2. Fantasies of deafness is the subtitle of Nelson & Berens' article, which is discussed in this section.
3. "Identifications belong to the imaginary; they are phantasmic efforts of alignment, loyalty, ambiguous and cross-corporeal cohabitation" (Judith Butler, 1993, p. 105).
4. The play between NESB and Deaf is not taken up by the authors, nor necessarily positioned as a positive resource. Rather, both constitute a newly complicated problem of governance in the chapter.

References

Anderson, B. (1991). *Imagined Communities: reflections on the origins and spread of nationalism.* London: Verso.

Arnold, Thomas. (1872). *The Education of the Deaf and Dumb: An exposition and a review of the French and German systems.* London: Elliot Stock.

Butler, Judith. (1993). *Bodies that matter: On the discursive limits of "sex."* New York: Routledge.

Bauman, H. & Dirksen, L. (1997). Toward a poetics of vision, space, and the body: sign language and literary theory. In *The Disability Studies Reader,* 315–331. New York: Routledge.

Baynton, Douglas. (2001). Disability and the justification of inequality in American history. In *The new disability history: American perspectives,* edited by Paul Longmore and Lauri Umansky, 33–57. New York: New York University Press.

———(1997). A silent exile on this earth: The metaphoric construction of deafness in the nineteenth century. In *The disability studies reader* (pp. 128–152). New York: Routledge.

Branson, Jan & Miller, Don. (1998). Achieving human rights educating deaf immigrant students from non-English-speaking families in Australia. In *Issues unresolved* (pp. 88–100). Washington, D.C.: Gallaudet University Press.

Brueggeman, B. (1999). Lend me your ear: the rhetorical construction of deafness. Washington D.C.: Gallaudet University Press.

Danforth, S. & Rhodes, W. (199). Deconstructing disability: a philosophy for inclusion. *Remedial and Special Education, 18* (6), 357–366.

Davis, Lennard. (1997). Universalizing marginality: how Europe became deaf in the eighteenth century. In L. Davis (Ed.) *The Disability studies reader* (pp. 110–127). New York: Routledge.

Davis, Lennard. (1995). *Enforcing normalcy: Disability, deafness, and the body.* London: Verso.

Dean, Mitchell & Hindess, Barry. (1998). Introduction: Government, liberalism, and society. In *Governing Australia: Studies in contemporary rationalities of government* (pp. 1–19). Cambridge: Cambridge University Press.

Derrida, Jacques. (1967/1997). *Of grammatology.* Gayatri Spivak trans. Baltimore: Johns Hopkins University Press.

Edwards, R. A. R. (2001). Speech has an extraordinary humanizing power: Horace Mann and the problem of nineteenth century American deaf education. In P. Longmore & L. Umansky (Eds.), *The new disability history: American perspectives* (pp. 58–82). New York: New York University Press.

Eghigian, Greg. (2000). *Making security social: Disability, insurance, and the birth of the social entitlement state in Germany.* Ann Arbor: University of Michigan Press.

Hill, M. (1884). *Object lessons on object pictures forming an elementary reading and language book for deaf children for use with the coloured illustrations.* London: A. N. Meyers & Co.

Lane, Harlan. (1997). Constructions of deafness. In L. Davis (Ed.), *The Disability Studies Reader* (pp. 153–171). New York: Routledge.

Longmore, Paul & Umanksy, Lauri, (Eds). (2001). *The new disability history: American perspectives.* New York: New York University Press.

Nelson, J. & Berens, B. (1997). Spoken Daggers, Deaf Ears, and Silent Mouths: Fantasies of deafness in early modern England. In *The Disability studies reader* (pp. 52–74). New York: Routledge.

Rogers, Linda & Swadener, Beth Blue. (2001). Introduction. In L. Rogers & B. B. Swadener (Eds.), *Semiotics & dis/ability* (pp. 1–18). Albany: SUNY.

Seashore, C. (1919). *The psychology of musical talent.* Boston: Silver, Burdett, and Company.

Spivak, G. Can the subaltern speak? In B. Ashcroft, G. Griffiths, & H. Tiffin (Eds.), (1995). *The post-colonial studies reader* (pp. 24–28). London: Routledge.

Wrigley, Owen. (1996). *The politics of deafness.* Washington, D.C.: Gallaudet University Press.

CHAPTER THIRTEEN

THE WEB, ANTIRACISM, EDUCATION, AND THE STATE IN SWEDEN

WHY HERE? WHY NOW?[1]

Camilla Hällgren and Gaby Weiner

This chapter explores a specific curricular development in Sweden (SWEDKID, an antiracist website for young people 12 years of age and upwards) as an exemplar of new patterns of governing in the late twentieth and early twenty-first centuries, in relation to historically and culturally specificity on the one hand, and global relations on the other (see introduction to this book). It illuminates recent shifts in Swedish discourses surrounding immigration, information technology, schooling, and welfare to show how global, state and educational policies, economies, and shifts in patterns of governance, impact on locally situated practices.

It is argued that while Sweden has an impressive history of promoting social democratic values and policies and indeed, in recent years, combating racism and xenophobia has been a "burning" issue for many Swedish politicians and educationists, this has not produced the required shifts at the level of the child. Policy documents, conferences, and media coverage proliferate in Sweden, each demanding a stronger engagement with democracy, more time to "talk about how we are acting towards each other" (Wärnersson, 1999), and more commitment from schools in consolidating democracy and fighting racism. In a speech in 2001 marking the

UN international day of struggle against racism, Prime Minister Göran Persson emphasized the economic as well as human cost of racism. "It is tremendously wasteful. Discrimination leads to the loss of people's willingness to work and to participate in the building of society. It leads to the loss of creativity and strength, knowledge and experience. It is a loss for everybody—economically, but above all it is a loss for humanity" (Persson, 2001).

Other Swedish politicians have followed suit. For example, Ingegerd Wärnersson, the then minister for schools, declared in a speech marking the introduction of a national values project in 1999, that the right to respect is fundamental to motivating learning.

> Everybody has the right to be treated with respect. That is how we grow as human beings; that is, how we gain the inclination to learn. Tolerance, democracy, solidarity and equal rights are fundamental values that should permeate schools. They are beautiful words—but what is their actual meaning, and how are they used in every day life? There is not enough discussion about these issues. I am convinced that if we want better schools, we need to find time to talk about how we behave towards each other. (Wärnersson, 1999)

Britta Lejon, the then democracy minister, in a lecture for the Democracy Gala in 2001, argued for a more active stance on racism. "We have to show that intolerance, racism and xenophobia can't be tolerated. We have to find the courage to say no—to make a statement, clearly take the part of those who stand up for justice and equal rights. Because those values are fundamental to democracy" (Lejon, 2001).

Yet, some have noted a deep divide between rhetoric and practice. For example, the Rumanian-born journalist Ana Maria Narti recently commented: "So many speeches against racism, so many promises about improving solidarity among all of the country's inhabitants, so many integration plans, so much money appropriated to strengthen the segregated areas of the large metropolitan areas! All the same, everyday life is increasingly harsh. No speech or plan has put a stop to the obvious and unfortunately continuous worsening of the social climate" (Narti, quoted in Pred, 2000, p. 267).

As researchers involved with a European Union project linking antiracism, technology, and education, we have noted the high level of interest, support, and publicity for the project in Sweden compared with other countries involved (Spain, Italy, UK). We suggest that one reason why the project has attracted such interest in Sweden—from ministers, national agencies, private sector companies, the media, our university, etc.—is that it is seen as a practical and feasible means of combating racism among young people that has been missing from other initiatives. An added advantage is that it utilizes the Internet, which has become a key secular symbol in Sweden's struggle for economic and social advantage. A further advantage is that it is a means by which the state can maintain its influence over values ("värdegrund" in Swedish) at a time of policy devolution and fragmentation (Bauman, 1998).

In this article we first offer a brief introduction to the project SWEDKID and its role in both engaging and disrupting Swedish youth cultures and values. We draw on the work of Castells, Pred, and Bauman, in seeking to understand the relationship between the Net, education, and anti/racism in Sweden. We apply this analysis to examine to what extent Swedish policies on racism, IT, and education are models of discursively organized educational reforms in a culture of performativity. According to Ball (2001, p. 3), "performativity is a technology, a culture and a mode of regulation that employs judgements, comparisons and displays as means of control, attrition and change." "Who controls the field of judgement," Ball argues, "is crucial." We conclude the chapter with a reflection on how these policies, discourses, and practices contribute to current understandings of the Swedish child, and to governance, technology, and pedagogy.

SWEDKID

SWEDKID is part of a European Union–funded project, EUROKID, which was involved from 2000 to 2003 in the research, design, implementation, and evaluation of the website, as a pedagogical tool to challenge racist and antidemocratic ideas among young people. EUROKID's main aim was to create and deliver on-line, national home-language websites in four countries (Spain, Italy, Britain, Sweden), with a fifth linked Europe-wide website to follow. At the time of writing (November 2002) three websites are on-line (Britkid, Spanishkid, and SWEDKID) and Italy has dropped out of the project. The Swedish part of the project, SWEDKID, has four parts:

- Development of a specific form of on-line teaching and learning resource, which addresses antiracism and multicultural issues and which is freely available to Swedish schools, and to anyone with access to the Internet.
- Investigation of young people's experiences of racism in Sweden, to inform web content.
- Research on the process and outcomes of a specific form of web-development, and its pedagogical application and outcomes.
- Exploration of theoretical and research frameworks for the use of website technology in addressing antiracism and other values issues in schools, youth cultures, and teacher education.

SWEDKID involves the presentation in web form of evidence and discussion on a range of experiences concerning racism, ethnicity, and identity. The project has an empirical base: in-depth interviews with young people in Sweden, from various minority and majority ethnic groups. The virtuality of the Net, a notably youth-oriented medium, is used to problematize and "trouble" young people's experiences and perspectives regarding what it is to be "truly" or "newly" Swedish. In so doing, the website seeks to illuminate, challenge, and intervene in the processes of racialization of Swedish society and culture.

Visitors to the website, which went on-line in October 2002 (at www.swedkid.nu), are encouraged to interact with the characters as a means of reflecting on their own ethnicity, identity and approaches to racism and antiracism. They can enter the website by choosing one of eleven characters of varying ethnicities and identities (majority, minority, and hybrid) drawn from Swedish society (see website illustration below). Once a character has been chosen, an overview sketch is provided of the character's family, culture, religious or other beliefs, interests, problems, etc., drawn from the interview data. The website-user is directed to other parts of the website where he/she is able to access semifictional discussions (based on incidents and experiences of interviewees) between characters illustrating the impact of racism on their daily lives.

Access is also provided to a wide range of linked websites and information sources. The importance of discursive terminology and definitions is recognized in the creation of a glossary of terms used on the website. Overall, the linguistic and iconographic components of the website have a twin function: to attract, sustain, and challenge young people; and more difficult perhaps, to inform and stimulate the actions of teachers.

Certainly, within days of going on-line, SWEDKID gained much positive attention in the national and local press but also detailed, critical analyses from nationalist and racist groups such as the so-called Swedish Resistance Movement. Thus the latter's webportal immediately put out the following message: "Swedkid isn't only about how 'prejudice' should be handled. 'Racial integration' is a typical feature. Soon, this kind of propaganda will be forced on yet more Swedish pupils" (our translation, Patriot.nu, 2002, p. 1).

Terminology

We take the position that though language is the common factor in any analysis of social organizations, power, and individual consciousness, it is also a domain in which subjectivity and social order are defined, contested,

and constructed (Weedon, 1987). For example, in this paper (and in the website) we refer to terms such as racism, antiracism, and xenophobia (as well as the softer "multiculturalism" or "inter-culturalism") to highlight and symbolize the "tough" rhetorical stance that Sweden has taken on these issues as a key actor in the European Union. The language of the SWEDKID dialogues (or scenarios) is largely drawn from the phraseology of the interviewees; yet attempts are also made to induct website-users into the language of meaning-makers such as politicians, researchers, and activists through the provision of a glossary of key concepts and terms. In encouraging the use of a range of literacies (linguistic, aesthetic, and technological) within the website, we nevertheless acknowledge that terminology, phraseology, and interpretation constitute contested systems of understanding, having different discursive meanings at different times, among different social groups and in different cultures.

The attraction of the Internet is that it is as a global phenomenon that, we suggest, disrupts national and generational constraints and assumptions. The concept of "globalization" has thus been key to project thinking. Generally, the term global refers to trends toward worldwide rather than national or local economic and cultural change. As Castells (2000) points out, however, globalization encompasses a number of factors including the collapse of the old, post–World War II order; the loss of power of the nation-state to the transnational corporation; and the use of communication technology to transact business (and process knowledge) across national and continental boundaries. Globalization is seen by some as a new freedom but by others as the "new world disorder" where "no one seems now to be in control" (Bauman, 1998, pp. 59 & 58). Unlike global diplomacy, which in the past sustained and preserved national boundaries and entitlements, globalization does the opposite. It blurs boundaries and, it is argued, imposes one-for-all ideologies. "For everybody . . . 'globalization' is the intractable fate of the world, an irreversible process; it is also a process which affects us all in the same measure and in the same way. We are all being 'globalized'—and being 'globalized' means much the same to all who 'globalized' are" (Bauman, 1998, p. vii).

We suggest, however, that the experience and impact of globalization takes different forms in different contexts. As Castells (2000) points out, policy decisions within the European Union are directly and indirectly generated by globalising forces.

> For European governments, the Maastricht Treaty, committing them to economic convergence, and true unification by 1999, was their specific form of adopting globalisation. It was perceived as the only way for each government to compete in a world increasingly dominated by American technology, Asian manufacturing, and global financial flows which had wiped out European monetary stability in 1992. Engaging global competition from the strength of the European Union appeared to be the only chance of saving European autonomy, while prospering in the new world. (Castells, 2000, p. 143)

Thus, the conception of globalization adopted for this paper is not that it is outside of, or exclusionary to, nationhood, local events, or individual identities. Rather that globalization insinuates itself in various ways at national and local levels, which in turn respond in specific and situated ways.

Understanding Generalities and Specifics

The Information Age. From Manuel Castells (1996, 1997, 1998, 2000) we can see how technology has transformed society and culture and how technology itself is transformed by shifting materialities and identities at global, national, and local levels. Of particular significance is the way in which culture is seen to act as a filter through which globalization flows, mediated by local, provincial, and parochial politics and conditions.

Castells claims that the transformation of consciousness and identity at the turn of the twenty-first century is due to a variety of factors, not least economic globalization and the technological revolution. Castells' angle on a world out of control is to show it as a bricolage (or collage) of the defining features of our times. They are summarized below (but not reduced substantially in number) because, for us, they are markers or signifiers of the post, postmodern world of young people in today's Sweden. We interpret the post, postmodern world as one where no one seems to be in control, and where contradiction and dissonance are commonplace and comprehensible.

For Castells, globalization involves:

- technological revolution
- restructuring of capitalism with increased flexibility, networking, and decentralization; increased individualization and diversification at work
- economic globalization and interdependence; the reworking of the relationship between the economy, state, and society
- state intervention to deregulate markets and dismantle welfare states, and legitimacy crisis of political systems
- decline in power of organized labor
- reshaping of the world's economic blocks—rise of Asian Pacific, consolidation of EU, decline in third world and former Soviet states
- increased global economic competition and organized crime
- new technology as means to satisfy hitherto unachievable, illicit, and taboo desires
- decline of patriarchy, emergence of gender as contested domain, and redefinition of relationships between women, men, and children. Increased participation of women in the workforce, though in discriminatory circumstances
- spread of environmental consciousness, and as a counter, its co-option politically
- tendency of social movements toward fragmentation, single-issue orientation, and the ephemeral
- crisis of individual and collective identity—especially social, cultural, religious, and ethnic identity, search for new identities and resurgence of older identities and fundamentalisms
- resurgence of nationalism, racism, and xenophobia; social fragmentation, social exclusion, breakdown in communication
- emergence of millenarism through various guises and gurus, e.g., technology gurus, new age prophets, political and religious extremists (summarized from Castells, 1996, pp. 1–3).

Castells points to the emergence simultaneously of the apparent seamless-ness of the network society and worldwide shifts toward increased inequality, nationalism, racism, and xenophobia. Thus, while offering the promise of limitless opportunity, global networks operate to exclude:

> global networks of instrumental changes selectively switch on and off individuals, groups, regions and even countries according to their relevance in fulfilling the goals processed in the network, in a relentless flow of strategic

> decisions. It follows a fundamental split between abstract, universal instrumentalism, and historically rooted, particularistic identities. Our societies are increasingly structured around a bipolar opposition between the Net and the Self. (Castells, 1996, p. 3, original emphasis)

Racisms in Sweden. As a complement to Castells' work, Pred (2000) links global pressures to local situated practices at the level of the nation-state—in this case, Sweden. He identifies, during the 1990s in Sweden, the intensification of cultural racism, proliferation of negative racial stereotypes, and spatial segregation of the "non-Swedish"—"even" in such a modern, prosperous, Westernized, strongly social democratic country as Sweden (the point of the title of Pred's book, *Even in Sweden*). Global economic restructuring, Pred argues, has generated experiences that have lent themselves to cultural reworkings as distinctive expressions of racism. Pred captures the zeitgeist in his introduction:

> [Writing] *Even in Sweden* has been anything but easy. . . . I have borne the intense discomfort of bearing witness to an immense tragedy, of observing good intentions coming completely apart, of seeing what was once arguably the world's most generous refugee policy, what was once a remarkably humane and altruistic response to cruelties committed abroad, become translated at home into the cruelties of pronounced housing segregation, extreme labor-market discrimination, almost total (de facto) social apartheid, and frequently encountered bureaucratic paternalism. (Pred, 2000, p. xii)

The "tragedy" that Pred recognizes is the collapse of the altruistic (or paternalistic) state. "Racisms" are defined by Pred (2000, p. xiv) as a constellation of "relations, practices and discourses," and as unavoidable in present-day Sweden. Like Castells and others, Pred argues that, although intensified by globalization processes, racism is shaped and produced locally, involving ordinary people. Thus, he writes that it is: "through participation in particular *locally situated practices*—that individuals and groups become racialized, that migrants, refugees and minorities have their racialization again and again reinforced, regardless of the differences in their biographical background or the diversity of their previous social experiences and subjective positions" (Pred, 2000, p. 18).

Developing Practice

Identification of "locally situated practices" (Pred, 2000; Lave and Wenger, 1991; Säljö, 2000), and problematization of the "ordinary" are key tasks for

SWEDKID, in challenging racist perspectives and practices among young people. However, there have been relatively few analyses of how to interrupt racist practices at the levels of the school, classroom, and individual (e.g., Lodenius & Wingborg, 1999; Lahdenperä, 1997);[2] even though a range of policy analysts and educational researchers across Europe have charted the development of racism and the perspectives of politicians, education policymakers, school practitioners, and students (e.g., Troyna & Carrington, 1990; Frankrijker, 1996; Osler & Vincent, 2002).

Other groups of researchers (feminist, action researchers, etc.) have been relatively more attentive to the importance of addressing the practices and values of students and teachers in chasing the elusive goal of educational equity (e.g., Walkerdine, 1990; Berge & Vé, 2000). From them, we have learned about the dangers of stereotyping and the need to pay attention to imbalances in the knowledge configurations embedded in pedagogical assumptions and messages, whether in conventional or virtual texts.

Our perception of the website, then, is not as a utopian solution to society's ills; but rather as a means of encouraging meaningful social action, and transformative politics among young people who may be more tuned into the message of SWEDKID through its medium of virtuality. We now turn to Sweden's specific positioning in relation to three key policy areas involving SWEDKID: racism and antiracism, information technology, and education.

Sweden, Racism, and Antiracism

Sweden has long been a multiethnic society despite the presumption in the 1960s of "one language, one race and no religion" (Andrae-Telin & Elgqvist-Saltzman, 1987, p. 4). Often forgotten in today's discussions are the number of Swedish-born minorities—Sami (Laplanders), Swedish Finns, Roma, Tornedalors, and Jews—each of which has made significant contributions to Swedish society and culture (Regeringens proposition, 1998/99). Today, of Sweden's nine million inhabitants, approximately 10 percent (over 900,000) were born abroad. Of these, 40 percent have lived in Sweden for 20 years or more. An additional 700,000 have at least one parent from abroad (Regeringens proposition, 1997/98). At present, a quarter of all children in Sweden have such a background—excluding national minorities (SCB, 2000).

How are we to understand the rise in concern about racism and xenophobia in Sweden recently? Is Sweden especially racist? Is there a racism lurking, as it were, waiting to ignite? Oddly perhaps, to understand Swedish approaches to racism "within," it is necessary to consider its history of peaceful coexistence, going back nearly 200 years. Sweden was a powerful Nordic imperial power from the seventeenth until the early twentieth century, when its empire came to an end with the granting of independence to Norway in 1905. Sweden's espousal of neutrality from then onwards meant that it was able to avoid the worst excesses of the two twentieth-century world wars. It was thus able to contribute to, and benefit from, the reconstruction of Europe in the immediate postwar period, and consequently, achieve a higher living standard than other European countries. There was little challenge or disruption to the pockets of Nazism and Fascism that developed in Sweden in the 1930s, as in other countries (Kaplan, 2001). Moreover, those growing up in Sweden after 1945 prospered from a welfare state, social cohesion, and a level of prosperity that attracted the admiration of the rest of the world, even as Sweden's shady compromises with Nazi Germany were forgotten.

Thus Sweden was shielded during the 1950s, 1960s, and 1970s from self-scrutiny regarding racism that was forced on many European countries. Discrimination such as that against African Americans in the United States or that of the apartheid system in South Africa seemed remote; the fight against racism was mainly an international, not a national, matter. Sweden had its share of right-wing extremist and nationalist groups, but there were too few to be visibly influential. Immigration controls were instituted only in 1967, at the start of a downturn in the Swedish economy when need for foreign labor power decreased. In the 1970s, along with the burgeoning of Sweden's economy once more, came economic migrants, mostly from neighboring Nordic countries, especially from Finland—resulting in nearly half of immigrant workers in Sweden being white and of Finnish origin. In the 1980s and 1990s, however, new migrants came to Sweden, largely refugees and asylum seekers from Eastern Europe and the developing world, many of whom were people of color (Kaplan, 2001).

One consequence of this change in immigration patterns was a heightened visibility of racist groups opposed to what became known as the "new" Swedes. For example, in the 1980s, the first known, active neo-nazi group, Bevara Sverige Svenskt, (BSS) meaning "Keep Sweden Swedish," coordinated the burning of crosses in different parts of the country. Several other small extreme-right political parties also came into existence from the mid-1980 onwards, such as: Sweden Democrats (Sverigedemokraktaterna, SD); Sjöbo Party (Sjöbopartiet); Progressive Party (Framstegspartiet); and New Democracy (Ny Demokrati). Consequently, racism became of national concern for the first time in the postwar period (Lange, Lööw, Bruchfeld, & Hedlund, 1997).

Yet, there was little concerted action in response to repeated requests from municipalities, schools, and trade unions for guidance, advice, and support on how to fight racism, despite some piecemeal interventions (e.g., small action research projects supported by the National Agency for Education [Skolverket]). There was (and still is) no systematic means of gathering or distributing information and research relating to racism (and antiracism) in Sweden. This has led, according to the National Office of Integration (Integrationsverket), to a general lack of knowledge about the nature of racism, how it is characterized, or what motivates perpetrators.

Following this knowledge vacuum, "common-sense" assumptions prevail. Thus, 48 percent of adults in Sweden believe in the existence of distinctive human races, each of which can be differentiated according to skin color, culture, and religion, and also, in some cases, by personality, looks, temperament, and treatment of women, according to a survey of 1,000 Swedes of 15 years and above in 2001 (Integrationsverket, 2001). A larger study in 1997 focused on the incidence of racism, and attitudes toward democracy, of nearly 8,000 school students (between 6th and 9th grade). The study found that 3 percent (233) of the students had, during the previous 12 months, been exposed at least once to a violent, racist, ethnic or politically-related incident; 7 percent had suffered from verbal threats; and 13 percent reported unfair treatment because of their ethnic background. Racist or ethnic abuse had been experienced by 23 percent of the students, while 17 percent had been contacted by racist organizations and 11 percent admitted to reading racist magazines at least once. The study also showed high levels of "conscious" racism; for example, 11 percent considered "rasblandning" (racial integration) against the laws of nature; 12 percent expressed the view that Jews have too much influence in the world today; and 29 percent claimed that there is too much emphasis in Sweden on the evils of Nazism and the Holocaust. Even more disturbingly, over a third (34 percent) agreed or partly agreed with the statement that non-European immigrants should return home (Lange et al., 1997).

Racist beliefs thus underpin the values of a significant minority of young people in Sweden, despite the public outpourings of concern about racism and despite the fact that several antiracist campaigns have targeted schools and young people. For example, in 1992, a number of government agencies were jointly given the task of trying to influence young people's perspectives on, and behavior toward, immigration, immigrants, and ethnic minority groups. They came to the conclusion that it is difficult to change perspectives by information alone—an appeal to values and feelings is also important. This point has been taken up in recent governmental emphasis on values among schools students (viz. the comments of the Schools Minister earlier in this chapter), but so far this work is at the stage of fact-gathering and mapping rather than intervention or practice.

Another strategy has been to develop government action plans outlining what has been achieved so far, and what needs to be done, in the fight against racism, xenophobia, etc. Here, once again, the rhetoric flows. Schools are described as important institutions for getting in touch with young people and with unique possibilities to create among young people an understanding of democracy and respect for human dignity (Regeringens skrivelse, 2000/01). Likewise, various regulatory frameworks for schools (e.g., Utbildningsdepartementet, 1985) and the curriculum (e.g., Lpo 1994; Lpfö 98) require that schools incorporate issues of democracy, respect, tolerance, and equality in their work. Yet there are few suggestions about how these changes of attitude may be achieved in practice.

Sweden and Information Technology

Sweden is a networked society. In 2002, 82 percent of Swedes had access to a computer, either at home, at work, or at school, and 72 percent had access to the Internet. In the same year, nearly all schools (96 percent) had a connection to the Internet, which is nearly 16 percent higher than the average for the European Union (MMS, 2002; Statens institut för kommunikationsanalys, 2002). The computer also has high symbolic value in Sweden, perhaps more so than in other countries. It is seen as a guarantee for the future—in a sense, as Sweden's saviour in the global marketplace.

In an investigation into the impact of information technology (IT) on Swedish schools over the last decade, Riis (2000) found a high level of confidence in the potential of IT despite the gap between the expectations of what IT can achieve educationally, and the reality of how it works at the level of the classroom. In particular, IT is seen as the key to changing the way schools and teachers work, and thus, as a means of ensuring "higher quality" teaching and learning.

Riis (2000) argues that for some, IT is viewed as one solution to recent cutbacks in school funding, which have led to increased class sizes, and a subsequent lack of attention to individual student needs. It provides, on the one hand, a means by which information and knowledge can be disseminated swiftly to schools and, on the other, the possibility of customizing curricula to individual requirements. Politicians have a yet more idealistic view, envisioning IT as providing possibilities for a fundamental transformation of the Swedish school (Skolverket, 1998b). As the then Minister for Schools explained: "No part of the everyday lives of children and adults remains unaffected by Information Technology. Working life has in the course of a few years been dramatically transformed as a consequence of IT. Today the flow of information and data is much greater in scope and accessibility than ever before" (Wärnersson, 2000).

The most important government initiative in recent years has been the project IT in Schools (ITIS), a four-year program involving in-service programs for 70,000 teachers. Launched in 1999, it constitutes the biggest ever investment in school development and in-service education in Sweden (1.7 billion Swedish crowns or 150 million US dollars). As an inducement to join the project, teachers ranging from preschool teachers to adult educators are promised personal computers, connections to the Internet, and personal e-mail addresses (see http://www.itis.gov.se/english/index.html). ITIS constitutes an enormous investment and therefore trust in the possibilities of IT, yet little has been suggested about how learning will change as a consequence of the project. Indeed, there is little indication of any shift in practice since the research overview carried out by Bergman in 1997, which showed a dearth of practical examples of the uses to which IT can be put.

We can see, then, that IT has not only achieved great symbolic value in Sweden but has become an increasingly expensive item in the national education budget. Significantly, more attention has been paid to making sure that the technology is in place and that potential users have appropriate technical skills, than in addressing key pedagogical questions such as what contribution IT can make to present-day schooling or what transformative possibilities it offers. So far, IT take-up has been low in schools. So, an additional attraction of SWEDKID is its evident strategy of encouraging computer literacy as a way to achieving its main goals regarding anti/racism.

Sweden and Education

In Sweden, as in other countries, confidence in schools has plummeted. A recent survey of the attitudes of pupils, parents, teachers, and the wider public indicates that schools are seen as failing to provide pupils with adequate knowledge and skills (Skolverket, 1998a). The school climate has become tougher, with teachers reporting substantial increases in their workload. Stress levels have risen, with nearly a third of teachers reporting increased problems with violence, bullying, and racism. Further, three quarters of teachers consider that they do not have sufficient skills and knowledge to deal with pupils of different ethnic backgrounds and cultures (Skolverket, 1998a).

Meanwhile, as highlighted by Castells, the legitimacy of politicians has been called into question. 85 percent of teachers maintain that local government is not sufficiently competent to take responsibility for education (i.e., in planning and putting in place nationally set goals for student achievement). Forty percent of teachers consider the formal school plan to have little significance for their daily work. Moreover, work aimed at promoting democracy, it is claimed, is neither rewarded nor evaluated. Schools are viewed as not being sufficiently adjusted to increased cultural diversity among students, thus exacerbating exclusion of certain minority groups (Skolverket, 2000). Likewise, a survey of how racism and xenophobia is treated among teacher educators, found much confusion about how teachers should be supported in working with diversity (Vinterek and Gustavsson, 1998).

As in other countries, the authority of teachers as professional arbiters of school outcomes has also been placed in doubt. There is a perception that standards are falling in schools. For example, recently for the first time, a significant number of students in Sweden failed to graduate from compulsory school at the age of 16. In 2000, a quarter of 16-year-old school dropouts failed one of the three core subjects (Swedish, Mathematics, English), and 15 percent, two or more. Notably, over a third (38 percent) of such students came from immigrant backgrounds (Skolverket, 2001). Sweden also compares badly with other countries in how minorities fare in education. For example, an OECD survey of 256,000 students in 32 countries shows Swedish students with an immigrant background as the poorest performing of all participating countries (OECD, 2000).

Why Here? Why Now?

In exploring why the Internet, antiracism, and education are seen as such a potent combination in today's Sweden, we have focused in this chapter on aspects of globalization and how they have been acted out and managed at a specific historical and cultural conjuncture. Our analysis is threefold, focusing on: why Swedish institutions have been so favorably disposed to SWEDKID; its exemplification and value as a pedagogical innovation; and its implications for Sweden's future.

The support of IT (and SWEDKID) in Sweden is an example of a discursively organized reform that seeks to address the uncertainties of the network society in an era of globalization, at the same time as dealing with national concerns regarding global competition and the need for a stable society. As a small country, with a high standard of living and welfare, newly incorporated into the European Union, Swedish policymakers and industrialists have been aware of the global pressures that seek to undermine the country's economic gains, made in the post–World War II period. Sweden's international reputation remains one of a unique combination of market economy and strong public sector with "one of the most comprehensive and generous systems of welfare provision in Europe and the world" (Gould, 1996, p. 91). The predominance of social democratic policies and popular support for a large welfare state, however, are at odds with the move of capital out of Sweden, and a weakening national currency. In this sense, Sweden is both a modern and postmodern state—a nation in transition from centralized collectivism to competitive individualism. Its policymakers and citizens are confronted with accommodation to certain aspects of the new order (e.g., restructuring of capitalism, increased flexibility, networking, decentralization, technological change); yet resistance to others (e.g., downsizing of welfare, resurgence of nationalism, racism and xenophobia, social fragmentation, social exclusion). The appeal of SWEDKID is that it is both symbolic and pragmatic in Sweden's attempt to smooth the fissures between the modern, postmodern, and even post, postmodern.

Sweden's positioning (or branding) in the international (global) arena is as a nation or people concerned with the promotion of human rights, and indeed it has consistently supported freedom fighters, for example in South Africa in the apartheid era. Yet, its failure to support and protect its minorities at home has been profoundly shocking to some, for example Pred, as we have seen. So, it seems, there can be deep discrepancies between different levels and pressures of policymaking. Kaplan (2001, p. 2) argues that the interface between the global, national, and local is both fluid and interactive: ". . . . no action, event or process in the field of racism anti-Semitism or violence is truly local. What happens in one place has repercussions far beyond city limits or national borders. . . . Yet if it is true that events at the local level are inextricably intertwined with national, regional and international developments, and vice versa, it is equally true that local contexts and conditions are of key importance."

We can thus see support of SWEDKID as a recognition of the wider dilemma that now faces Sweden—as one of the most prosperous, most networked societies internationally, which, nevertheless, continues to segregate and exclude many of its citizens.

As we have seen, the Internet, symbolically and materially, is able to offer cultural reworkings of social phenomena and meanings. Knowledge can be utilized in different ways on the web—and this novelty and diversity appeals to young people—although knowledge is itself mediated by the understandings and values of individual and group "readers." The web is also a *practical* medium, which can be activated for progressive purposes (such as in the case of global environmentalists and SWEDKID) or for counter-progressive purposes (as in the case of Sweden's ultra-racist groups and their initial response to SWEDKID mentioned earlier). As Castells said recently, the Internet is a tool for helping us to think locally and act globally:

> The network society is not an abstraction, it is a fact. And the Internet is the key media in that society. When society transforms itself into networks, society needs the structure of the Internet. People are using the technology to get what they want; they do not follow the technology. People in the network society think locally, act globally, and Internet is the tool. To understand our society we have to create a new methodology. (Castells 2001)

Castell thus argues that it is people who use technology, not the other way around. The inclination of today's youth (and perhaps others, too) is to start with the self, the body and identity, and then reach out to the global. SWEDKID's utilization of the person in the form of characterization, as an entry into more general debates about racism, builds on this. The Net, however, is also a means of gaining access to information, products, and entertainment. Castells' reference to a new methodology refers to this process.

As creators and researchers of the SWEDKID website, our main concern has been to raise the consciousness of Sweden's post, postmodern generation, and to engage young people in exploring their identities and ethnicities as fluid entities. Reducing the level of racism in Swedish society is a more structural aim. However, given the instabilities and rapid changes in youth cultures and unintended and unforeseen outcomes of any innovative undertaking (Giddens, 1991), website investment is a risky business. Connecting to SWEDKID is a voluntary undertaking, not a compulsory unit for all Swedish 12-year-olds. We cannot at this stage (a couple of months on from the launch), foresee how popular SWEDKID will be with young people (or their teachers), or any specific impact or outcomes, however promising initial press coverage or usage might be.

Education and the Swedish Child

To summarize the arguments of this chapter, we have surmised that SWEDKID is "hot" in Sweden at present because it conjures up a number of possibilities, simultaneously:

- for engaging with a new methodology, and with the seemingly unlimited potential of the Net;

- for addressing desires for the future, involving trust, democracy, and most important, prosperity; and
- for offering practical and symbolically powerful responses to present crises of political legitimacy and governance.

We are less optimistic, however, about what kind of Swedish child is being produced by the different sets of global, national, and local discourses that we have outlined in this chapter. We have noted the high level of rhetoric concerning democracy and rights in Sweden, which is nevertheless paralleled by the fear that Sweden's postwar economic position and, especially, its levels of prosperity, are under threat. It could be argued, indeed, that such rhetoric provides an obstacle to reflective and conscious change, say on the part of teachers, because the fear of being thought of as racist denies them a secure environment in which they can open up about their inadequacies and difficulties.

We have also outlined trends in Swedish society pointing to increased racism, concerns about educational failure, and continuing belief in the potential of information technology to solve social problems. We suggest that the form of child produced by such discourses and trends is framed more by anxieties arising from neoliberalism, social disorder, and the global marketplace than by democratic ideals. We argue that rather than celebrating the potential of childhood and youth, current discourses surrounding anti/racism, information technology and education produce the Swedish child at the turn of the twenty-first century as one to be feared, for example:

- as a potentially "out of control" racist;
- as a potentially under-performing immigrant
- as unable to behave toward others in an "orderly" fashion
- as unable to take up the dual challenges of the network age and smart capitalism
- as unable to deliver Sweden's desires for the future.

Following this, we suggest that SWEDKID has been important symbolically for Sweden; as an intervention in what is seen by many as the downward slide of Sweden, economically and culturally. In an era of performativity, SWEDKID is able to activate the symbolism of the Net, in order to support state strategies to maintain social order and raise academic achievement among Sweden's youth. Thus, SWEDKID is perceived as a means by which the national and local state can *show* that they are actively addressing Sweden's current problems.

As educators and researchers, we are differently positioned with regard both to the child and to the possibility of change. We are tentative about what can be achieved by information technology, given its human origins and social and cultural limitations. On the other hand, we are hopeful of the potential of education—and of the development of electronic pedagogies—in engaging young people in the exploration and discussion of their personal values for the benefit of the wider community. Nevertheless, whoever we may be—whether pupils, politicians, policymakers, teachers, or researchers—we are inevitably part of new and changing patterns of governance, technology, and pedagogy that position us both similarly and differently to previous generations, to other countries and continents, and to each other. SWEDKID's emphasis on the person suggests that in an era where it feels as if no one is in control, taking responsibility for the quality and ethics of our personal practice is, perhaps, the most that any of us can aspire to!

Notes

1. Revised version of paper presented at the conference "Governing Patterns of the Child, Education and the Welfare State," Norsjö, Sweden, 18–20 May 2001
2. There is a large literature on teaching "multiculturalism" in the United States; however, this is not easy to translate to the situated practices and understandings of educators within the European Union.

References

Andrae-Telin, A. & Elgqvist-Saltzman, I. (1987). *Side by side in classrooms and at work—Ideology and reality in Swedish educational policy and practice.* Arbetsrapport nr 47, Umeå, Seweden: Pedagogiska Institutionen, Umeå Universitet.

Ball, S. J. (2001). The teacher's soul and the terrors of performativity. *Voice of the Research Students' Society, 38.* Retrieved November 25, 2002, from http://www.ioe.ac.uk/rss/voice/voice38.htm

Bauman, Z. (1998). *Globalization: The human consequences.* Cambridge: Polity Press.

Bergman, M. (1997). *När IT kom till skolan. Det stora IT projektet om IT användning och hur det hela började för en skola,* [When IT came to school. The great project about using IT and how it all started in a school.]. Tema T Arebetsnotat 174, Linköping: Linköpings Universitet.

Berge, B-M., & Vé, H. (2000). *Action research for gender equity.* Buckingham: Open University Press.

Castells, M. (1996). *Vol. 1: The rise of the network society.* Oxford: Blackwell.

———(1997). *Vol. 2: The power of identity.* Oxford: Blackwell.

———(1998). *Vol. 3: End of millennium.* Oxford: Blackwell.

———(2000). *Vol. 1: The rise of the network society* (2nd ed.). Oxford: Blackwell.

———(2001) Speech at Umeå Forum, Umeå University, Sweden, April 2001.

Frankrijker, R. L. (1996). Expert beliefs about teacher training in multicultural perspective: The multicultural competent teacher in the Netherlands. In M. O. Valente, A. Bárrios, A. Gasper, & V. D. Teodoro (Eds.), *Selected papers from the 18th Conference of the Association for Teacher Education in Europe: Teacher training and values education* (pp. 479–496). University of Lisbon. ATEE.

Giddens, A. (1991). *Modernity and Self-Identity: Self and society in the late modern age.* Cambridge: Polity Press.

Gould, A. (1996). Sweden: The last bastion of social democracy. In V. George & P. Taylor-Gooby (Eds.), *European welfare policy* (pp.72–94). London: Macmillan.

Integrationsverket. (2000). *Manskliga Rättigheter Rasism Etnisk Diskriminering* [Human rights, Racism, Ethnic discrimination]. Stockholm, Sweden: SIFO.

IT in Schools (ITiS) (2002). http://www.itis.gov.se/english/index.html Retrieved December 19, 2002, from http://www.swedkid.nu

Kaplan, J. (2001). *Racism, anti-Semitism and violence: The local studies perspective.* Sweden: Stockholm International Forum Combating Intolerance.

Lahdenperä, P. (1997). *Invandrabakrund eller skolsvårigheter? En textanalytisk studie av åtgärdsprogram för elever med invandrarbagrund* [Immigrant background or school difficulties? A text analysis study of action programmes for pupils with immigrant background]. Sweden: Stockholm University.

Lange, A., Lööw, H., Bruchfeld, S., & Hedlund, E. (1997). *Utsatthet för etnisk och politiskt relaterat våld hot m m, spridning av rasistisk och antirasistisk propaganda samt attityder till demokrati m m bland skolelever* [Exposure of ethnical and political related violence and attitudes to democracy etc. among school pupil(s)]. Centrum för invandringsforskning, Stockholm: Stockholms Universitet.

Lave, J., & Wenger, E. (1991). *Situated learning: Legitimate peripheral participation.* New York: Cambridge University Press.

Lejon, B. (2001, February 20). Speech to The Student Democracy Gala, Gothenberg, Sweden.

Lodenius, A., & Wingborg, M. (1999). *Svenskarna först? Handbok mot rasism och Främlingsfientlighet* [The Swedes first? A Handbook against racism and xenophobia]. Stockholm: Atlas.

Lpo 94. (1994). *Läroplan för det obligatoriska skolväsendet och de frivilliga skolformerna Lpo 1994* [The Curriculum for the compulsory school and the non-compulsory school system Lpo 94]. Stockholm: Utbildningsdepartementet.

Lpfö 98. (1998). *Läroplan för det obligatoriska skolväsendet, förskoleklasses och fritidshemmet* [Curriculum for the compulsory school, the pre-school class and the after school centre Lpo 94]. Stockholm: Utbildningsdepartementet.

Mediamätning i Skandinavien, (MMS). (2002). *Tillgång till datorer och Internet November 2001 Oktober 2002* [Access to computers and the Internet November 2001 October 2002]. Retrieved November 24, 2002, from http://www.mms.se/

Organization for Economic Development and Cooperation. (OECD). (2000). *Measuring student knowledge and skills: The PISA 2000 assessment of reading, mathematical and scientific literacy.* Paris: OECD Publication Service.

Osler, A. & Vincent, K. (2002). *Citizenship and the challenge of global education*. Stoke on Trent, UK: Trentham Books.

Patriot.nu. (2002). *Ny skolmetod för indoktrinering* [New school method of indoctrination]. Retrieved October 21, 2002, from http://www.patriot.nu/

Persson, G. (2001, March 21). Speech on United Nation's International Day Against Racism and Race Discrimination. Stockholm, Sweden.

Pred, A. (2000). *Even in Sweden: Racisms, racialized spaces and the popular geographical imagination*. Berkeley, CA: University of California Press.

Regeringens proposition. (1997/1998). *Sverige, framtiden och mångfalden från invandringspolitik till integrationspolitik* [Sweden, future and diversity from immigration policy to integration policy]. (No. 16). Stockholm: Regeringen.

Regeringens proposition. (1998/1999). *Nationella minoriteter i Sverige* [National minorities in Sweden]. (No. 143) Stockholm, Sweden: Regeringen.

Regeringens Skrivelse. (2000/2001). *En Nationell handlingsplan mot rasism, främlingsfientlighet, homofobi och diskriminering* [A national action plan against racism, xenophobia and discrimination]. (No. 59). Stockholm: Regeringen.

Riis, U. (2000). *IT i skolan mellan vision och praktiken forskningsöversikt* [IT in school between vision and practice: A research overview]. Stockholm: Skolverket.

Säljö, R. (2000). *Lärande i Praktiken ett sociokulturellt perspektiv* [Learning in practice: A sociocultural perspective]. Stockholm: Prisma.

Statistiska Centralbyrån (SCB). (2000). *Barn och deras familjer* [Children and their families]. Stockholm: SCB.

Skolverket. (1998a). *Who believes in our schools? Attitudes to the Swedish school in 1997: A summary.* Stockholm: Skolverket.

———(1998b). "*(utvecklingen beror då inte på användningen av datorer." IT-användningen i den svenska skolan* ["development is not dependent on the use of computers." The use of IT in Swedish schools]. Stockholm: Skolverket.

———(2000). *En fördjupad studie om värdegrunden* [An in-depth study of social values]. Stockholm: Skolverket.

———(2001). *Rapport 206: Beskrivande data om barnomsorg och skola och vuxenutbildning* [Descriptive data about child care system, school and adult education]. Stockholm: Skolverket.

Statens institut för kommunikationsanalys. (2002). *Facts about information and communications technology in Sweden 2002*. Stockholm: SIKA.

SWEDKID. Retrieved December 19, 2002, from http://www.swedkid.nu

Troyna, B., & Carrington, B. (1990). *Education, racism and reform*. London: Routledge.

Utbildningsdepartementet. (1985). *Skollagen* [School law]. (No. 1100). Stockholm, Sweden: Utbildningsdepartementet.

Vinterek, M., & Gustavsson, S. (1998). *Interkulturella frågor i lärarutbildningarna med särskild inriktning mot rasism och främlingsfientlighet* [Intercultural questions in teacher education with main focus on racism and xenophobia]. Sweden: Umeå Universitet.

Walkerdine, V. (1990). *School girl fictions.* London: Verso.

Wärnersson, I. (1999). Public speech marking the introduction of the National Values Project. Stockholm, Sweden.

———(2000). *National Action programme for ICT in Schools.* Retrieved December 19, 2002, from http://www.itis.gov.se/english/about_national_programme.html

Weedon, C. (1987). *Feminist practice and poststructuralist theory.* Oxford: Basil Blackwell.

Startpage of SWEDKID:

CONTRIBUTORS

BERNADETTE BAKER is an Associate Professor in the Department of Curriculum and Instruction, University of Wisconsin-Madison (USA). She is the author of *In Perpetual Motion: theories of power, educational history, and the child.* She is founding co-Chair of the Foucault and Education, as well as the Post-Colonial Education Special Interest Groups of the American Education Research Association.

MARIANNE N. BLOCH is a Professor in the Department of Curriculum and Instruction and in the Department of Human Development and Family Studies, University of Wisconsin - Madison (USA). Her recent publications include: *Woman and Education in Sub-saharan Africa* (1998) (co-edited with J Beoku-Betts, & B. Tabachnick), and *Partnership and the State: The paradoxes of governing schools, children and the family* (co-edited with B. Franklin, and T. S. Popkewitz), to be published by Palgrave-MacMillan Press.

GAILE S. CANNELLA is a Professor of Educational Psychology at Texas A&M University (USA). She is the editor of the Childhood and Cultural Studies section of the *Journal of Curriculum Theorizing.* Dr. Cannella's books include: *Deconstructing Early Childhood Education; Embracing Identities in Early Childhood Education;* and *Kidworld: Childhood Studies, Global Perspectives, and Education.*

LOÏC CHALMEL is Professor of History of Education, University of Rouen, France. He is the author of many publications in the field of history of education, including, most recently, *La petite école dans l'école. Origine piétiste-morave de l'école maternelle française* (Paris, Berne: Peter Lang 1996/2000).

GUNILLA DAHLBERG is a Professor of Education at the Stockholm Institute of Education, and Professor in the Reggio Emilia Institute there. Her recent publications include many chapters and articles, as well as two co-authored volumes: Dahlberg, G., Moss, P., and Pence, A. *Beyond Quality in Early Childhood and Care,* and Hultquist, K. and Dahlberg, G. *Governing the Child in the New Millenium* (Routledge Press).

MIRIAM DAVID is Director of the Graduate School of Social Sciences and Professor of Education at Keele University in England. Her research has focused on families, gender, education, and public policy. Recent publications include *Personal and Political Feminisms; Sociology and Family; Feminist theory: How gender has been involved in family-school choice* (co-authored with A. Stambach)(forthcoming in SIGNS); and, with Madeline Arnot and Gaby Weiner, *Closing the Gender Gap: Post-war education and social change.*

INÉS DUSSEL is Director of the Department of Educational Research of the Latin American School for the Social Sciences (FLACSO)/Argentina and Assistant Professor at the Universidad de San Andres, Buenos Aires, Argentina. Her research interests include educational reform, curriculum, and histories of the body. Her recent publications include: *Curriculum, Humanismo y Democracia en la Enseñanza Media (1863–1920)* (1997) [Curriculum, Humanism and Democracy in Secondary Schooling (1863–1920)], and Dussel, I., Tiramonti, G., & Birgin, A., Decentralization and recentralization in the Argentine educational reform: Reshaping educational policies in the '90s. In T. Popkewitz (Ed.), *Educational knowledge: Changing relationships between the state, civil society, and the educational community* (2000).

CAMILLA HÄLLGREN is a doctoral student in the research program focused on Teachers' Work at Umeå University. Her doctoral studies involve the conceptualization, design, development, implementation, dissemination, and evaluation of the website SWEDKID, which is aimed at challenging racist and anti-democratic ideas among young people.

KERSTIN HOLMLUND is Senior Lecturer of Social Science and Swedish Language in the Faculty of Teacher Education at Umeå University (Sweden). Her research focuses on the history of early education, child care, and women's work and the development of social policies related to child care in Sweden. Recent publications include: *Låt barnen komma till oss: förskollärarna och kampen om småbarns-institutionerna 1854–1968.* (Let the children come to us: pre-school teachers and their struggle for the childcare institutions). (1996) (Umeå: Umeå University, Borea Bokförlag), and "Kindergartens for the poor and kindergartens for the rich: two directions for early childhood institutions in Sweden (1854–1930). *History of Education* (1999), vol 28 no 2.

INGEBORG MOQVIST is Senior Lecturer in the Faculty of Education at the University of Vaxjo (Sweden), and a senior advisor to the Swedish society for the protection of families and children against abuse. Her publications include *Den kompletterade familjen. Föräldraskap, forstran och förändring i en svensk småstad* [The Augmented family]. (1997). (Department of Education, Umeå University, Sweden).

THOMAS S. POPKEWITZ is a Professor in the Department of Curriculum & Instruction, Chair of the Committee on International Education for the School of Education, University of Wisconsin - Madison (USA). His studies explore the systems of reason that order teaching, teacher education, and the educational sciences. His most recent books include: *A Political Sociology of Educational reform: Power/knowledge in teaching, teacher education, and research* (1991); *Struggling for the Soul: The politics of education and the construction of the teacher* (1998); *Educational Knowledge: Changing relationships between the state, civil society, and the educational community* (2000); and *Cultural History and Education: Critical studies on knowledge and schooling* (edited with B. Franklin & M. Pereyra) (2002).

BETH BLUE SWADENER is Professor of Education at Arizona State University. Her research focuses on early childhood education, multicultural education, and international education. Her recent publications include: (with S. Kessler) *Reconceptualizing the Curriculum in Early Childhood Education* (1992); (with S. Lubeck) *Deconstructing the Discourse of "At Risk": children and families "At Promise"* (1995); Swadener, B. B., Kabiru, M., & Njenga, A. (2000), *Does the Village Still Raise the Child?: A collaborative study of changing childrearing,* and an in-press edited volume by Kagenda Matua and Beth Blue Swadener, *Decolonizing Research in Cross-cultural Contexts: Critical personal narratives* (Albany, NY: State University of New York Press).

PATRICK WACHIRA is a Kenyan scholar who is completing his Ph.D. in Curriculum & Instruction at Kent State University's College & Graduate School of Education in the USA. Patrick holds a bachelors degree from the University of Nairobi, Kenya. His research and scholarly interests include equity in education, mathematics education, and educational policy in and out of Africa.

GABY WEINER is professor of teacher education and research at Umeå University in Sweden. She moved there from her post as professor of educational research at South Bank University, London in 1998. She has written and edited numerous books and reports on social justice, equal opportunities, and gender and is co-editor, with Kathleen Weiler, of The Open University Press series "Feminist Educational Thinking". Her publications include: *Feminisms in Education: An introduction; School Effectiveness for Whom? Challenges to School Effectiveness and School Improvement* (with R. Slee and S. Tomlinson); *and Closing the Gender Gap: Postwar educational and social change* (with M. Arnot and M. David).

INDEX